Volume 15

Quantum Electronics

PART B

Edited by

C. L. TANG

*School of Electrical Engineering
and the Materials Science Center
Cornell University
Ithaca, New York*

D0101170

1979

ACADEMIC PRESS · New York   San Francisco   London

A Subsidiary of Harcourt Brace Jovanovich, Publishers

621.366
Q1
v. 2

ACADEMIC PRESS, INC.
111 Fifth Avenue, New York, New York 10003

United Kingdom Edition published by
ACADEMIC PRESS, INC. (LONDON) LTD.
24/28 Oval Road, London NW1 7DX

Library of Congress Cataloging in Publication Data

Main entry under title:

Quantum electronics.

    (Methods of experimental physics ; v. 15)
    Includes bibliographical references.
    1. Lasers. 2. Quantum electronics. I. Tang, Chung
Liang, Date   II. Marton, Ladislaus Laszlo,
Date   III. Series.
TA1675.Q36   621.36'6   79-14369
ISBN 0-12-475955-6 (v. 15B)

PRINTED IN THE UNITED STATES OF AMERICA

79 80 81 82   9 8 7 6 5 4 3 2 1

Methods of
Experimental Physics

VOLUME 15

QUANTUM ELECTRONICS

PART B

METHODS OF EXPERIMENTAL PHYSICS:

L. Marton and C. Marton, *Editors-in-Chief*

CONTENTS

6. Color Center Lasers
   by LINN F. MOLLENAUER

CONTRIBUTORS

Numbers in parentheses indicate the pages on which the authors' contributions begin.

J. E. BJORKHOLM, *Bell Telephone Laboratories, Holmdel, New Jersey 07733* (232)

TERRILL A. COOL, *School of Applied and Engineering Physics, Cornell University, Ithaca, New York 14853* (95)

E. P. IPPEN, *Bell Telephone Laboratories, Holmdel, New Jersey 07733* (185)

P. F. LIAO, *Bell Telephone Laboratories, Holmdel, New Jersey 07733* (232)

T. MANUCCIA, *Naval Research Laboratory, Washington, D.C. 20375* (55)

LINN F. MOLLENAUER, *Bell Telephone Laboratories, Holmdel, New Jersey 07733* (1)

S. K. SEARLES, *Naval Research Laboratory, Laser Physics Branch, Washington, D.C. 20375* (86)

C. V. SHANK, *Bell Telephone Laboratories, Holmdel, New Jersey 07733* (185)

Y. R. SHEN, *Physics Department, University of California, Berkeley, California 94720* (249)

J. J. WYNNE, *IBM/T. J. Watson Research Center, P.O. Box 218, Yorktown Heights, New York 10598* (210)

JAMES T. YARDLEY,* *Department of Chemistry, University of Illinois, Urbana, Illinois 61801* (269)

F. ZERNIKE, *North-American Philips, Briarcliff Manor, New York 10510* (143)

* Present address: Allied Chemical Corporation, Corporate Research Center, P. O. Box 1021R, Morristown, New Jersey 07960.

FOREWORD

Professor C. L. Tang has organized a splendid and broad-based volume on experimental methods in quantum electronics. Because the laser seems to have penetrated nearly every aspect of scientific research, this volume should prove useful to a wide range of researchers. Lasers of various types are discussed, and their principles and characteristics for the experimenter are explored. Each author is quite well known in the area from which he reports, and numerous valuable insights are presented in the succeeding chapters.

With the exception of that on classical methods, the other volumes of "Methods of Experimental Physics" are devoted primarily to a particular field of study. While this volume too is devoted to a particular field, it is also true that one may regard it as dealing with an experimental tool.

We extend our thanks to the volume editor and authors for their efforts.

L. MARTON
C. MARTON

PREFACE

Quantum electronics was the name adopted for the first conference on maser and related physics over twenty years ago. With the advent of the laser shortly afterward, the field has since experienced an explosive growth. Lasers and related devices are now used in many branches of science and technology and have made possible the development of numerous new experimental techniques in physics and chemistry.

There are many excellent review articles and books on numerous specialized topics of quantum electronics and laser physics. The present volume of the "Methods of Experimental Physics" series was prepared with the intent of providing a concise reference or guidebook for investigators and advanced students who wish to do research in the field of quantum electronics or to use laser-type devices and related techniques in their own specialties. As with many of the earlier volumes of this series, it quickly became obvious that to give adequate coverage to the main topics planned, a single volume would be too restrictive. This led to the present two parts of the volume: Volume 15A and Volume 15B. Even with two volumes, a choice still had to be made on whether to emphasize either lasers and related devices as sources of coherent radiation from the IR to the UV or the numerous of experimental techniques that make use of lasers. Although new lasers and related generators of coherent radiation are still being developed, this part of the field appears to have reached a certain degree of maturity. Several basic types of lasers and related devices have emerged and will likely remain as prototypes of sources of coherent radiation from the far IR to the near UV for some time. It is now timely to review and summarize the basic facts and principles of these sources. Therefore, the bulk of Volume 15 deals with these lasers and devices. However, new experimental techniques using lasers are still being rapidly developed. Therefore it is premature as well as extremely difficult to give a comprehensive review that will not become obsolete in a relatively short time. This of course is not to say that laser-related experimental techniques have not already produced important results in physics and chemistry; far from it. A number of such applications have been chosen here as examples of the power and unique potential of laser-related experimental techniques. The time will soon come when a more complete review will be needed.

It has been an honor and a pleasure for me to work with the authors of this volume.   Each is an authority and has contributed much to the original research in his specialty.   Part 1, however, is an exception.   The editor failed in his search for anyone willing to take time off from his busy research to collect and record some of the more basic material.   Time was running out and the task unhappily fell upon himself.   In addition, as is inevitable with edited works, there is a slight unevenness in the latest literature reviewed in the various articles.

Finally, it is a privilege indeed for me to have worked with Bill Marton, Editor-in-Chief of the "Methods of Experimental Physics" series.   I deeply regret that it had not been possible to publish this volume before his passing.   I can only hope that this volume has lived up to the high standards of the rest of the volumes of this series.

CONTENTS OF VOLUME 15, PART A

CONTRIBUTORS TO VOLUME 15, PART A

WILLIAM B. BRIDGES, *California Institute of Technology, Pasadena, California 91125*

HENRY KRESSEL, *RCA Laboratories, Princeton, New Jersey 08540*

OTIS GRANVILLE PETERSON, *Allied Chemical Corporation, Morristown, New Jersey 07960*

C. L. TANG, *School of Electrical Engineering and The Materials Science Center, Cornell University, Ithaca, New York 14853*

M. J. WEBER, *Lawrence Livermore Laboratory, University of California, Livermore, California 94550*

Methods of
Experimental Physics

VOLUME 15

QUANTUM ELECTRONICS

PART B

6. COLOR CENTER LASERS*

6.1. Introduction

Certain color centers in the alkali halides can be used to make efficient, optically pumped, broadly tunable lasers for the near infrared.[1-3] Although extended coverage may be possible in the future, the presently known laser-active center types allow operation over the wavelength range $0.8 \leqslant \lambda \leqslant 3.3$ μm. This region is of fundamental importance to molecular spectroscopy, chemical dynamics, pollution detection, fiber optic communications, and the physics of narrow-bandgap semiconductors. It is in terms of this special tuning range that the color center lasers have their greatest advantage: on a scale of increasing wavelength, they take over just where the organic dyes fail.

Like their dye-laser counterparts, the color center lasers can be operated cw as well as in a pulsed mode. In the cw mode, pump powers at threshold are quite modest—often as low as 20 mW.[2] In pulsed operation, outputs on the order of 10 kW have been obtained,[3] and it is reasonable to expect this figure can be scaled up by at least an order of magnitude. In neither mode has there been any sign of bleaching or aging effects, as long as the crystals were properly cooled.

Frequency definition of the color center lasers is excellent, with performance in this respect promising to equal or exceed that attained with the best dye lasers, in either cw or pulsed operation. In particular, there are no fundamental obstacles to the production of an oscillator–amplifier combination, capable of ~100 kW output pulses, continuously tunable over the range 3000–4000 cm^{-1}, whose frequency width is essentially Fourier-transform limited. Such frequency definition would represent many orders of magnitude improvement over the best performance attainable from the only other high-power tunable sources in the near infrared, the parametric oscillators. Furthermore, by using the stimulated Raman effect, the output of such a pulsed color center laser could be efficiently

[1] L. F. Mollenauer and D. H. Olson, *Appl. Phys. Lett.* **24,** 386 (1974).

[2] L. F. Mollenauer and D. H. Olson, *J. Appl. Phys.* **46,** 3109 (1975).

[3] G. C. Bjorklund, L. F. Mollenauer, and W. J. Tomlinson, *Appl. Phys. Lett.* **29,** 116 (1976).

* Part 6 is by Linn F. Mollenauer.

METHODS OF EXPERIMENTAL PHYSICS, VOL. 15B

translated down, in a single step, to just about any other frequency range of interest to molecular physics.[4] Thus, one would have a source at once unique in its capabilities, and attractive in terms of its relative simplicity and economy.

The usefulness of color centers to quantum electronics has been further enhanced through the recent discovery[5] of a highly efficient photochromic process for the production of the desired centers. Initiated by a two-photon absorption of a high-powered, UV laser beam, the process features a controllable and essentially unlimited penetration depth, and allows the spatial distribution of the created centers to be precisely varied on a microscopic scale. For example, by interfering the UV laser beam with itself, one can write thick gratings of any desired period greater than half the UV wavelength. Gratings of color centers written in that manner have already been used to make successful distributed-feedback lasers.[3] In a more general sense, the two-photon coloration process should allow the creation of a host of integrated optics devices.

It should be noted that laser action was demonstrated nearly 12 years ago in a flash-lamp-pumped rod containing $F_A(II)$ centers.[6] Until recently, however, the potential of color centers for useful laser action has been ignored, largely due to a lack of communication between workers in the field of color center physics and those in quantum electronics. The primary purpose of the work described above and in the following has been to explore and demonstrate that potential. In view of the interest shown by intended users, especially on the part of chemists, and in view of the fact that the technology of the longer-wavelength color center lasers is by now well developed, the near future should see rather extensive use of these. We hope this work may also help to rekindle interest in the basic physics of color centers, a field perhaps too early declared mature, in view of the many fundamental questions raised by the search for laser-active centers and for better ways to create them.

6.2. Some Pertinent Color Center Physics

6.2.1. General

To facilitate understanding among readers who are not experts on color centers, we summarize here the pertinent physics of those F-like centers

---

[4] See, for example, W. Schmidt and W. Appt, *Z. Naturforsch., Teil A* **27**, 1373 (1972). The use of H_2 is specifically discussed there, but other molecules, such as N_2 and O_2, should work as well.

[5] L. F. Mollenauer, G. C. Bjorklund, and W. J. Tomlinson, *Phys. Rev. Lett.* **35**, 1662 (1975).

[6] B. Fritz and E. Menke, *Solid State Commun.* **3**, 61 (1965).

that might be considered for laser action.[7] Both here and in Chapter 6.4, we show how some of these are suitable for laser action, while others are not.

The known laser-active centers (F_A and F_2^+) are shown in the right-hand column of Fig. 1, with their progenitors (F and F_2) to the left. All are based on a simple anion (halide ion) vacancy in an alkali halide crystal having the rock salt structure. The ordinary F center consists of a single electron trapped at such a vacancy, when the immediate surroundings are essentially unperturbed. On the other hand, if one of the six immediately surrounding metal ions is foreign, say a Li^+ in a potassium halide, one has an F_A center. Two F centers adjacent along a [110] axis constitute the F_2, and, of course, the F_2^+ is its singly ionized counterpart. The processes by which these centers may be created will be discussed in Chapter 6.3.

6.2.2. The F Center

Although not itself laser active, the ordinary F center is something of an archetype to the other varieties of center; as such, it should be discussed and understood first. The optical pumping cycle of the F center is shown in Fig. 2. It consists of four steps: excitation, relaxation, luminescence, and relaxation back to the normal configuration. The fundamental absorption arises from transitions from the s-like ground state to the p-like first excited state of the square-well formed by the vacancy. The oscil-

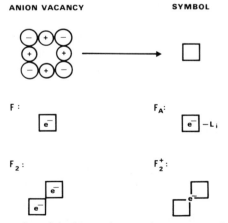

FIG. 1. Basic configuration of the known laser-active centers and their progenitors (see text).

[7] For a comprehensive review of color centers in alkali halides, see W. B. Fowler, *in* "Physics of Color Centers" (W. B. Fowler, ed.), Chapter 2. Academic Press, New York, 1968).

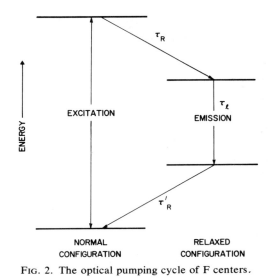

FIG. 2. The optical pumping cycle of F centers.

lator strength f associated with this transition is near unity. Relaxation consists of a simple expansion of the vacancy, and of a corresponding adjustment in the electronic wavefunction.

However, when the system reaches its relaxed excited state (RES) the associated electronic wavefunction becomes spatially very diffuse. As a result of the poor overlap between this wavefunction and that of the ground state of the relaxed system, the oscillator strength of the luminescence band is quite small ($f \sim 0.01$). Also, the RES is energetically quite shallow, and hence there exists the possibility of self-absorption into the conduction band of the emitted photons. As we shall detail in Chapter 6.4, these two properties taken together make rather unlikely the attainment of a net optical gain.

6.2.3. F_A(II) Centers

The F_A centers[8] can be classified according to their relaxation behavior. Those of type I retain the single vacancy, and their behavior in this respect is almost indistinguishable from that just described for the ordinary F center. Hence the type-I F_A centers are also not suitable for laser action. However, in the other (and rarer) class are the F_A centers of type II. These relax to a double-well configuration following optical excita-

---

[8] For a more extensive treatment of F_A centers, see F. Lüty, *in* "Physics of Color Centers" (W. B. Fowler, ed.), Chapter 3. Academic Press, New York, 1968.

tion. We shall show theoretically, and our experiments have verified, that the type-II F_A centers are eminently suited for tunable laser action.

The vacancy and double-well configurations of the lattice for $F_A(II)$ centers are shown in Fig. 3; also shown there are the associated energy levels. Note that the relaxed $F_A(II)$ center is a radically different system from its unrelaxed counterpart. The relaxed system is somewhat analogous to the H_2^+ molecular ion (in this case with an additional negative charge between the two attractive centers). The wavefunctions for the excited and ground states of the relaxed center are made up of anitsymmetric and symmetric combinations, respectively, of a single-well s-state. The resultant oscillator strength for an electric dipole transition is quite large ($f \sim 0.2$–0.35).

Two other important features of $F_A(II)$ behavior are implicit in Fig. 3. First (in the normal configuration), the presence of the foreign ion causes p_z orbitals to be distinguished from p_x and p_y, such that the absorption exhibits a characteristic splitting. That is, transitions to the p_z orbitals re-

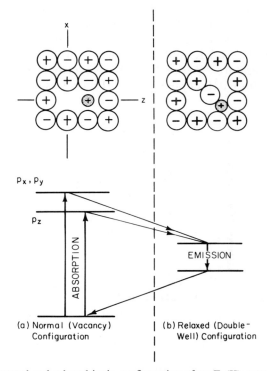

FIG. 3. Top: normal and relaxed ionic configuration of an $F_A(II)$ center. Bottom: associated energy levels of the center.

sult in a longer-wavelength band that is often well resolved from the main band. As far as laser applications are concerned, this extra band greatly increases the probability of overlap with a convenient pump source.

It should also be noted from Fig. 3 that there is a 50% probability that the halide ion separating the two wells will move into the original F_A center vacancy upon completion of the optical pumping cycle. When such occurs, the F_A center axis will have switched orientation by 90°. Such reorientation effects must be taken into account in the design of $F_A(II)$ lasers. Pumping of the longer-wavelength band with light propagating along a [100] axis bleaches that band for the pump light itself. A similar effect occurs when pumping the short-wavelength band with light polarized along a [100] axis. A practical configuration for avoiding these effects will be discussed in Chapter 6.6.

The luminescence bands associated with most of the presently known type-II centers are shown in Fig. 4.[8,9] Our experiments have so far involved only the F_A centers in KCl:Li and RbCl:Li; and satisfactory laser action has been obtained with both. (For further details, see Chapter 6.6). Taken together, these two hosts allow continuous coverage of the range $2.5 \leqslant \lambda \leqslant 3.3$ μm ($3000 \leqslant \tilde{\nu} \leqslant 4000$ cm^{-1}).

The quantum efficiency η of $F_A(II)$ center luminescence in KCl:Li is about 50% for $T = 1.6$ K and decreases approximately linearly with increasing temperature[10,11] (see Fig. 5). Nevertheless, laser action has been obtained with $F_A(II)$ centers in KCl:Li for T as large as 200 K. The

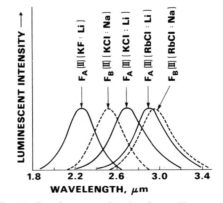

FIG. 4. Luminescence bands of type-II centers.

[9] L. F. Mollenauer, B. Hatch, D. H. Olson, and H. Guggenheim, *Phys. Rev. B* **12**, 731 (1975).

[10] G. Gramm, *Phys. Lett.* **8**, 157 (1964).

[11] E. Hammonds and L. F. Mollenauer, unpublished.

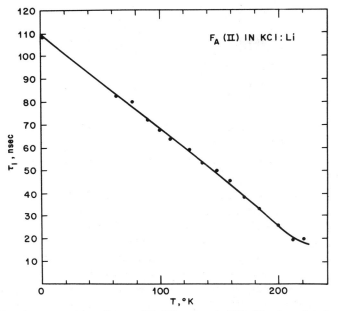

FIG. 5. Luminescence decay time τ_l of $F_A(II)$ centers in KCl: Li. Associated quantum efficiency is about 50% at $T = 0$ (see text).

mechanisms for the nonradiative decay are not yet completely understood. Behavior of type-II centers in other hosts ought to be similar, although $\eta(T)$ has not yet been measured for these. The only exceptions are the $F_A(II)$ centers in the hosts KF:Li and KF:Na,[9] for which the measured efficiencies are quite low, even at $T = 1.6$ K.

6.2.4. $F_B(II)$ Centers

The $F_B(II)$ centers referred to in Fig. 4 are quite similar to the $F_A(II)$, except that they involve two foreign metal ions, instead of the single foreign ion of the F_A center.[12] F_B centers are obtained in substantial quantities when the dopant concentration represents at least several percent of all metal ions. The ionic configuration of the $F_B(II)$ center is shown in Fig. 6. (A type-I F_B center is formed when the foreign ions lie along a common [100] axis.)

As predicted earlier,[2] cw laser action with $F_B(II)$ centers has indeed been obtained recently.[12a] The particular hosts were KCl:Na and

[12] N. Nishimaki, Y. Matsusaka, and Y. Doi, *J. Phys. Soc. Jpn.* **33**, 424 (1972).
[12a] G. Litfin, R. Beigang, and H. Welling, *Appl. Phys. Lett.* **31**, 381 (1977).

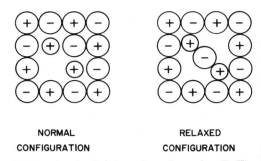

NORMAL   RELAXED
CONFIGURATION   CONFIGURATION

FIG. 6. Normal and relaxed configurations of an $F_B(II)$ center.

RbCl:Na, with tuning ranges of 2.2–2.65 and 2.5–2.9 μm, respectively. Pump sources were an argon ion laser for the KCl:Na (477–530 nm) and a krypton-ion laser (647–676 nm) for the RbCl:Na.

One of the potential difficulties with the $F_B(II)$ centers is that they are always accompanied by a certain population of $F_A(I)$ and $F_B(I)$ centers. Furthermore, the absorption bands of all three center types tend to overlap rather strongly.[12] In an earlier work,[2] we had suggested that this problem could be circumvented, at least in part, by taking advantage of the orientational bleaching effects[8,12] possible with $F_A(I)$ and $F_B(I)$ centers. In particular, we had suggested aligning the pump polarization vector in a [100] plane. Litfin et al.[12a] have done just that, and claim that the scheme works well. However, we wonder if the relatively low slope efficiencies they observed (~2%) might not be due in part to a residue of unbleached $F_A(I)$ and $F_B(I)$ centers. That is, the orientational bleaching effects are never 100% complete, due to imperfect resolution of the pump bands, and optically pumped type-I centers are known to produce an absorption in the region of the type-II luminescence bands (see Park and Faust,[12b] sect. 4.1).

At one time we had hoped that it would be possible to create $F_B(II)$ centers in a heavier alkali-halide lattice, in order to push the cw laser tuning range well into the $\lambda > 3$ μm region. (A band centered about $\lambda \sim$ 3.3 μm would be especially useful to molecular spectroscopy, since such wavelengths correspond to the fundamental bond-stretch frequency of most C–H bonds.) However, recent experiments in this laboratory with KBr:Na, RbBr:Na, and RbBr:K have so far shown no indication of type-II luminescence bands. It is especially puzzling that the host RbBr:Na apparently does not allow for a type-II center, since there is no problem with relative ion sizes, and since it would seem to be so closely

[12b] K. Park and W. L. Faust, *Phys. Rev. Lett.* **17**, 137 (1966).

analogous to the host KCl:Na. Clearly, a careful and thorough examination of the question of type-II vs. type-I formation needs to be carried out, both experimentally and theoretically. A simple but promising theoretical approach has been taken recently by Kung and Vail.[12c]

6.2.5. F_2^+ Centers

The F_2^+ center consists of a single electron shared by two vacancies, as already indicated in Fig. 1. The arrangement shown there applies to both emission and absorption. For the F_2^+, relaxation consists merely in a slight increase in separation of the two vacancies. The corresponding Stokes shifts are then quite small, usually just enough to prevent significant overlap of the absorption and emission bands. This behavior contrasts sharply with that of the $F_A(II)$ and $F_B(II)$ centers, where as already indicated, relaxation involves both a radical change in ionic configuration and a Stokes shift ratio of about 5 to 1 between absorption and emission energies.

On the other hand, in their relaxed configurations, the F_2^+ and type-II centers are quite similar. The model of an H_2^+ molecular ion embedded in a dielectric continuum seems to fit the F_2^+ particularly well. Good quantitative fit has been obtained between that model and the observed spectrum in quite a number of hosts.[13] Figure 7 shows the F_2^+ energy level diagram. The levels there are named after their molecular ion counterparts.

There are two strong transitions: the $1S\sigma_g \rightarrow 2P\sigma_u$ in the infrared, and the $1S\sigma_g \rightarrow 2P\pi_u$ in the visible. The emission of the visible transition is strongly quenched at all but very low temperatures ($T \leqslant 50$ K) by a nonradiative transition to the $2P\sigma_u$ level.[13] For this and other reasons[14] the visible transition is not suitable for laser action, at least not in any practical way.

However, very practical laser action can indeed be obtained with the infrared transition, as has recently been demonstrated experimentally in this laboratory.[15] With the host crystal NaF, the laser was continuously tunable over the range $0.885 \leqslant \lambda \leqslant 1.00$ μm, the pump source was a

---

[12c] A. Y. S. Kung and J. M. Vail, *Phys. Status Solidi* B **79**, 663 (1977).

[13] M. A. Aegerter and F. Lüty, *Phys. Status Solidi* **43**, 244 (1971).

[14] Severe overlap of the higher energy pump band with that of the F centers could lead to the production of F' centers, through a well-known tunneling effect. (The F' is a vacancy containing two electrons. Also, it is essentially impossible to have F_2^+ centers without a substantial accompanying population of F centers.) Unfortunately, the F' centers would absorb strongly photons of the higher emission band. For example, see R. S. Crandell and M. Mikkor, *Phys. Rev.* **138**, A1247 (1965).

[15] L. F. Mollenauer, *Opt. Lett.* **1**, 164 (1977).

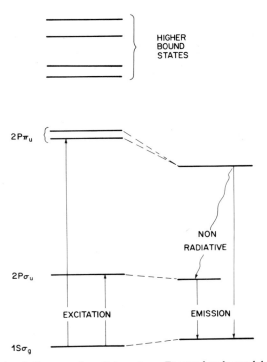

FIG. 7. Energy level diagram of an F_2^+ center. Energy levels are labeled according to their counterparts on the H_2^+ molecular ion model (see text).

krypton-ion laser operating at 0.7525 μm, and threshold pump power near band center was ~40 mW into the crystal. We have also observed cw laser action in the host KCl, in a band near 1.69 μm, pumped with a Nd:YAG laser operating at 1.34 μm. In both cases, there were no bleaching or fading effects even after many hours of operation at 79 K.[15a]

The emission and absorption bands of the infrared transition in NaF are shown in Fig. 8. The reciprocal of the pump powers required at threshold are shown there also, as a function of laser wavelength, for operation at 79 K. The relatively poor performance on the short-wavelength side of the band was due to a rapid fall-off of mirror reflectivities for $\lambda \gtrsim 0.9$ μm. Nevertheless, it was possible to tune the laser over a good fraction of the luminescence band with only a very modest maximum available pump power (~150 mW). (The laser cavity was essentially the same as that described in Section 6.6.1 for our experiments involving F_A(II) centers except, of course, for an appropriate change in mirror coatings, and for the substitution of a quartz prism as the tuning element.)

[15a] However, see Section 6.8.3.

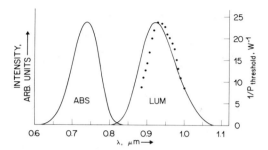
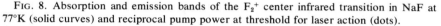

FIG. 8. Absorption and emission bands of the F_2^+ center infrared transition in NaF at 77°K (solid curves) and reciprocal pump power at threshold for laser action (dots).

Even before that direct demonstration, there was strong evidence that the infrared transition possessed nearly ideal properties for laser action. First, from the small Stokes shift, it could reasonably be assumed that the emission had essentially the same high oscillator strength ($f \sim 0.2$) as the absorption. Second, there was strong evidence for a 100% quantum efficiency, even though that quantity had not been measured absolutely. That is, the relative quantum efficiency was known to be temperature independent,[13] and the measured luminescence decay times[15] (~ 31 and ~ 80 nsec in NaF and KCl, respectively) correspond exactly to the calculated radiative decay times. Third, from the calculated energies of the H_2^+ molecular ion,[16] the splitting between the lowest-lying even-parity excited state ($2S\sigma_g$) and $2P\sigma_u$, could be predicted as being much too large to allow for self-absorption at the emission energy. Based on the earlier quantitative successes of the molecular-ion model, this prediction could be made with a rather high degree of confidence. Finally, there was no serious threat of absorption from a species foreign to the F_2^+. For example, in KCl, the only such interfering species known is a peculiar variety of the F_3^+ center,[17] whose density can be held to negligible levels through proper handling of the crystal.

It should be noted that the light-to-light conversion efficiencies of lasers using F_2^+ centers is potentially very high. Slope efficiencies of $\sim 80\%$ are theoretically possible, due to the small Stokes shift, and due to the fact that for pumping with linearly polarized light, all excited centers can radiate efficiently into a laser mode polarized in the same plane. This is in contrast to the situation obtained with $F_A(II)$ centers, where maximum theoretical slope efficiencies are of the order of 10% (see Section 6.6.6).

Last, but by no means least, in the list of advantages of the F_2^+ is its

[16] D. R. Bates, K. Ledsham, and A. L. Stewart, *Philos. Trans. R. Soc. London, Ser. A* **246,** 215 (1953).
[17] I. Schneider, *Solid State Commun.* **4,** 91 (1966).

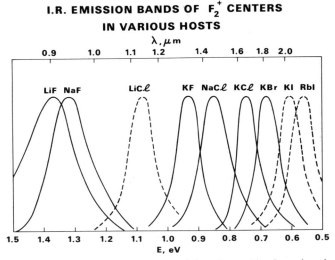

FIG. 9. Luminescence bands of the F_2^+ center infrared transition in various hosts. Solid curves are measured experimentally, dashed are predicted from H_2^+ molecular ion model (see text).

near universality: in principle, at least, the F_2^+ can be created in any alkali halide host. Figure 9 shows the infrared luminescence bands in a number of hosts[13,18-22]; it is evident from these that the F_2^+ should allow for continuous coverage of a large and important tuning range.

6.2.6. Systems with a Pair of Spins

From time to time, it has been suggested that various two-electron centers (such as the F_2 or F_3^+) or an electron–hole system (the self-trapped exciton) might be useful for tunable lasers. Unfortunately, laser action with such systems having a pair of spins is often precluded by absorption from the triplet state. Furthermore, the particular systems mentioned above present another serious problem: as recent experiments have indicated (see Section 6.8.1), intense pumping of the fundamental transition ionizes the F_2 and F_3^+ centers through a two-step process. Thus, these centers are readily bleached. This bleaching probably explains the lack of gain in early experiments with the F_2 center in KCl.

[18] I. Schneider and M. N. Kabler, *Phys. Lett.* **17**, 213 (1965).

[19] I. Schneider and C. E. Baily, *Solid State Commun.* **7**, 657 (1969).

[20] J. Nahum, *Phys. Rev.* **174**, 1000 (1968).

[21] J. Nahum and D. A. Wiegand, *Phys. Rev.* **154**, 817 (1967).

[22] K. Konrad and T. J. Newbert, *J. Chem. Phys.* **47**, 4946 (1967).

Also, recently several groups have reported[22a] laser action with F_2 center in LiF, but in all cases the laser action could be sustained for only a few pulses. Once again, this fading of the laser action is undoubtedly due to the two-step photoionization. The prognosis for sustained pulsed or cw laser action with F_2 centers is not very good.

The self-trapped exciton produces two luminescence bands, one typically in the hard UV, due to decay of singlet states, and the other at lower energies, due to decay of triplets.[23] Of course, since the ground state is of spin 0, only the singlet band is of sufficient oscillator strength to allow for significant gain cross section. Even then, the singlet emission is strongly quenched at all but very low temperatures,[24,25] i.e., the temperature is required to be well below 77 K. Furthermore, when self-trapped excitons are created, the formation of the triplet state is favored, and the triplet system exhibits a strong absorption that overlaps the singlet emission band.[26] Although it has been shown that, at least in some cases, the triplets can be converted back to singlets with the use of an intense infrared laser pulse,[27] a device simultaneously requiring electron beam and laser pumps, as well as liquid He for cooling, would hardly be of much practical interest.

6.3. Processes for Color Center Formation

6.3.1. Apologia

The creation of precisely controlled densities of the desired centers in crystals of high optical quality is a vital part of the color center laser art. Unfortunately, much of the lore pertinent to the generation of color centers lies scattered throughout the literature of over four decades, and to the best of our knowledge, has never been brought together in one

[22a] Yu, L. Gusev, S. N. Konoplin, and S. I. Marennikov, *Sov. J. Quantum Electron.* (*Engl. Transl.* **7**, 1157 (1977); R. W. Boyd and K. Teegarten, *IEEE J. Quantum Electron.*, *1978* p. 697 (1978); G. Litfin, R. Beigang, and H. Welling, private communication.

[23] M. N. Kabler and D. A. Patterson, *Phys. Rev. Lett.* **19**, 652 (1967).

[24] M. Ikezawa and T. Kojima, *J. Phys. Soc. Jpn.* **27**, 1551 (1969).

[25] For this reason, the supposedly line-narrowed emission in KBr at 77°K attributed to the self-trapped exciton [E. L. Fink, *Appl. Phys. Lett.* **7**, 103 (1965)] must in fact have some other explanation, such as an impurity effect. Furthermore, the emission wavelength listed there is not that of the self-trapped exciton singlet band (see Kabler and Patterson[23] and Ikezawa and Kojima[24]).

[26] R. T. Williams and M. N. Kabler, *Phys. Rev.* **9**, 1897 (1974).

[27] R. T. Williams, private communication.

place. Furthermore, the known techniques, although straightforward in principle, involve a number of subtleties. Hence, the cookbook approach is no substitute for a genuine understanding of the underlying physics. Finally, the demands of the laser art have stimulated the creation of some new and elegant variations on the older processes. Therefore, in the following, we summarize the "well-known" methods and associated physics, as well as present the new techniques.

6.3.2. Additive Coloration

To make any of the color centers discussed in the previous section, ordinary F centers are created first, usually through the process of additive coloration, or by subjecting the crystal to radiation damage. In general, additive coloration is preferred for the stability of its end product, but the radiation damage process has proven vital for efficient F_2^+ center production.

A crystal containing F centers is chemically equivalent to a perfect crystal plus a stoichiometric excess of the alkali metal; hence, additive coloration involves bringing the crystal into equilibrium with a bath of the alkali vapor. The equilibrium density N_0 of F centers is determined by a simple solubility equation,

$$N_0 = \alpha N', \tag{6.3.1}$$

where N' is the metal vapor density, and the solubility constant α is only weakly temperature dependent.[28] For KCl colored in K vapor at 600°C, $\alpha \cong 2.3$.

However, the approach to equilibrium is determined by the diffusion of F centers. For a thin slab, the concentration builds up as

$N(x, t) =$

$$N_0 \left\{ 1 - \frac{4}{\pi} \sum_{m=0}^{\infty} \frac{1}{(2m + 1)} \left[\sin \frac{(2m + 1)\pi x}{l} \right] \exp \left[- \frac{D(2m + 1)^2 \pi t}{l^2} \right] \right\},$$

$$\tag{6.3.2}$$

where x is the distance in from one surface of the slab, l is its thickness, and D is the diffusion constant. After a while, only the leading term in the sum is important, and the solution reduces to

$$N(x, t) \cong N_0 \left\{ 1 - \frac{4}{\pi} \left(\sin \frac{\pi x}{l} \right) \exp \left[- \frac{D\pi^2 t}{l^2} \right] \right\}, \tag{6.3.3}$$

[28] H. Rögener, Ann. Phys. (Leipzig) [5] 29, 386 (1937).

from which we can extract a characteristic diffusion time,

$$\tau_D = l^2/\pi^2 D. \qquad (6.3.4)$$

The diffusion constant for F centers involves a large activation energy, and is thought to vary as[29]

$$D(T) = AT^{2/3} \exp(-E_0/KT). \qquad (6.3.5)$$

However, since E_0 is typically on the order of 1 eV, the empirically determined behavior of $D(T)$ can be fit equally well, over the range of interest, to the following:

$$D(T) = D_0 \exp(-T_0/T). \qquad (6.3.6)$$

For example, in KCl, $D_0 = 1.22 \times 10^2 \mathrm{cm^2/sec}$ and $T_0 \cong 14,430$ K. Unfortunately, $D(T)$ has been measured in only a very few substances.[30]

The temperature range used for additive coloration is bounded on the high side by the melting point of the crystal, and on the low side by a temperature at which the formation of colloids is favored. For example, in KCl, the melting point is 768°C, and the colloids begin to form for $T \leqslant$ 400°C.[31] In practice, one tends to use the lowest temperature for which the coloration time is conveniently short. For example, for a slab of KCl 2 mm thick, $D = 8 \times 10^{-6} \mathrm{cm^2/sec}$ at 600°C, and $\tau_D = 8.4$ min. In this case, an exposure to the vapor of 30 min or more duration should result in a coloration sufficiently uniform for most purposes.

An apparatus[32] for additive coloration, allowing for precise control of the F center density and for very convenient loading and unloading of crystals, is shown in Fig. 10. It is based on the principle of the heat pipe. Liquid metal is confined entirely to the wick (made of fine stainless steel mesh). The dividing line between the region of pure metal vapor and the N_2 buffer gas occurs at that level on the wick where the temperature is at the dew point. Obviously, such an arrangement allows for very precise control of the metal vapor pressure, through control of that of the buffer gas, since pressure equilibrium must exist, and since the separation of the gases is complete everywhere save perhaps for a very narrow region at the border between them.

The operating cycle is as follows: With the ball valve closed, the crystal container is loaded into the air lock space and the air there replaced with N_2 of the correct pressure. The valve is then opened and the crystal low-

[29] N. F. Mott and R. W. Gurney, "Electronic Processes in Ionic Crystals," 2nd ed., p. 143. Dover, New York, 1940.
[30] O. Strsiw, *Z. Tech. Phys.* **11**, 343 (1935).
[31] A. B. Scott and W. A. Smith, *Phys. Rev.* **83**, 982 (1951).
[32] L. F. Mollenauer, *Rev. Sci. Instrum.* **49**, 809 (1978).

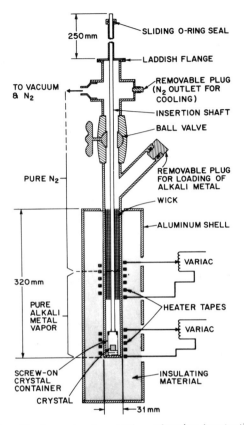

FIG. 10. Apparatus for additive coloration (see text).

ered into the coloration zone. A set of baffles in the crystal container prevents condensation of vapor onto the crystal while the container temperature rises through the dew point. For removal of the sample, this procedure is simply reversed. While it is still in the air lock space, the sample is cooled rapidly to room temperature by a stream of N_2 gas. Such handling has made it possible to color an optically polished sample with little or no loss of surface quality. This has obviously facilitated the preparation of laser-quality crystals.

6.3.3. F Center Aggregation

The formation of F_A, F_B, or F_2 centers results from a simple aggregation process than can be described[33] as follows: First, thermal ionization

[33] H. Härtel and F. Lüty, Z. Phys. 177, 369 (1964).

of optically excited F centers results in the formation of pairs of F' centers and empty vacancies. (The F' center consists of a pair of electrons trapped at a vacancy.) At sufficiently high temperatures ($T \geqslant -50°C$), the empty vacancies wander through the lattice until they meet either an F center or a foreign metal ion. Recapture of an electron (from optically ionized F' centers) by the vacancy then leads to formation of F_2 centers in the first instance, or to formation of F_A (or F_B) centers in the second (see Fig. 11).

If the foreign metal ion concentration is several orders of magnitude greater than that of the F centers, an essentially complete conversion can be carried out, with F_A centers as the exclusive end product. However, the creation of F_2 centers cannot be carried to completion without an accompanying creation of higher aggregates, such as F_3 or F_4. Thus, the optimum conversion to F_2 will necessarily involve a finite residue of F centers.

6.3.4. The U Center

The U center is an H^- ion trapped at an anion vacancy. U centers absorb only in the hard UV and are very stable. They are important to the laser art in at least two ways: They form the basis of the two-photon coloration process mentioned in the introduction, and they allow for the efficient formation of electron traps useful in the creation of F_2^+ centers.

U centers are formed by baking a crystal already containing F centers in an atmosphere of H_2:

$$F + \tfrac{1}{2}H_2 \xrightarrow{\;550°C,\;100\ atm\;} U \qquad\qquad (6.3.7)$$

During conversion, the H_2 concentration varies linearly from a value at the crystal surface determined by the external H_2 density N_{H_2} and the sol-

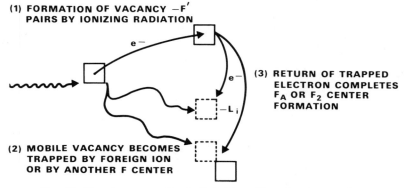

(1) FORMATION OF VACANCY $-$F'
PAIRS BY IONIZING RADIATION

$e-$

(3) RETURN OF TRAPPED
ELECTRON COMPLETES
F_A OR F_2 CENTER
FORMATION

$e-$

$-L_i$

(2) MOBILE VACANCY BECOMES
TRAPPED BY FOREIGN ION
OR BY ANOTHER F CENTER

FIG. 11. F center aggregation to form F_A or F_2 centers (see text).

ubility constant α_{H_2} to zero at the boundary between converted and un-converted regions. Thus the concentration gradient is inversely propor-tional to the thickness x of the converted region. It can easily be shown[34] that x increases with time as

$$x = (2D_{H_2}\alpha_{H_2}N_{H_2}t/N_F)^{1/2}, \qquad (6.3.8)$$

where N_F is the F center concentration in the unconverted region. A spa-tially uniform distribution of U centers can be obtained only to the extent that the rate of F center diffusion is negligible compared to the rate of H_2 diffusion. For this reason, the hydrogen pressure is made as large as practicably possible. It is also helpful in this regard that D_{H_2} is orders of magnitude greater than D_F, and its temperature dependence much less ($D_{H_2} \sim 4 \times 10^{-4} \text{cm}^2/\text{sec}$ in KBr at 550°C). Measured values of α_{H_2} are $\sim 2 \times 10^{-4}$ in KCl[35] and $\sim 4 \times 10^{-5}$ in KBr[34] at 550°C.

The usefulness of the reaction of Eq. (7) is based on the fact that it can be reversed at lower temperatures through excitation of the U center, either by pumping the UV absorption band, or with X rays, or with two-photon excitation. For sufficiently low temperatures (in KCl, $T \leqslant$ 200°K), the U centers are converted into pairs of empty vacancies and in-terstitial H^- ions.[36] At higher temperatures, the reaction goes all the way to the formation of F centers and H_2.[37]

6.3.5. Dynamics of the Two-Photon Coloration Process

The two-photon coloration[5] is based on the U to F conversion process described in Section 6.3.4. If a high-intensity beam of photon energy

$$h\nu \geqslant \tfrac{1}{2}Eg \qquad (6.3.9)$$

is incident on the crystal, where Eg is the alkali halide bandgap, the resul-tant two-photon absorption produces electron–hole pairs. These may recombine at a U center, at an F center, or at a halide ion. In the first in-stance, the desired U to F conversion is obtained with 100% quantum effi-ciency. The reverse process results, with efficiency η, from recombina-tion at an F center. Thus, the time averaged behavior of the coloration process is governed by the equation

$$\frac{dN_F}{dt} = \frac{\sigma_U N_U - \eta\sigma_F N_F}{\sigma_U N_U + \sigma_F N_F + \sigma_L N_L}, \qquad (6.3.10)$$

where N is the number of electron–hole pairs generated per cubic centi-meter per second; N_F, N_U, and N_L are the F center, U center, and

[34] R. Hilsch, *Ann. Phys. (Leipzig)* [5] **29**, 407 (1937).
[35] L. F. Mollenauer, unpublished.
[36] B. Fritz, *J. Phys. Chem. Solids* **23**, 375 (1961).
[37] M. Ueta, M. Hirai, and H. Watanabe, *J. Phys. Soc. Jpn.* **15**, 593 (1960).

halide-ion densities, respectively; and σ_U, σ_F, and σ_L represent the relative rate coefficients for electron–hole annihilation at U centers, F centers, and halide ions, respectively. Equation (6.3.10) can also be used to describe the U to F conversion by X rays, where a shower of electron–hole pairs is produced for each absorbed photon.

It is evident from Eq. (6.3.10) that the two-photon coloration is efficient as long as N_U is large with respect to the quantities βN_F and γ, where $\beta \equiv \sigma_F/\sigma_U$, and $\gamma \equiv \sigma_L N_L/\sigma_U$. In KCl, $\beta \sim 8$ and $\gamma \sim 4 \times 10^{16} \mathrm{cm}^{-3}$ (see Mollenauer *et al.*[5]).

Strictly speaking, Eq. (6.3.10) is not quite complete. In the first place, it neglects the creation of F centers from decay of self-trapped excitons.[38] However, the F centers created in this way are largely unstable, and, in any event, the term to be added would be important only in the limit of very low U center concentration. Second, to the right-hand side of Eq. (6.3.10) should be added a term $- \sigma_F' \eta N_F I/\hbar\omega$, where σ_F' is the single-photon absorption cross section and I the mean intensity. However, when the pulse intensities are high enough ($I_{\mathrm{pulse}} \gtrsim 10 \ \mathrm{MW/cm^2}$), the rate for the two-photon process is large enough that the single-photon process can be neglected.[5]

There is one more single-photon effect that is of importance when the coloration is carried out at temperatures near 300 K. Photoionization of the F centers by the UV beam initiates the aggregation process described in Section 6.3.3. In a crystal heavily doped with foreign ions, and where F_A centers are desired, the aggregation is an additional benefit of the two-photon coloration process.

Figure 12 shows a typical experimentally determined curve[5] of coloration versus two-photon exposure. It corresponds closely to the solution of Eq. (6.3.10) as modified by the single-photon term. Following an initial steep rise whose slope is essentially independent of N_U for $N_U \gtrsim 1 \times 10^{17} \mathrm{cm}^{-3}$, the curve bends over and finally saturates at an optical density determined by the initial value of N_U and by the mean intensity I of the UV pulses.

To complete characterization of the two-photon coloration process, the corresponding absorption cross section is needed. A careful measurement[5] of the initial slope of the coloration curve for a crystal containing a large density of U centers ($N_U > 1 \times 10^{18} \mathrm{cm}^{-3}$), when combined with measurements of the UV beam, yielded the value

$$\sigma_2 \cong 7 \times 10^{-50} \quad (\mathrm{cm^4 sec/photon})/\mathrm{KCl \ molecule}$$

for the two-photon absorption cross section at $\lambda = 266$ nm (the fourth harmonic of a Nd:YAG laser). Or in terms that are more convenient for

[38] J. N. Bradford, R. T. Williams, and W. L. Faust, *Phys. Rev. Lett.* **35**, 300 (1975).

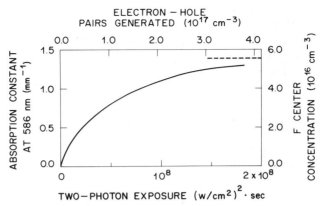

FIG. 12. Induced absorption at 586 nm, as a function of two-photon exposure, for a 0.63-mm-thick Li-doped KCl crystal initially containing 1.3×10^{17} U centers/cm³. The dashed line represents the best-fitting saturation coloration. Approximately 10^4 laser pulses were required to produce this curve.

experimental work, the corresponding induced absorption has the coefficient

$$\alpha_2 \cong 1.5 \times 10^{-3} \quad \text{cm}^{-1}/(\text{MW}/\text{cm}^2).$$

Thus it can be seen that even for intensities in the range of tens of MW/cm², samples of millimeter or centimeter dimensions will be optically thin.

6.3.6. F_2^+ Formation

There are several known processes for the production of F_2^+ centers. By far the most practical and satisfactory was the technique we used for the NaF laser (see Section 6.2.5), for which a brief description follows: shaped, annealed, and polished crystals ($1.9 \times 10 \times 10$ mm) were sealed in a single layer of Al foil ($\sim 12 \ \mu$m thick) in a dry N₂ atmosphere, and exposed to a 1 MeV, 1.6 μA/cm² e⁻ beam for 15 min on each 10×10 mm side while cooled to T $\sim$ 170–200 K by a stream of dry N₂ gas. This process creates high densities of both empty vacancies and F centers. The process of F_2^+ formation[20] is then completed upon warming the crystals to room temperature, where the vacancies become mobile and attach themselves to F centers. In practice, the packets were submerged in liquid N₂ immediately following e⁻ beam exposure and held there until needed. Warming to room temperature was carried out in a dry box and under a safe light ($\lambda \geqslant 0.55 \ \mu$m). The process of loading the crystals into the laser dewar provided the necessary time ($\sim$5–10 min) for the above-mentioned aggregation. Since the F_2^+ centers in NaF have a finite lifetime at room temperature (half-life ~ 6 hr), the crystals were cooled to operating temperature soon after the loading process was complete. Measured absorption coefficient at the peak of the absorption band was ~ 24 cm^{-1}, more than an order of magnitude greater than the values that had been obtained previously with X rays.[20,22]

When F_2^+ centers are formed as above, the excess charge is trapped as F' centers (vacancies containing 2e$^-$) or possibly also at various impurity or defect sites. (The safelight was employed to avoid discharging of the F' centers.[22]) The finite lifetime of the F_2^+ centers at room temperature is due to deionization through slow release of this trapped charge, rather than to any intrinsic instability of the center proper. In NaF, for temperatures below about $-40°C$, the charge leakage becomes negligible, making the center lifetimes indefinitely long. (If more stable traps can be found, as by the inclusion of certain impurities, the centers may become stable and usable as laser media at room temperature.)

The process we used to create F_2^+ centers in KCl[39] was much more complex, and gave less satisfactory results. We do not recommend it for laser work, and include the following description mostly for the record. In a crystal containing a high density of U centers ($N_U \sim 1 \times 10^{18}$cm^{-3}), a fraction of these are converted to F centers (by X-raying the crystal; see Section 6.3.5); alternatively, a second additive coloration can be performed. (The ratio of U to F center diffusion is sufficiently small that the U center distribution will be essentially unaffected if the additive coloration time is carefully chosen.) Next, the F centers are converted to F_2 centers by way of the aggregation process indicated in Section 6.3.3. Then the crystal is cooled to $T = 77$ K or lower, and the U centers are excited with hard UV, thereby creating a high density of electron traps (anion vacancies) (see Section 6.3.5). Finally, the F_2 centers are ionized to F_2^+ by using a wavelength in the near UV ($\lambda \sim 4000$ Å for KCl), chosen to maximize the ratio of F_2 to F center ionization rates.[40]

The aggregation is probably the most critical step in the above. To obtain a sufficiently uniform product, the wavelength of the exciting light should lie at one extreme edge of the F band, preferably on the long-wavelength side. (Light corresponding to the short-wavelength edge of the band tends to partially destroy the produced F_2 centers.[41]) Also, it should be noted that the ratio of F_2 to higher aggregates seems to be enhanced by use of the lowest temperature for which the aggregation is not

[39] M. A. Aegerter and F. Lüty, *Int. Conf. Color Cent. Ionic Cryst., 1971* Abstract 47 (1971).

[40] M. A. Aegerter and F. Lüty, *Phys. Status Solidi* **43**, 227 (1971).

[41] C. Z. van Doorn, *Philips Res. Rep., Suppl.* **4** (1962).

too slow (in KCl, $T \sim -10$ or $-20°C$). If the aggregation is halted at the optimum, over two-thirds of the initial F center population can be converted into F_2 centers, before the production of F_3 centers becomes significant. Obviously, it is very helpful in the achievement of this optimum to be able to monitor the pertinent absorption bands during the aggregation process itself.

Optical excitation of the U band also presents problems. At the densities required, the absorption at the peak of the U band is very high (typically ~ 300 cm^{-1}), such that the UV beam must "eat" its way through the crystal. The difficulty here is that one of the end products, the H$^-$ interstitial ion, also absorbs in the region of the U band. As a consequence, the penetration of the UV beam slows down rapidly. (Absorption by the impurity OH, often present in alkali halide crystals, also compounds the effect.[42]) Furthermore, prolonged exposure of the already converted region to the UV beam tends to destroy the desired centers. Hence, the process is truly efficient only for relatively thin layers, and one must take account of this in the design of the laser crystal geometry: The laser beam path must lie near and parallel to the edge through which the conversion light passes.

The principal disadvantage of the above technique lies in the fact that the density of F_2^+ centers produced is marginal for use in a laser cavity with a highly focused mode. In such a case, one typically needs an absorption coefficient at the pump wavelength on the order of 30 cm^{-1} (to produce an o.d. ~ 1 in a crystal 1 or 2 mm thick). So far, under very carefully controlled conditions, the best results have been several times below that value. Thus, there is no margin for error. Another severe disadvantage of the method lies in the fact that, if the crystal is allowed to warm up to room temperature, the traps are destroyed in seconds.[36] Hence it is necessary to create them *in situ*, thereby greatly increasing the complexity and awkwardness of the laser cavity mount.

6.4. Optical Gain

In this section, we shall develop some useful formulas for the optical gain, or amplification that can be attained in a four-level system. Not only will these formulas prove useful in the description of working lasers, but also they will allow us to conclude the general discussion, begun in Chapter 6.2, of the suitability of various center types for practical laser action.

[42] H. W. Etzel and D. A. Patterson, *Phys. Rev.* **112**, 1112 (1958).

6.4.1. The Problem of Self-Absorption

As already indicated in Chapter 6.2, the optical pumping cycle of F and F-like centers consists of four steps: excitation, relaxation, luminescence, and relaxation back to the ground state of the normal configuration (see Fig. 2). Since both relaxation times τ_R and τ_R' are very short (τ_R and $\tau_R' \leqslant 10^{-12}$ sec) with respect to the luminescence decay time τ_l, the only populations of any significance are N and N^*, i.e., those of the ground and relaxed–excited states, respectively. Thus, population inversion of the luminescence levels is obtained at any finite pump rate, and in this limited sense, any F or F-like center forms an ideal four-level system.

But one must also consider the possibility of absorption at the luminescence frequency. That is, in the Beer's law expression for the net optical gain,

$$G \equiv I_{\text{out}}/I_{\text{in}} = e^{\alpha z},$$

where z is the gain path length, the net gain coefficient is the sum of two terms,

$$\alpha = \alpha_g - \alpha_l,$$

where α_g is the coefficient calculated for the inverted luminescence levels alone, and where α_l is the coefficient of absorption loss. Of course, only when $\alpha_g > \alpha_l$ can there be a net gain.

Even when the unexcited crystal is perfectly transparent at the luminescence frequency, there sometimes exists the possibility that the optical pumping may create a new and absorbing species. Such occurs, for example, with F and $F_A(I)$ centers. As already indicated in Chapter 6.2, one new species is the RES itself, since absorption at the luminescence frequency from the RES into the conduction band is a distinct possibility. Such absorption from the RES of the F center has been measured for slightly lower photon energies, and it has been found to be quite large.[12b] For the case of optically pumped F and $F_A(I)$ centers, whenever $N \geqslant 10^6 - 10^{17}/\text{cm}^3$, F' centers are formed in significant quantities.[43] These constitute yet another absorbing species, since the F' absorption band[44] severely overlaps the F luminescence band.

The above problems are nonexistent for the $F_A(II)$ centers and for the infrared transition of the F_2^+. In both cases, the separation between the RES and the lower edge of the conduction band is greater than the luminescence photon energy, and there is no overlap between the F' absorption and the luminescence band. In the case of the F_2^+, we have already

[43] F. Porret and F. Lüty, *Phys. Rev. Lett.* **26**, 843 (1971).
[44] R. S. Crandall and M. Mikkor, *Phys. Rev.* **138**, A1247 (1965).

pointed out in Section 6.2.5 that there are no bound–bound transitions that would lead to self-absorption.

6.4.2. Some Useful Gain Formulas

For a four-level system, with Gaussian luminescence band of full width at the half-power points $\delta\nu$, the gain coefficient at the bank peak α_0 can be calculated from the well-known formula

$$\alpha_0 = \frac{N^*\lambda_0^2\eta}{8\pi n^2\tau_l}\frac{1}{1.07\ \delta\nu},\qquad (6.4.1)$$

where λ_0 is the wavelength at the band center, n is the host index, η is the quantum efficiency of luminescence, τ_l is the measured luminescence decay time, and N^*, as defined above, is the population in the relaxed excited state.

The gain formula (6.4.1) represents an extremely useful form, and one that affords maximum insight. In the first place, it contains only experimentally measurable quantities, with the sole exception of N^*. Furthermore, in general, N^* is not limited by the pump, since extremely high intensity sources are now available. Instead, the limit on N^*, if any, is created by other factors, such as a rapid increase of interaction among centers with increasing concentration.[45] From this point of view, it is more revealing to display N^* directly, rather than to express it in terms of a pump intensity.

In keeping with the above philosophy, let us compare the gain coefficients possible with various color center types for a fixed value of N^*. In the tabulation below, we show values of α_0 calculated from Eq. (6.4.1) for the F and F_2^+ centers in KCl, and for the $F_A(II)$ center in KCl:Li. In each case, $n = 1.49$ and $T = 77$ K. Note that α_0 is two orders of magnitude smaller for the F center than for the $F_A(II)$ center. This difference reflects the combined effects of an approximately 22 times smaller oscillator strength and a roughly 4 times larger $\delta\nu$ for the F center.

Quantity	F	F_2^+	$F_A(II)$	Units	Reference
λ_0	1.0	1.68	2.7	μm	8, 13, 46
$\tau_{l/\eta}$	600	80	200	nsec	10, 11, 46
δ_ν	6.3	1.69	1.45	10^{13}Hz	8, 13, 47
α_0	0.044	3.5	4.2	cm^{-1}	

[45] D. Frölich and H. Mahr, *Phys. Rev.* **148**, 868 (1966).
[46] L. F. Stiles, M. P. Fontana, and D. B. Fitchen, *Phys. Rev. B* **2**, 2077 (1970).
[47] W. Gebhardt and H. Kuhnert, *Phys. Status Solidi* **14**, 157 (1966).

Although it is not as fundamental as Eq. (6.4.1), we shall have use for an equation that expresses the gain as a function of pump intensity. Since the pump rate out of the ground state is equal to the photon absorption rate, we may write

$$NW = \beta I / E_p, \qquad (6.4.2)$$

where W is the pump rate, β is the absorption coefficient at the pump wavelength, E_p is the pump photon energy, and I is the pump beam intensity. We may then write

$$N^* = NW\tau_l = \beta(I/E_p)\tau_l. \qquad (6.4.3)$$

Finally, substituting (6.4.3) into (6.4.1), we obtain

$$\alpha_0 = \frac{1}{8\pi} \frac{\lambda^2}{n^2} \frac{\eta}{(1.07 \, \delta\nu)} \frac{\beta I}{E_p}. \qquad (6.4.4)$$

For a coaxially pumped system, where one can assume a variation of the gain coefficient of the form

$$\alpha = \alpha_0 e^{-\beta z}, \qquad (6.4.5)$$

and if furthermore, the gain path is long with respect to β^{-1}, then it is easy to show that

$$G = e^{\alpha_0/\beta}. \qquad (6.4.6)$$

Equations (6.4.4) and (6.4.5) may be combined to yield

$$\ln(G) = \frac{1}{8\pi} \frac{\lambda^2}{n^2} \frac{\eta}{1.07 \, \delta\nu} \frac{I}{E_p}. \qquad (6.4.7)$$

6.5. Laser Cavities with a Highly Concentrated Modal Beam

In cw dye lasers, the modal beam is highly concentrated in the region of the dye cell. The beam is focused to a diffraction limited spot, called the beam waist, whose diameter is typically on the order of 10 μm. This small spot size allows the pump beam also to be highly concentrated, such that a power of, say, 1 W can produce a pump intensity in the region of the dye cell on the order of 10^6 W/cm^2. In addition to allowing for the maximum possible concentration of pump power, such tightly focused, coaxially pumped cavities are energetically highly efficient, since practically all the incident pump power can be absorbed in a volume of the amplifying medium that is coincident with that swept out by the laser mode itself.

Such focused-beam cavity designs are also quite suitable for use in a cw color center laser. For the most part, one needs merely to substitute a thin slab of colored crystal for the dye cell. Otherwise, the technology that has been developed around such cavity designs can be taken over more or less wholesale.

One such cavity design is shown in Fig. 13. The amplifying crystal is located in the neighborhood of the beam waist, as shown. The angle ϕ is Brewster's angle, such that reflection loss at the crystal surfaces (for a mode whose electric field is in the plane of the paper) can be avoided without the necessity for antireflection coating them; this is a big advantage for the color center device, where the necessity to add antireflection coatings would pose difficulties in crystal handling. Another very important feature of this design is that the astigmatism induced by mirror M_2 can be made to exactly compensate for the astigmatism created by rotation of the crystal to Brewster's angle. This compensation can be accomplished through adjustment of the reflection angle 2θ. Without this compensation, it would be impossible to reduce the beam waist below a certain critical size, and still maintain a stable mode.

The mode itself[48] can be described as follows: The beam waist, of radius W_0, is located almost exactly at the center of curvature of mirror M_1, and just a little outside the focal length f of mirror M_2. When the output mirror is a flat (we shall make that assumption from now on), the modal beam has its largest diameter ($2W_1$) at mirror M_2, and slowly tapers down to a second beam waist at the output mirror. However, since one usually has $d_2 \gg 2W_1$, for all practical purposes, the mode consists of an essentially parallel beam in the region between M_2 and the output mirror. This region of parallel light constitutes yet another important feature of the cavity design, since it greatly facilitates the insertion of tuning elements and other intracavity devices.

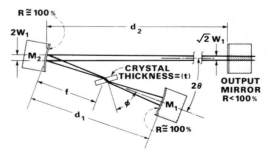

FIG. 13. A cavity suitable for cw tunable lasers and its mode pattern.

[48] For a good general review paper on the characteristics of laser modes, see H. Kogelnik and T. Li, *Appl. Opt.* **5**, 1550 (1966).

The cavity of Fig. 13 has been described quantitatively in great detail by Kogelnik *et al.*[49] The most important results of their paper can be summarized as follows: The mode is said to be stable as long as W_1 is of finite size. Let us define δ such that

$$d_1 \equiv r_1 + f + \delta, \tag{6.5.1}$$

where r_1 is the radius of curvature of M_1. A stable mode will be obtained as long as δ lies within the range

$$0 \leqslant \delta \leqslant f^2/d_2 \equiv 2S. \tag{6.5.2}$$

The quantity $2S$ is known as the *stability range*. The *confocal* parameter b is defined as the distance between points along the beam path where the mode diameter is $2^{1/2}$ times that obtaining at the beam waist. One always has

$$b = 2\pi W_0^2/\lambda. \tag{6.5.3}$$

When the cavity is adjusted to the middle of its stability range, one also has

$$b \cong f^2/d_2 = 2S. \tag{6.5.4}$$

By combining Eqs. (6.5.3) and (6.5.4), one can calculate W_0:

$$W_0 = f(\lambda/2\pi d_2)^{1/2}. \tag{6.5.5}$$

Yet another important quantity is the far-field angle ϕ, which can be calculated as

$$\phi = \lambda/\pi W_0 \tag{6.5.6}$$

and from which W_1 can be calculated as

$$W_1 \cong f\phi. \tag{6.5.7}$$

It can also be shown that when the cavity is adjusted to the middle of its stability range, the beam diameter at the output mirror has its smallest value, which is

$$\sqrt{2} W_1 = (4d_2\lambda/\pi)^{1/2}. \tag{6.5.8}$$

Finally, the astigmatic compensation mentioned above will be obtained whenever the relation

$$f \sin \theta \tan \theta = t(n^2 - 1)(n^2 + 1)^{1/2}/n^4 \tag{6.5.9}$$

[49] H. W. Kogelnik, E. P. Ippen, A. Dienes, and C. V. Shank, *IEEE J. Quantum Electron.* **qe-8,** 374 (1972).

is satisfied, and where t and n are the crystal thickness and index of refraction, respectively.

Kogelnik *et al.*[49] also discuss the nature of the mode shape inside the amplifying medium. Although in general the behavior is quite complicated, they find that for the case $S \gg t$, the beam waist area A is given by the expression

$$A \cong \pi n W_0^2, \qquad (6.5.10)$$

whereas for the case $S \ll t$, they obtain

$$A \cong \lambda t (n^2 - 1)[n^3(n^3 + 1)^{1/2}]^{-1}. \qquad (6.5.11)$$

6.6. Construction and Performance of a cw Laser Using $F_A(II)$ Centers

6.6.1. Basic Configuration

To date we have constructed and operated several cw lasers using $F_A(II)$ centers.[1,2] The latest version is shown in Fig. 14. The basic cavity configuration is that described in Chapter 6.5, with the following parameters: $r_1 = f = 25$ mm, $d_2 = 600$ mm, $t = 1.7$ mm, and $2\theta = 20°$. The arrangement of Fig. 14 was also used for the F_2^+ laser described in Chapter 6.2; only the mirror reflectivity peaks and the pump source were changed.

For operation with the $F_A(II)$ centers of KCl:Li and RbCl:Li, the pump source was a krypton ion laser operating at 6471 Å. Both input and output mirrors were of the multilayer dielectric type. Thus it was possible for the input mirror to be a high reflector ($R \sim 100\%$) for the wavelengths in the $\lambda = 2.7$ μm band, and at the same time to have rather high transmission at the pump wavelength. (Multilayer dielectric, gold and silver coatings have been used with about equal success for the intermediate mirror M_2 of Fig. 13.) A simple lens located immediately behind M_1 served to bring the pump beam to a focus at the crystal. Actually, it was also necessary to use a second, very weak lens located just outside the pump beam input window in order to achieve perfect focus, but this allowed for very convenient external adjustment of both the pump beam position as well as its focus.

As already indicated in Section 6.2.3, the crystal temperature should preferably be held below $T = 100$ K in order to have a high quantum efficiency of luminescence and a low pump power required for threshold. To obtain the necessary cooling, the crystal slab was held against a copper finger ($T \sim 77$ K) with a gentle spring clamp. No grease or other thermally conductive compound was used, in order not to strain or fracture the crystal by differential thermal contraction. Despite the rather poor thermal contact between the crystal and cold finger, cw pump inputs as high as ~600 mW could be tolerated without undue heating of the crystal.

The two opposing large faces of the crystal were [100] planes. The crystal was oriented such that the plane of the paper in Fig. 14 contained a 110 axis. Since the pump beam was also polarized in this plane, both sets of centers whose axes lay in the plane of the slab were pumped equally. Also, due to the Brewster's angle orientation, the laser beam traversed the crystal at a considerable angle (36°) to the slab normal, and hence centers whose axes lay along that direction were also pumped. In this way it was possible to avoid bleaching the crystal, even though the long wavelength absorption band was used for pumping. A second advantage of the [110] orientation was that it prevented strain-induced birefringence from affecting the laser mode, since such strains tended to lie along the [110] axis.

The entire cavity was surrounded by a vacuum enclosure, indicated by the dashed lines of Fig. 14. The vacuum was required for two reasons: first, to provide thermal insulation for the crystal, and second, to prevent atmospheric absorption (especially from H$_2$O) from interfering with laser operation. The cylindrical can surrounding the crystal and spherical mirror section was open at the top, and the vacuum seal was completed by a removable liquid nitrogen storage can.

The copper cold finger was referenced mechanically to the cavity support through some thin phenolic pieces, whose low thermal conductivity

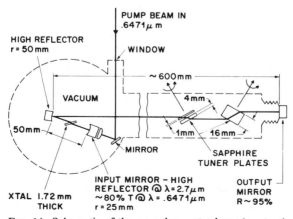

FIG. 14. Schematic of the cw color center laser (see text).

and shape allowed for a heat leak of only a few tens of milliwatts, while at the same time providing a very rigid mount. On the other hand, excellent thermal contact was made to the cold finger by a small flexible metal bellows, attached to the bottom of the nitrogen can. Aided by 1 atm differential pressure, the bellows pressed down on top of the finger with ~ 70 N. By direct measurement, cold finger temperatures were only 1 or 2 degrees higher than the boiling point of liquid N_2.

After roughing with an external pump, the vacuum in the can was maintained by an internal cryopump: A metal tube, sealed at the top and normally open at the bottom to the vacuum, extended up into the liquid N_2 storage space and was filled with Linde molecular sieve. To prevent damage to the mirrors from swirling dust particles of the sieve during the roughing operation, a valve operated by a bimetallic strip separated the (pre-evacuated) sieve storage space from the main vacuum chamber. Only when the liquid N_2 was added would the valve open, connecting the two evacuated spaces. Before this valve was added, we experienced considerable damage to a number of mirror coatings and to crystal surfaces from the "dust storm" effect.

Although not shown in Fig. 14, we have found it convenient to add a Brewster's angle window of CaF_2 to separate the vacuum space of the main can from the arm containing the tuner and output mirror. With provision for separate evacuation of the output arm, experimental changes in the tuner and output mirror could be made in rapid succession without the need for warming up or otherwise disturbing the laser crystal. Also, the output arm could be filled with various gases for intracavity absorption experiments.

6.6.2. Tuning Elements

To date a number of tuning schemes have been used for color center lasers, with varying degrees of success: prisms, gratings, and birefringence plates. For example, we have found a simple fused-silica prism (with both surfaces at Brewster's angle to the beam) to make a very convenient and efficient tuner for the F_2^+ laser in the 1 μm region.[15] Prisms of sapphire have been used[50] for the $F_A(II)$ and $F_B(II)$ center lasers, although the rather low dispersion of most acceptable materials, and the relatively high cost of many of these, makes the prism somewhat less attractive for the 3 μm region.

A truly superior tuning element for the lasers using type-II centers can be made from one of the very high-efficiency gratings that are now commercially available for the 3 μm region, by using it in retroreflection as a

[50] R. Beigang, G. Litfin, and H. Welling, *Opt. Commun.* **22**, 269 (1977).

replacement for the flat output mirror of Figs. 13 and 14. For example, one of the commercial gratings,[50a] blazed at 2.8 μm and having 457.8 grooves/mm, has an efficiency profile that is nearly flat from 2.4 to 3.5 μm. Measured efficiency (first order) is approximately 95%, and since only first- and zeroth-order reflections are possible, almost all of the remaining 5% of the incident energy appears in zeroth order. (These figures have been checked in actual $F_A(II)$ center laser operation with the grating.[50b]) The zeroth-order reflection provides for a very convenient output coupling, especially when a small mirror, parallel to and rotated with the grating, is used to maintain a constant direction of the output beam. Other advantages of the grating include low-cost, high-frequency selectivity, and the fact that a very precise and linear wavelength readout is possible with a simple sine-bar drive. And when the grating is combined with a single intracavity etalon, very stable single-frequency operation can and has been obtained easily.[50b] (See Section 6.6.6.)

In our earliest experiments on $F_A(II)$ center lasers, we used a set of sapphire birefringence plates for tuning. Although we no longer recommend these, due to their high cost and relatively poor performance, a description follows for the sake of the record. (The data of Fig. 16 were obtained with these.) The plates were oriented at Brewster's angle with respect to the beam. The optic axis of each lay in the plane of the plate. A simple mechanical linkage allowed all three plates to be rotated simultaneously about their normal axes. In this way, it was possible to tune the KCl: Li laser smoothly and continuously over the full range $2.5 \leqslant \lambda \leqslant 2.9$ μm with rotation of a single knob. Although three plates are shown in Fig. 14, in most experiments to be described only the thinner two (1 and 4 mm) were used.

Space does not permit an extensive discussion here of the birefringence tuner, but a rather complete description has been given recently in the literature.[51] The essential principle of the tuner is this: In general, a linearly polarized mode is made elliptically polarized by the plates, and a large reflection loss results. However, there exists one wavelength for which the linear mode polarization is unaffected, and there is no loss; laser action then occurs at this favored wavelength. One can easily show that the lasing wavelength varies with plate rotation angle θ_0 as

$$\lambda = \lambda_p(1 - \cos^2 \phi \cos^2 \theta_0), \tag{6.6.1}$$

where ϕ is Brewster's angle and λ_p a constant determined by the plate thickness and birefringence constants. For sapphire, $\phi = 60°$, or

[50a] Jobin Yvon, type 1-1-L19, or Bausch and Lomb No. 2691.

[50b] K. German, private communication.

[51] A. L. Bloom, *J. Opt. Soc. Am.* **64,** 447 (1974).

$\cos^2\phi = 0.25$, such that a practical tuning range somewhat less than 25% wide is possible with a single set of plates.

6.6.3. Alignment

The cavity was prealigned with the aid of a $\lambda = 5682$ Å krypton ion laser line, for which wavelength the mirrors were rather good reflectors. A dummy (transparent) crystal was inserted into the cavity for this purpose. With the aid of this visible light, the limits of the cavity stability range could be found easily and rapidly, and the mirrors set into proper angular adjustment. Usually no further adjustment was required for obtaining laser action after such prealignment, except perhaps for a slight touch-up of the output mirror and pump beam input steering lens. (The output mirror was mounted on a flexible metal bellows, to allow for angular adjustment during laser operation.)

The alignment procedure was straightforward and consisted of the following steps: (1) The input beam was brought to focus at the center of curvature of the input mirror, M_1. This was accomplished by adjusting the lateral position and focus of the input steering lens until the reflection from M_1 formed a parallel beam that exactly retraced the path of the input beam. When this adjustment had been correctly made, threshold of the pump laser was considerably lowered. (2) The intermediate mirror was adjusted to center the beam on the output mirror. (3) The multiple spot pattern formed at the output was observed on a white card. If the spots were well defined, they were made to coalesce into a single spot through angular adjustment of the output mirror. If the secondary spots were fuzzy, then the cavity spacing d_1 was outside the stability range; in that case, appropriate change was made in d_1 (through translation of M_2 with a calibrated screw), and steps (2) and (3) were then repeated. The limits of the stability range could be established in that way after a few trials, and d_1 set to the middle of the range.

6.6.4. Crystal Preparation

Slabs of slightly greater than the desired final thickness were cleaved out of raw crystal stock. (The cleaving takes place along [100] faces.) After rough grinding to the desired thickness, the slabs were carefully cleaned in absolute ethanol to remove all traces of grit. They were then polished on cotton cloth laps using Linde B (a very fine Al_2O_3 powder) and absolute ethanol. The first and most important lap was made by stretching a well-laundered cotton handkerchief over an ~ 15 cm diam. flat glass disk. The finishing lap was of a commercial "cotton velvet" material and was used only briefly to remove the smaller scratches. With care, this process will produce a good optical surface, i.e., one with low dig and scratch numbers, although the cloth laps do not allow for a high degree of surface flatness. Fortunately, when the crystal is used at the mode focus, a high degree of flatness is not required. (Where an optically flat surface is required, a pitch lap can be used, although more effort is required to make this work well.)

The $F_A(II)$ centers have been prepared in two ways: In our earlier experiments, crystals containing $\sim 1 \times 10^{18}$ U centers/cm³ were cut, polished, and then X-rayed at room temperature to reconvert a fraction of the U centers to F centers (see Section 6.3.4). At the time, this represented the only known sure way to create a colored crystal of high optical quality. However, with development of the particular additive coloration apparatus shown in Fig. 10 (See Section 6.3.2), recently we have preferred to directly color a previously polished crystal. This latter technique has the advantages of simplicity, precise control over F center concentration, and greater chemical stability of the F centers. (The crystal contains no H_2; hence the F centers cannot revert back to U centers.)

Polished and colored crystals are kept in a dry box and in the dark until needed. Immediately prior to use, crystals are loaded into the laser and cooled to -30 to $-10°C$. The F centers are then converted into $F_A(II)$ centers by optically pumping the F band for nearly 1 hr with a microscope lamp (see Section 6.3.3). (It has been found necessary to obtain very complete conversion, since optically excited F centers are rather strong absorbers in the $F_A(II)$ luminescence band.) Finally, the crystal is cooled to 77 K for laser operation.

6.6.5. $F_A(II)$ Center Stability

As already indicated in the introduction, there is no detectable bleaching of the centers in normal operation of the laser, as long as the crystal temperature is held well below 200 K. However, there is a stability problem associated with long-term storage: when the crystals are kept in the dark at room temperature, the $F_A(II)$ centers slowly dissociate (in KCl with a time constant on the order of 24 hr) into F centers and isolated Li ions.[8] We have shown experimentally that the F_A centers may be regenerated a number of times with little apparent degradation, through repetition of the aggregation process; on the other hand, it is not known if such recycling can be carried out *ad infinitum*. Fortunately, a large activation energy is involved, such that the dissociation becomes negligible with cooling of the crystal just a few tens of degrees below zero centigrade. Thus, a small and inexpensive thermoelectric cooler will suffice to maintain the F_A centers between periods of laser operation. As an alternative, it is not hard to obtain long liquid N_2 storage times; in our laser, a storage can of 1.6 liter capacity has a hold-time in excess of 48 hr.

6.6.6. Performance

In order to measure the gain capabilities of the KCl:Li laser, the usual high-reflectivity output mirror was replaced with one having $R = 50\%$. In this way the unknown intracavity losses were then overwhelmed by the huge output transmission loss. In this experiment, threshold for laser action (at band center) occurred at an input pump power P to the crystal of 130 mW. At threshold, the gain (for a double pass through the crystal) just compensates for the total cavity loss. Thus a single pass gain of $2^{1/2} = \exp(0.346)$ was implied. This measured gain is to be compared with that calculable from Eq. (6.4.7).

To make that calculation, first we must estimate the pump intensity I. From Eqs. (6.5.4) and (6.5.5) and the cavity parameters listed above, we obtain $b = 1.02$ mm and $W_0 \cong 20\ \mu$m. Since the conditions for Eq. (6.5.11) are better satisfied than those for Eq. (6.5.10), we estimate the mode beam cross section from Eq. (6.5.11) as $A = 0.83 \times 10^{-5}$cm². By making the further assumption that the pump beam was well matched in size to the modal beam, we may estimate I from the modal area and the threshold pump power of 130 mW. Then from Eq. (6.4.7), we finally calculate $\ln G = 0.39$, in excellent agreement with the measured value.

When a high reflectivity output mirror ($T = 1.6\%$) was used, the pump power at threshold dropped to 14 mW. Since $\ln G$ is directly proportional to the pump power, a value of $\ln G = 0.037$ is implied by extrapolation from the previous experiment. Hence the total cavity loss was $2 \times 0.037 = 7.4\%$, of which 1.6% represents output coupling, and the remaining 5.6% is the intracavity loss. Since we were not able to measure the individual mirror reflectivities, an exact accounting of the intracavity loss is impossible.

The maximum energetic efficiency of the KCl:Li laser should be about 10%, since the ratio of pump to luminescence photon energies is 5, and only about one-half the centers will be oriented such that they can radiate into a linearly polarized laser mode. This figure must be further multiplied by the ratio of output mirror transmission to total cavity loss. Thus, with $T = 1.6\%$ and the loss of 7.4% cited above, our laser should have had an energetic efficiency, when operated far above threshold, of approximately $0.1 \times 1.6/7.4 = 2.2\%$. Figure 15 is a graph of laser output power as a function of pump input for operation with the $T = 1.6\%$

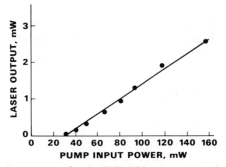

FIG. 15. Output vs. input power for the KCl:Li laser; output mirror transmission 1.6%, total cavity loss 7.4%.

output mirror. The behavior shown there is consistent with the above efficiency estimate.[52]

In Fig. 16, the reciprocal of the pump power required at threshold, P_{th}^{-1}, is plotted as a function of laser wavelength. Ideally, this curve should have the same shape as the luminescence band. However, the deviation between the two curves is real. In fact, the laser often could not be made to function for $\lambda \geq 2.8$ μm, and the data of Fig. 16 represent the best attainable behavior. The origin of this sporadic effect on the tuning is not known. It could possibly be attributed to the existence of an internal ab-

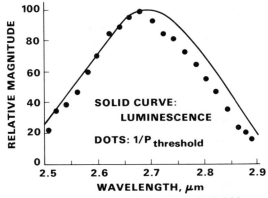

FIG. 16. Tuning characteristics of the KCl:Li laser.

[52] Recently, G. Litfin, R. Beigang, and H. Welling have obtained a slope efficiency of 7.7% in a KCl:Li laser by using a rather output coupling ($T \sim 30\%$) to overwhelm internal cavity losses. With 1.2W of pump power, this yielded an output of 85 mW (G. Litfin, R. Beigang, and H. Welling, *1977 CLEA* Abstract 7.7, and G. Litfin, private communication.)

sorption in the crystal, as we speculated earlier, but no known potential absorbing species should behave that way. A more likely explanation may lie in terms of dichroism induced in the crystal by the pumping, or in terms of a slight mutual angular misalignment of the optic axes of the birefringent tuner plates; those who use birefringence tuners with organic dye lasers have reported that such alignment is extremely critical.

Figure 17 shows the spectral purity of the laser output, as measured by a high-resolution interferometer, when the laser frequency is controlled by a pair (1 and 4 mm thick) of birefringence plates alone, and with the laser operating far above threshold. Each of the peaks of Fig. 17 represents operation on a single-cavity mode. Separation between the peaks is determined by a spatial hole-burning effect. The requirement that the standing wave peaks of one lasing mode correspond to the standing wave nulls of the next leads to the formula

$$\Delta\tilde{\nu}_m \sim (2m + 1)/4D, \qquad m = 0, \pm 1, \ldots , \qquad (6.6.2)$$

where $\Delta\tilde{\nu}$ is the separation in cm^{-1} between peaks and D is the distance between mirror M_1 and the crystal. The formula is approximate only, since the spacings must also correspond to a multiple of the laser resonator mode spacing. In our laser, $D = 2.5$ cm, implying $\Delta\tilde{\nu}_0 \sim 0.1$ cm^{-1}, in essential agreement with the measured $\Delta\tilde{\nu}_0 \sim 0.095$ cm^{-1}. Although Fig. 17 shows just two modes, sometimes three appear; in that case, two of these must be in strong competition with each other. It has been shown[53] that the intensities of the competing pair alternate at frequency c $\Delta\tilde{\nu}$.

It is possible to obtain single-mode operation with the addition of just one intracavity etalon.[50,50b] (In Beigang et al.,[50a] linewidths of less than 250 kHz are reported.) The free spectral range of the etalon should be twice the $\Delta\nu_0$ of Eq. (6.6.2); in that case, with the etalon tuned to one mode, all modes of the spatially complementary set [the $\Delta\tilde{\nu}_m$ given by Eq.

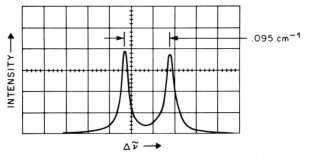

FIG. 17. Spectral purity of the KCl:Li laser output with the birefringence tuner alone.

[53] I. V. Hertel, W. Muller, and W. Stoll, *IEEE J. Quantum Electron.* **qe-13**, 6 (1977).

(6.6.2)] will lie at the etalon transmission nulls. In such single-mode operation, frequency definition of the laser output is determined by the stability of the cavity. In this respect, the color center laser, with its static amplifying medium, has a fundamental advantage over the dye lasers, where index and thickness fluctuations of the dye stream modulate the cavity length.

In summary, we have shown that performance of the cw $F_A(II)$ center laser is excellent in essentially all aspects, including low threshold, wide tuning range, and good frequency definition. Perhaps the relative ease with which such performance can be obtained should be emphasized. The large gains possible allow for easy cavity alignment and initiation of laser action. Although the crystals must be prepared with care for the best performance, again the high gain capabilities of the $F_A(II)$ center allow a rather large margin for imperfection in this regard. The requirements for cooling with liquid N_2 is no more complicated or expensive than that involved with cooling of most detectors to be used in conjunction with the laser. From the operator's point of view, the quiet liquid N_2 can is much easier to live with than a screaming dye pump and the constant potential for accidental dye spills.

6.7. Generation and Performance of a Distributed-Feedback Laser Using $F_A(II)$ Centers

6.7.1. General

For certain applications, such as pollution monitoring, or the selective stimulation of a given species in a chemical reaction, it would be desirable to have a very inexpensive laser that would be fixed-tuned to a predetermined frequency, such as a prominent absorption line of a given molecular species. Figure 18 shows such a device.[3] It consists simply of a slab of KCl:Li containing a grating of $F_A(II)$ centers, and it makes use of the principle of distributed feedback (DFB).[54] That is, if either the index n or the gain coefficient α is modulated spatially with period d, there will be strong feedback at those wavelengths that satisfy the Bragg condition

$$m\lambda = 2d, \qquad (6.7.1)$$

where m is an integer and λ is the wavelength in the medium. For the device of Fig. 18, the feedback is principally due to modulation of α.

[54] H. Kogelnik and C. V. Shank, *J. Appl. Phys.* **43**, 2327 (1972).

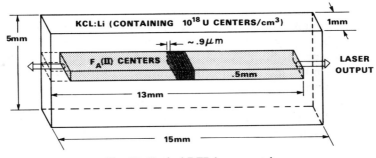

Fig. 18. Typical DFB laser crystal.

6.7.2. Creation of the Grating

The $F_A(II)$ centers were produced from U centers by using the room temperature, two-photon coloration process described in Section 6.3.5. A specially designed interferometric technique[55] was used to write the gratings. That technique (for details, see Appendix) permits the piecewise generation of precisely registered gratings of arbitrary size by sequential exposure of small segments of the photosensitive medium. The concentration of laser pulse energy thus afforded is an important advantage for the two-photon coloration process, since for constant laser power the exposure time necessary to produce a certain concentration of $F_A(II)$ centers is proportional to the square of the beam spot area. Furthermore, when the intensity becomes too low, the maximum center concentration is limited by the single-photon initiated back process (see Section 6.3.5).

The two interfering beams of spatially filtered 266 nm radiation (the fourth harmonic of a pulsed Nd:YAG laser, fourth harmonic energy 0.25 mJ, width 8 nsec), focused to a spot ~ 0.55 mm in diameter, were incident on the crystal samples through the large front face. The interfering beams were slowly translated longitudinally to produce 0.5 mm wide $\times$ 13 mm long gratings. Each grating exposure took approximately 10 min, involved 6000 pulses of 266 nm radiation, and produced a spatially modulated distribution of $F_A(II)$ centers with average population of $5 \times 10^{16}/cm^3$. To determine the periods of the gratings, aluminized flats were exposed in exactly the same way as the crystals. (The laser intensity is high enough to remove some of the aluminum in the bright fringes of the interference pattern.) Measurement of the autocollimation angles of the resulting aluminum gratings allowed the grating periods to be determined with an accuracy of about 0.3%.

[55] L. F. Mollenauer and W. J. Tomlinson, *Appl. Opt.* **16**, 555 (1977).

6.7.3. Performance of the DFB Laser

The DFB samples were placed in a liquid-N_2 immersion dewar and either longitudinally or transversely pumped with 532 nm pulses (10 nsec pulse width, 10 pulses/sec, pulse energy $\leqslant 2$ mJ − the second harmonic of the Nd : YAG laser system.) The pump pulses were circularly polarized to avoid the orientational bleaching effect described in Sections 6.2.3 and 6.6.1. (A linear polarization along a 110 axis would have worked equally as well.)

Figure 19 shows the output spectra of three different DFB laser samples whose measured periods of spatial modulation, as corrected for thermal contraction and change of index of the host from 300 to 77 K, were 0.8804, 0.9185, and 0.9483 μm. The measured wavelengths of the DFB laser outputs indicated in Fig. 19 are in excellent agreement with the values 2.595, 2.708, and 2.796 μm calculated from the corrected spatial periods and the index ($n = 1.474$) of KCl at 77 K. The linewidth of the

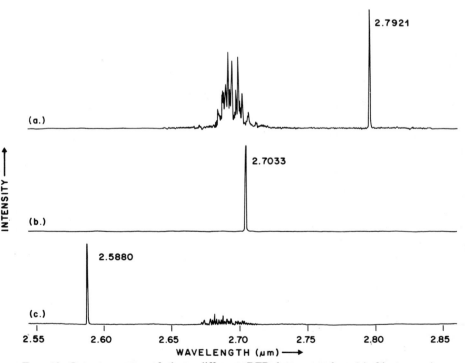

FIG. 19. Output spectra of three different DFB laser samples: (a) fringe spacing 0.9483 μm; (b) fringe spacing 0.9185 μm; (c) fringe spacing 0.8804 μm.

DFB output was in each case determined to be narrower than the 0.2 nm resolution limit of the analyzing grating spectrometer.

Two of the three DFB laser samples exhibited some spurious laser action over a 0.05 μm range centered about the peak of the gain line at 2.8 μm. The spectra of these spurious emissions typically exhibited a regular structure whose spacings corresponded closely to the axial mode spacings of the laser crystal or the dewar windows. It should be possible to eliminate all such spurious laser emissions by preparing the crystal sample with nonparallel end faces and by mounting the crystal in such a way that the laser emission is incident at an oblique angle to the dewar windows.

In all cases laser action was achieved more efficiently and at lower thresholds with longitudinal than with transverse pumping. Figure 20 shows the relative output power as a function of absorbed pump power

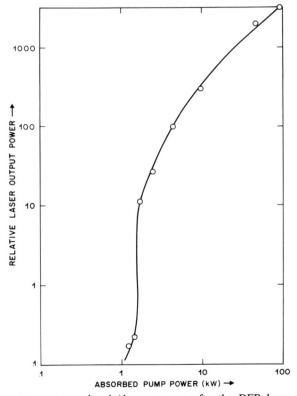

FIG. 20. Output power vs. absorbed pump power for the DFB laser with output at 2.7033 μm.

for the longitudinally pumped sample emitting at 2.7033 μm. A sharp threshold for laser action was observed for an absorbed pump power of 1.5 kW. The laser output spectrum was monitored near threshold and found still to consist only of the single sharp DFB line with no spurious laser emission. For operation well above threshold, measured energetic efficiency of the laser was 6.7%, for a quantum efficiency of 34%. (The output from both ends was included in determining that efficiency figure.)

As long as the DFB laser crystal was kept at liquid-N_2 temperature, absolutely no bleaching or other degradation of the laser performance was noticed, even after tens of hours of operation at pump intensities up to the dielectric breakdown threshold of the KCl crystal itself. This is one of the most significant observations to result from the DFB laser experiments. In the first place, it was not at all clear a priori that the $F_A(II)$ centers would not be destroyed by nonlinear optical effects occurring at such high pump power levels. Second, it has implications for more than the DFB laser itself; it allows the confident prediction, as in the introduction (Chapter 6.1), that a very high power pulsed tuned oscillator–amplifier combination can be practically realized using $F_A(II)$ centers.

6.8. Recent Developments in F_2^+ Center Lasers

6.8.1. Enhanced Production of F_2^+ Centers through Use of a Two-Step Photoionization and Extrinsic Electron Traps

A simple process for the production of F_2^+ centers through radiation damage has already been described in Section 6.3.6. Although this technique, as described there, is sufficient in the hosts LiF and NaF, in general there is an additional problem to be overcome: In general, the (thermally unstable) F' centers formed during the radiation damage process release large densities of electrons as the crystal is warmed to room temperature, thereby deionizing the F_2^+ centers. However, the F_2^+ centers can be regained through the use of a highly selective, two-step photoionization process, used in conjunction with divalent metal ions as extrinsic traps. This discovery[56] has greatly extended the number of host crystals from which viable F_2^+ center lasers can be made.

The need for an alternative to the ordinary (single-step) photoionization process is based upon the following: (1) By any of the presently known methods of creating F_2 or F_2^+ centers, these aggregate centers are consid-

---

[56] L. F. Mollenauer, D. M. Bloom, and H. J. Guggenheim, *Appl. Phys. Lett.* **33**, 506 (1978).

erably outnumbered by accompanying F centers; (2) the ground states of the F and F_2 centers lie at approximately the same level below the conduction band, as illustrated in Fig. 21, and furthermore, the absorption cross sections and ionization efficiencies of these two center types are comparable in the photoionization wavelength range.[40] As shown in Mollenauer *et al.*[56] through some simple rate equations, the fractional ionization of F_2 centers obtained under such single-step ionization is rather low, unless the density of external traps is larger than the sum of all ionizable species. But such a large density of traps is not usually attainable.

The way out of this dilemma is to provide a highly selective way to ionize the F_2 centers, such that no positively ionized species, other than the F_2^+ centers, are produced. The selectivity of the two-step process is based upon the fact that excitation to the first bound state of the F_2 center requires considerably lower energy than the corresponding transition of the F center, as also illustrated in Fig. 21. Furthermore, the fundamental excitation energy of the F_2 center is slightly greater than half the minimum single-photon ionization energy. Thus, a source tuned to the fundamental F_2 transition provides for a two-step ionization, whose rate is many orders of magnitude greater than the nonresonant two-photon ionization rate of the F center. In fact, the latter rate is truly negligible.

This new, two-step photoionization process was first demonstrated[56] in NaF containing approximately 100 ppm of Mn^{2+} as the traps. The second harmonic of a pulsed Nd:YAG laser was used as the two-step ionization source, since the corresponding wavelength ($\lambda = 532$ nm) lies within the

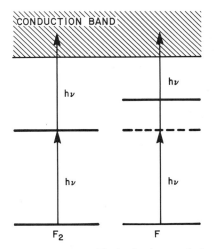

FIG. 21. How a light source resonant with the fundamental absorption of the F_2 center ionizes that center, but leaves F centers essentially unaffected (see text).

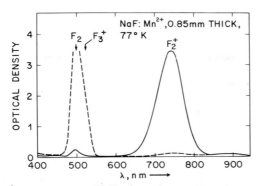

FIG. 22. Absorption spectrum at 77 K of an electron-beam-irradiated, 0.85 mm thick crystal of NaF : Mn²⁺. Dashed curve: Initial curve; peak of combined F_2–F_3^+ band has OD of about 6. Solid curve: After two-step photoionization; F_2^+ center density $\sim 8 \times 10^{17}$/cm³ (see text).

fundamental absorption band of the F_2 center in NaF.   Figure 22 shows the absorption spectra of an electron-beam-irradiated crystal, both before and after exposure to a few dozen intense ($I \sim 10$ MW/cm²) pulses of the second harmonic.   (To prepare the "before" curve, the F_2^+ band had been reduced through brief exposure of the crystal to UV light.)   Several light polarization directions were used, such that all F_2 centers would be affected.   From the figure, it is clear that a nearly complete conversion of F_2 to F_2^+ has taken place.

The experiment yielded an additional gain: A significant part of the original absorption band near $\lambda \cong 500$ nm is due to F_3^+ centers, and that, too, all but disappeared.   In fact, the number of F_2 centers indicated by the band height cannot alone account for the number of F_2^+ centers gained; it must be that the F_3^+ centers were also converted, at least in part, into F_2^+ centers.

The two-step photoionization can be carried out efficiently with a cw laser as well, since the extremely high intensities listed above are not truly necessary.   This was first demonstrated[56] in the host KF, doped with Pd²⁺.   In KF, the 648 nm line of a krypton ion laser provides a nearly perfect match to the F_2 band.   First, it was verified that a very large F_2^+ band could be obtained by pumping with the unfocused ($A \sim 0.01$ cm²), 4 W beam of the laser.   Characteristic times for the bleaching of the F_2 band were then measured.   The results are shown on a log–log plot in Fig. 23. The slope of the line drawn through the points is exactly two, verifying that the process involves pairs of photons.   It should also be noted that nearly complete conversion of the centers was obtained, independently of the rate at which it was carried out.

The cw photoionization experiments have a number of interesting im-

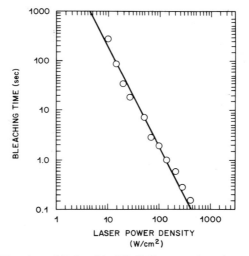

FIG. 23. Bleaching time of F_2 band in KF:Pd^{2+} vs. cw laser intensity (see text).

plications. First, when focused, the laser beam could have a power as low as a few milliwatts. Second, a small flashlamp or high-intensity arc could probably also serve as well; only a long-pass filter would be needed to avoid direct pumping of the F band. Third, the two-step photoionization makes it very difficult to operate an F_2 center laser,[22a] and the implications for an F_3^+ center laser may be even more fatal. One can probably generalize from the above, that the most durable centers under the intense pumping required for laser action will be those, like the F_2^+, whose first excited states (following relaxation) lie deep below the continuum, or at least well below the halfway mark to it.

The choice of certain divalent transition metal ions as dopants to make the necessary electron traps was based to a great extent on a hint provided by the work of Umbach and Paus[57] on the formation of F_Z centers. That is, Umbauch and Paus[57] found that divalent transition metal ions would *not* allow for the formation of F_Z centers, implying that an electron trapped by such an ion lies deeper below the continuum than the F center ground state. Based on that work, and also in part on the theoretical calculations of Simonetti and McClure,[58] the dopants Mn^{2+}, Cr^{2+}, and Ni^{2+} were tried in NaF, at the level of 30–100 ppm. All three seemed to work well as electron traps. However, in the host KF, Mn^{2+} produced no ef-

[57] K. H. Umbach and H. J. Plaus, *1971 Int. Conf. Color Cent. Ionic Cryst., 1971* Abstract 77 (1971).

[58] J. Simonetti and D. S. McClure, *1977 Int. Conf. Defects Insulating Cryst., 1977* Abstract 394 (1977).

fect.   It was then pointed out[59] to us that the divalent metal ions aggregate if the ion size is small enough relative to the host lattice.   The larger ion Pd^{2+} (the analog to Ni^{2+}) does not present this problem.

To summarize, the formula for creating a high density of F_2^+ centers is as follows: (1) Incorporate a sufficient number of divalent transition metal ions, of a size that fits well into the lattice, as electron traps.   (2) With the crystal cooled sufficiently to prevent vacancy diffusion, create anion vacancies and F centers by massive radiation damage, as from an electron beam.   (3) Temporarily warm the crystal to room temperature to allow for the formation of F_2 and F_2^+ centers.   (4) Cool the crystal to laser-operating temperature, and complete the F_2 to F_2^+ conversion with an *in situ,* two-step photoionization.

Although experiments have so far involved only the fluorides, there would appear to be no fundamental obstacle to the extension of this scheme to the heavier alkali halides.   Should this indeed prove possible, then the presently achieved tuning range (with the hosts LiF, NaF, and KF, $0.82 \leq \lambda \leq 1.5$ μm) can be extended to well beyond 2 μm (see Fig. 9).

6.8.2. F_2^+ Center Lasers in the Hosts KF and LiF

The two-step photoionization described in Section 6.8.1 has made it possible to create viable F_2^+ center lasers with the host KF.[60]   The absorption and luminescence bands of the laser transition in KF are shown in Fig. 24.   The laser was continuously tunable over the range $1.26 \leq \lambda \leq 1.48$ μm.   The pump source was a small, cw Nd:YAG laser operating at 1.064 μm.   The laser cavity and tuning arrangement are shown in Fig. 25.

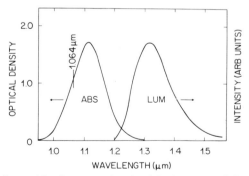

FIG. 24. Absorption and luminescence bands of the F_2^+ center infrared transition in KF; the optical density scale refers to a crystal thickness of 1.7 mm (see text).

[59] D. S. McClure, private communication.
[60] L. F. Mollenauer, D. M. Bloom, and A. M. DelGaudio, *Opt. Lett.* **3,** 48 (1978).

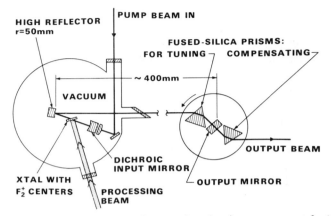

FIG. 25. Schematic of the F_2^+ center laser cavity, showing arrangement for *in situ* processing of crystal (see text).

In the laser experiments related here, so-called high-purity KF crystals were used. In that case, there were not sufficient electron traps to allow for complete F_2 to F_2^+ conversion. Nevertheless, it was possible to achieve a ~35% absorption of the pump light (at $\lambda = 1.064~\mu m$) in a single pass through the laser crystals.

Another difficulty encountered with the undoped crystals was the presence of a small, but significant absorption band in the region of the F_2^+ center luminescence. We discovered that this band could be eliminated by a second *in situ* treatment of our laser crystals with the combined fundamental (at $1.064~\mu m$) and second harmonic of a pulsed Nd:YAG laser. For the laser experiments on KF described below, the crystals were so treated. Such treatment considerably improved the laser performance in the short-wavelength half of the luminescence band.

The reciprocal pump powers required for threshold of laser action are shown in Fig. 26, for three different degrees of output coupling ($T \sim 2, 9$, and 40% over most of the band); also included is a theoretical gain shape curve (obtained from multiplying the luminescence band of Fig. 24 by a factor proportional to λ^2.) Except for the extreme short wavelength range, where the tail of the F_2^+ pump band (combined perhaps with a residue of the absorption mentioned above) reduces the net gain, the experimental curves fit the theoretical gain shape reasonably well. The tuning prism exhibited a small absorption peaking at $\lambda \sim 1.38~\mu m$, due to the first overtone of OH; this accounts for the small dip evident in the uppermost tuning curve. Finally, we note that the reflectivities of *all* mirrors involved in the laser cavity fell off steeply for $\lambda \gtrsim 1.46~\mu m$; otherwise, laser action should have been obtainable out to at least $1.50~\mu m$.

By comparing the $T = 2\%$ and $T = 40\%$ curves of Fig. 26, an internal cavity loss of ~12% can be inferred. This loss is probably to be ac-

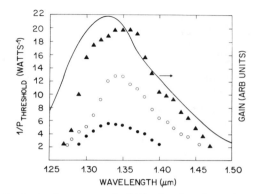

FIG. 26. Theoretical gain shape curve (solid line) and reciprocal pump powers required for laser threshold for various degrees of output coupling: $T \sim 2\%$ (triangles; $T \sim 9\%$ (open circles); $T = 40\%$ (dots).

counted for in terms of scattering at the crystal surfaces, of $\sim 3\%$ per pass per surface. Although KF is relatively hard as compared to the average alkali halide, it is also highly hygroscopic. With special techniques,[61] very good surface finishes are attainable in KF, but these were not implemented at the time of the experiments. Hence, in the future, considerable improvement is to be expected over the performance cited here.

The maximum output power was 40 mW near band center with the 9% output coupling mirror; this was for a pump power actually absorbed by the crystal of 160 mW. (The pump powers implicit in Fig. 26 refer to the power present at the entrance face of the crystal, not to absorbed power.) The slope efficiency measured in terms of absorbed power was 30%. When that actual slope efficiency is multiplied by the ratio of total loss to output coupling loss (20%/9%) a slope efficiency of $\sim 70\%$ is implied for a cavity with no internal loss. This figure compares rather well with the laser to pump photon energy ratio of $\sim 77\%$.

Following the laser experiments, the Pd^{2+}-doped KF crystals mentioned in Section 6.8.1 were grown. The large, clean F_2^+ absorption band shown in Fig. 24 was obtained with those doped crystals, where it was possible to obtain a complete F_2 to F_2^+ conversion. These new crystals should allow for a manifold increase in output power[61a] and a considerable reduction in pump power required at threshold over the laser performance

[61] An excellent polish can be produced on highly hygroscopic crystals such as KF and LiCl by using the combination of oelic acid and Linde A on a suitable lap. Evidently, the oelic acid is sufficiently polar to keep the abrasive particles dispersed, but does not attack the crystal itself (J. Simonetti, private communication).

[61a] With the Pd^{2+}-doped KF crystals, and with pump input powers of ~ 5 W, cw output powers of ~ 2 W have been obtained: The tuning range has been extended to 1.23 μm (L. F. Mollenauer and D. M. Bloom, unpublished).

cited above. Furthermore, there is no evidence of the interfering band
seen in the undoped crystals.

The potentially large light-to-light conversion efficiencies of F_2^+ center
lasers have in fact been largely realized in LiF.[62] Figure 27 shows laser
output versus pump input for operation near band center with the LiF,
F_2^+ center device. The slope efficiency of $\sim 60\%$ compares very favorably with the pump-to-output photon energy ratio of $\sim 70\%$. Figure
28 shows power output as a function of tuning. Note that power outputs
are on the order of 1 W over a large fraction of the tuning band. In the future, this sort of performance should be attainable from all F_2^+ center
lasers, regardless of host.

6.8.3. Mode-Locking of F_2^+ Center Lasers

Recent experiments with the host LiF have shown that the F_2^+ center
lasers are relatively easy to mode-lock by synchronous pumping.[62a] In

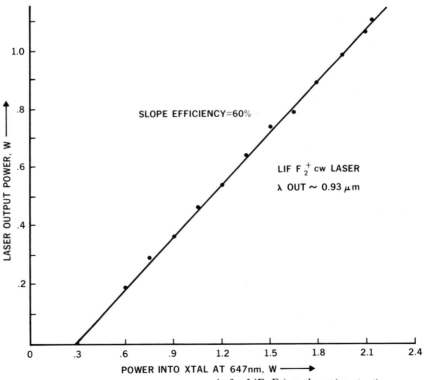

FIG. 27. Power out vs. pump power in for LiF, F_2^+ cw laser (see text).

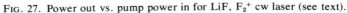

[62] L. F. Mollenauer, unpublished.

[62a] In recent experiments, the KF, F_2^+ center laser has yielded mode-locked output pulses
of ~ 5 psec over the tuning range $1.24 \leq \lambda \leq 1.44$ μm (L. F. Mollenauer and D. M. Bloom,
unpublished).

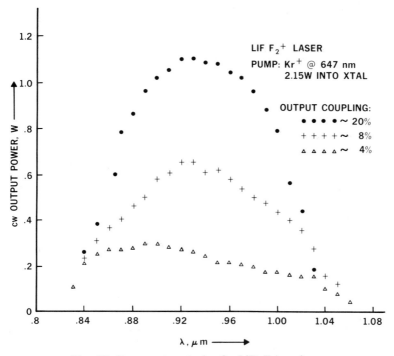

FIG. 28. Power out vs. tuning for LiF, F_2^+ cw laser.

those experiments, the pump source was a mode-locked krypton ion laser (pulse width ~ 100 psec) operating at 647 nm. The F_2^+ center laser cavity length was made exactly half that of the pump. Figure 29 shows an autocorrelation trace of the output pulses at ~ 0.9 μm: a pulse width of 4 psec can be inferred. In view of the large and strictly homogeneous luminescence bandwidth, the ultimate limit on pulse width with mode-locked F_2^+ center lasers should be well under 1 psec.

However, a very weak orientational bleaching effect was encountered in the mode-locking experiments, producing a fading of the laser output on a time scale of some tens of minutes. In order for F_2^+ centers to reorient, they must be excited to the $2P\pi_u$ or higher states; excitation to the $2P\sigma_u$ alone cannot produce a reorientation[13] (see Fig. 7). That is, the reorientation is made possible by the electronic energy released to the lattice during a nonradiative transition from $2P\pi_u$ to $2P\sigma_u$. (This transition is nonradiative at all but very low temperatures.[13]) In the laser experiment, excitation to the $2P\pi_u$ state was presumably by way of a weak, two-photon transition, excited by the pump light itself, from the relaxed $2P\sigma_u$ state. (The energy level spacings are apparently just right to allow

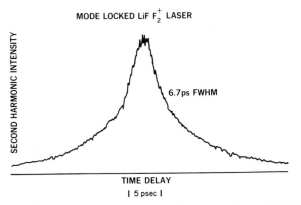

FIG. 29. Autocorrelation curve of output pulses from mode locked LiF F_2^+ center laser (see text).

for this.) It is hoped that this two-photon transition (and hence the orientational bleaching effect) can be made negligible through proper orientation of the crystal. That is, if the electric field of the pump beam is made to fall exactly along a [110] axis, then by orthogonality of the $2P\sigma_u$ and $2P\pi_u$ states, and the fact that the lobes of the $2P\sigma_u$ lie along the [110] axis, the net two-photon transition probability should become zero. The assumptions in the above, which need to be verified by experiment, are that the $2P\pi_u$ state is pure, and that the energy level spacings are indeed such that excited states other than the $2P\pi_u$ are not accessible by way of two pump photons.

6.8.4. Possible Future Developments

Although the F_2^+ centers are virtually indestructable when the crystal temperature is low enough ($T \leqslant 200$ K), at present they tend to fade away at room temperature, with a time constant of one day. This problem of inadequate ''shelf life'' represents the chief obstacle to widespread use of F_2^+ center lasers in laboratories not having facilities for their creation. The reasons for this fading are not yet entirely clear. It would seem to be due, at least in part, to a slow leakage of electrons back to the centers; in this case the cure would simply be to use the two-step photoionization *in situ* prior to use.

However, the fading may also be due to a tendency of the F_2^+ centers themselves to wander slowly at room temperature. If this is indeed the problem, then the cure may lie in associating the F_2^+ center with an impurity ion. For example, by associating the F_2^+ center with an alkali metal ion, the so-called M_A^+, i.e., $(F_2^+)_A$ centers are obtained.[63]   In addition to the possible stabilizing effect that M_A^+ formation may have, for a given host, both luminescence and absorption bands of the M_A^+ center show significant shifts from those of the isolated F_2^+ center.   This will undoubtedly prove useful, in some cases, in providing for a more convenient matching to available pump sources, and in providing useful shifts of the laser tuning bands.

6.A. Appendix: Technique for the Piecewise Interferometric Generation of Gratings

The technique is based upon a simple fact: For interferometers of a certain class, each and every illuminating ray is split into two rays that ultimately intersect on a common plane normal to the planes of interference fringes.   Thus, fringes are formed at any illuminated location on that plane, no matter what the size of the illuminating beam.   Small segments can be illuminated sequentially, and the resultant overall pattern will be identical to that which would be recorded if the full aperture were simultaneously illuminated with a plane wave.

A simple interferometer of the required type is shown in Fig. 30.   It consists of just two elements: beam splitter S and plane mirror M_1.   The split components of each incident ray intersect at points whose locus is the plane ab.   (The normal to ab is the bisector of the intersection angle ϕ.)   Mirror M_0 plays a vital role, even though it is not part of the interferometer proper.   When displaced as indicated by the dashed lines, M_0 serves to translate the beam parallel to itself (in what we arbitrarily define as the horizontal direction), thus successively illuminating portions of the interference pattern on the plane ab.   Clearly, similar means can be used to translate the input beam parallel to itself in the vertical direction (normal to the plane of the figure).   By combining horizontal and vertical translations, one can thus produce gratings of almost any aspect ratio or shape.

The accuracy of the gratings thus produced will depend on the accuracy with which the input beam can be translated parallel to itself (and, of course, on the accuracy of the interferometer elements).   It can be shown that a small deviation of the input beam through an angle α, in any direction, results in a change in the path difference of approximately α^2 times half of the original path difference.   The vital point is that here (as in most interferometers) there is no first-order dependence on α.   For a total path difference of 10 cm and a wavelength of 0.5 μm, the input beam must be

[63] I. Schneider, *Solid State Commun.* **9,** 45 (1971).

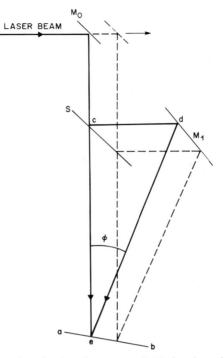

FIG. 30. Basic configuration of an interferometer suitable for piecewise generation of gratings.

rotated 3×10^{-3} rad to shift the interference pattern by one fringe. Mechanical translation stages capable of much better angular stability ($\alpha \leqslant 10^{-5}$ rad) are available commercially.[64]

It should be noted that the interferometer shown in Fig. 31, often used in the generation of gratings, is not suitable for the sequential illumination technique. The locus of points where the split ray components intersect each other is plane $a'b'$, parallel to the interference fringe planes and normal to the plane ab in which a grating is usually written. In general, the difference in the number of reflections suffered by the two interfering beams must be even (2 in Fig. 30) to permit the sequential writing of gratings. When it is odd, the behavior will be like that shown in Fig. 31.

The interferometer of Fig. 31 is often chosen because it can be adjusted to yield a zero path difference at the center of the grating. This minimizes

[64] Dover Instrument Corp., 19 Jones Road, Waltham, Massachusetts, 02154, Model 400-B.

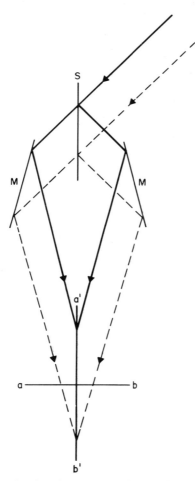

FIG. 31. Configuration of an interferometer that is not suitable for piecewise generation of gratings.

the requirements on both the coherence length of the source and the angular stability of the input beam. However, the addition of an extra pair of mirrors to the interferometer of Fig. 30 allows it to be similarly adjusted. One such modification is shown in Fig. 22. Note that the difference in the number of reflections suffered by the two beams is still even. (Figure 22 represents the interferometer actually used in the generation of the DFB laser gratings described in Chapter 6.7.)

The path difference in Fig. 32 can be analyzed as follows. Segment

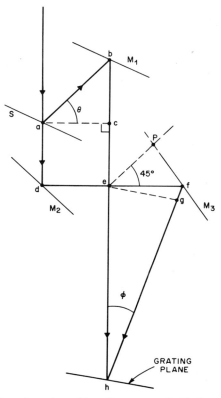

FIG. 32. Configuration of a balanced interferometer suitable for piecewise generation of gratings. Rotation of mirror M_3 about point P changes the period of the interference pattern while maintaining zero path difference at the center of the sample. (When M_3 is adjusted, the sample must also be repositioned, but none of the other elements of the interferometer need to be adjusted.)

$gh = eh$ by construction. Segments ad and ce are opposite sides of a rectangle. It is clear that adjustment of ac or of angle θ, or both, allows the remaining path segment abc to be made equal to the other remaining path segment $defg$. Furthermore, it is easy to demonstrate that path efg stays constant (equal to $2^{1/2}$ times the length of eP), as ϕ is adjusted by rotation of mirror M_3 about pivot point P. (P may lie anywhere along the dashed 45° line.) Hence the path difference need be adjusted to zero only once, and then it will remain so for all possible interference angles.

7. MOLECULAR LASERS

7.1. Molecular Infrared Lasers*

7.1.1. Introduction

This chapter is intended as an introduction to the basic theory, techniques, and overall strategy employed in the conception and development of new molecular infrared (IR) lasers. While remaining at the introductory level, this chapter attempts to outline the critical steps and concepts in the process of converting design goals into an experimental plan that will eventually yield an operable device. We shall attempt neither an exposition of a large number of existing devices nor a detailed account of the many theoretical and experimental procedures commonly encountered in contemporary laser physics, for these subjects are more than adequately covered by excellent texts and reviews. Rather, the reader will hopefully depart with some basic concepts, and order-of-magnitude estimates firmly in hand, a few potentially confusing areas (e.g., vibrational relaxation nomenclature) cleared up, and the knowledge of where and when to invoke the basic theoretical and experimental tools of the trade. In keeping with this objective, we have minimized the use of equations and detailed descriptions of experimental methodology. These can be found to almost any desired degree of complexity in the existing literature and their inclusion herein would defeat the uniqueness of this chapter. To direct the reader to these sources, an extensive bibliography is provided in which the first reference cited was chosen (if possible) to minimize the number of different sources to be consulted by the newcomer to the field.

The approach herein, where the development of new IR lasers is stressed, has been taken in order to reflect the current state of development of this field and the various uses to which new sources might be put. If operating wavelength is unimportant, brute-force cw and pulse power is available in the infrared using CO_2, CO, and HF/DF lasers. These sources have matured to the point where further improvements in their performance will not come from concepts that can be covered within the

---

* Chapter 7.1 is by T. Manuccia.

METHODS OF EXPERIMENTAL PHYSICS, VOL. 15B

limited scope of this chapter.   On the other hand, there are many regions
of the spectrum that are devoid of laser sources.   Devices operating in
these regions have great potential for applications in the areas of communications, long-range sensing (pollutants), and photochemistry, especially isotope separation.

For our purpose here, it is assumed that the reader has had an introductory exposure to general laser theory and the CO_2 laser at the level
of Yariv[1] or similar material in this and other volumes[2] of this series.
Basic concepts of molecular infrared spectroscopy such as can be found
in the same earlier volume of this series[2] or in Harmony[3] will also be
necessary.   While obviously desirable, the level of sophistication found
in the excellent works of Yariv,[4] Sargent et al.,[5] Steinfeld,[6] and Hertz-
berg[7-10] will not be required.

This chapter will have a broader but less detailed coverage than the
book of Willett[11] on electrical excitation, that of Anderson[12] on gas dynamic lasers, or the review of Chang and Wood[13] on high-pressure optically pumped CO_2 lasers.   It will mention some recently introduced device configurations such as high-pressure electron beam sustainer lasers
and gas flow lasers discussed primarily in the original research literature
and to some extent by Duley.[14] As coverage of the more conventional
device configurations for CO_2, CO, and other well-known lasing mole-

[1] A. Yariv, "Introduction to Optical Electronics." Holt, New York, 1971.

[2] D. Williams, ed., "Methods of Experimental Physics," Vol. 13B. Academic Press,
New York, 1976.

[3] M. D. Harmony, "Introduction to Molecular Energies and Spectra." Holt, New York,
1972.

[4] A. Yariv, "Quantum Electronics," 2nd ed. Wiley, New York, 1975.

[5] M. Sargent, III, M. O. Scully, and W. E. Lamb, Jr., "Laser Physics." Addison-
Wesley, Reading, Massachusetts, 1974.

[6] J. I. Steinfeld, "Molecules and Radiation." Harper, New York, 1974.

[7] G. Herzberg, "Spectra of Diatomic Molecules." Van Nostrand-Reinhold, New York,
1950.

[8] G. Herzberg, "Infrared and Raman Spectra of Polyatomic Molecules." Van Nostrand-
Reinhold, New York, 1945.

[9] G. Herzberg, "Electronic Spectra and Electronic Structure of Polyatomic Molecules."
Van Nostrand-Reinhold, New York, 1966.

[10] G. Herzberg, "The Spectra and Structures of Simple Free Radicals—An Introduction
to Molecular Spectroscopy." Cornell Univ. Press, Ithaca, New York, 1971.

[11] C. S. Willett, "Introduction to Gas Lasers: Population Inversion Mechanisms." Pergamon, Oxford, 1974.

[12] J. D. Anderson, Jr., "Gasdynamic Lasers." Academic Press, New York, 1976.

[13] T. Y. Chang and O. R. Wood, II, IEEE J. Quantum Electron. qe-13, 907 (1977).

[14] W. W. Duley, "CO_2 Lasers." Academic Press, New York, 1976.

cules appears in a variety of books and review articles,[15-25] the present chapter will make no attempt to cover these in detail or to cover topics closely related to IR molecular lasers such as chemical lasers,[26,27] laser measurements,[28] or excimer and other short ($\lambda \lesssim 1\ \mu$m) emission from molecules.[29] With few exceptions, there will be little discussion of the many applications of molecular lasers[30-38] or other optical devices and

[15] D. Williams, ed., "Methods of Experimental Physics," Vol. 13B, pp. 280-290. Academic Press, New York, 1976.

[16] E. R. Pike, ed., "High Power Gas Lasers—1975," Conf. Ser. No. 29. Inst. Phys., London, 1976.

[17] C. K. N. Patel, in "Lasers" (A. K. Levine, ed.), Vol. 2, p. 1. Arnold, London, 1968.

[18] P. K. Cheo, in "Lasers" (A. K. Levine and A. J. DeMaria, eds.), Vol. 3, p. 111. Dekker, New York, 1971.

[19] V. M. Fain and Ya. I. Khanin, "Quantum Electronics," Vols. I and II. MIT Press, Cambridge, Massachusetts, 1969.

[20] C. K. Rhodes and A. Szöke, in "Laser Handbook I" (F. T. Arecchi and E. O. Schultz-Dubois, eds.), p. 265. North-Holland Publ., Amsterdam, 1972.

[21] J. P. Goldsborough, in "Laser Handbook I" (F. T. Arecchi and E. O. Schulz-Dubois, eds.), p. 597. North-Holland Publ., Amsterdam, 1972.

[22] W. V. Smith and P. P. Sorokin, "The Laser." McGraw-Hill, New York, 1966.

[23] C. G. B. Garrett, "Gas Lasers." McGraw-Hill, New York, 1967.

[24] B. Lengyel, "Lasers." Wiley, New York, 1971.

[25] A. Maitland and M. H. Dunn, "Laser Physics." North-Holland Publ., Amsterdam, 1969.

[26] R. W. F. Gross and J. F. Bott, eds., "Handbook of Chemical Lasers." Wiley, New York, 1976.

[27] T. A. Cool, "Chemical Lasers," Part 8, this volume.

[28] H. G. Heard, "Laser Parameter Measurements Handbook." Wiley, New York, 1968.

[29] S. K. Searles, "Excimer Lasers," Chap. 7.2, this volume.

[30] J. Joussot-Dubien, ed., "Lasers in Physical Chemistry and Biophysics." Elsevier, Amsterdam, 1975.

[31] C. B. Moore, ed., "Chemical and Biochemical Applications of Lasers," Vols. 1, 2, and 3. Academic Press, New York, 1974, 1977, and 1978, resp.

[32] R. G. Brewer and A. Mooradian, eds., "Laser Spectroscopy." Plenum, New York, 1974.

[33] M. S. Feld, A. Javan, and N. A. Kurnit, eds., "Fundamental and Applied Laser Physics." Wiley, New York, 1973.

[34] T. R. Gilson and P. J. Hendra, "Laser Raman Spectroscopy." Wiley, New York, 1970.

[35] J. H. Eberly and P. Lambropoulos, eds., "Multiphoton Processes." Wiley, New York, 1978.

[36] J. F. Ready, "Effects of High-Power Laser Radiation." Academic Press, New York, 1971.

[37] S. F. Jacobs, M. Sargent, M. O. Scully, and C. T. Walker, eds., "Laser Photochemistry, Tunable Lasers, and Other Topics." Addison-Wesley, Reading, Massachusetts, 1976.

[38] M. L. Wolbarsht, ed., "Laser Applications in Medicine and Biology." Plenum, New York, 1971.

techniques often used to control the output of such lasers. Thus, topics such as modulators, deflectors, optical systems, detectors,[39] Q-switching, stable and unstable optical resonators, injection and mode locking and control, oscillator/amplifier configurations, and parasitic suppression were omitted because typically they do not arise in the conception and development of new lasers and because their operation is essentially identical whether applied to the infrared or other parts of the spectrum.

The body of the chapter will be organized into two major sections. The first will be a discussion of those aspects of basic physics, particularly spectroscopy and vibrational energy transfer, that are unique to lasers operating on IR molecular transitions and crucial in the conception of new lasers. Carbon dioxide is used where possible as an illustrative example, because of all molecular IR lasers, it is the most completely understood, because it is probably familiar to the widest number of readers, and because CO_2 lasers have been operated successfully in the widest variety of circumstances. This species thus represents an archetype against which most other lasers can be compared and contrasted.

The second major section deals with a few of the myriad device configurations available in this portion of spectrum. As these various options usually reflect differing excitation methods, some of the associated basic physics is covered here as well.

With the availability of items such as a collection of *Scientific American* articles on lasers,[40] computerized listings of 5200 distinct gas laser emission wavelengths,[41] and with 41 references to books containing substantial quantities of material on molecular lasers cited in this introductory section alone, it is obvious that this field is broad, reasonably well reviewed, and above all, massive. It is in the particular combination of subjects chosen that this review purports to be unique. The selection was guided by the attempt to avoid repetition of material and approach that can be found in traditional laser texts, but it was obviously little more than personal bias that finally dictated the small sampling of topics chosen and the emphasis given thereto.

7.1.2. Basic Physics

7.1.2.1. Molecular Spectroscopy.
Central to the design of new infrared molecular lasers is the determination of the energies of the vibra-

[39] W. L. Wolfe, ed., "Handbook of Military Infrared Technology." Office of Naval Research, US Govt. Printing Office, Washington, D.C., 1965.

[40] "Lasers and Light," a collection of Scientific American articles. Freeman, San Francisco, California, 1969.

[41] R. Beck, W. Englisch, and K. Gürs, "Table of Laser Lines in Gases and Vapors." Springer-Verlag, Berlin and New York, 1978.

tional–rotational states of molecules and the determination of the nature of the wavefunctions corresponding to states of interest. The first item determines the frequency of the transition while the second determines the oscillator strength and hence gain of the transition.

Because low-pressure molecular lasers are only discretely (i.e., line) tunable and because of the limitations of frequency-shifting technology, the exact operating frequency of a laser is often of paramount importance in those applications that use the high spectroscopic resolution these devices are capable of. For example, for a laser process to photochemically separate the isotopes of uranium, a 16 μm laser was needed whose frequency coincided with the desired absorption feature in UF_6 to an accuracy orders of magnitude better than existing spectroscopic data or theories could predict. The bulk of the data currently available[42–45] will usually allow the frequency of low-lying transitions in small molecules to be estimated to no better than ± 1 cm^{-1}, and only in special cases to better than ± 0.1 cm^{-1}. Even when spectroscopic constants for a species are well known and it is merely desired to calculate the transition frequencies for an isotopically labeled variant of the molecule, difficulties arise when accuracy of this degree is required because of perturbations in the spectra. Applications of simple formulas such as the "product rule"[8] will rarely yield numbers to better than ± 1 cm^{-1}.

In the 16 μm case alone, the effort expended by the laser community to overcome this one spectroscopic difficulty was massive. This reflects both the complexity of performing spectroscopy at the required level of absolute wavelength accuracy and the virtual impossibility of finding a coincidence between the set of narrow emission lines from a given low-pressure molecular laser and a specified, narrow absorption feature. If possible, efficient frequency-shifting technology[46,47] should be employed in such cases. At the least, any hypothesized spectral coincidence should be established as rapidly as possible by direct techniques such

[42] Extensive tables of spectroscopic constants are presented as appendices to Herzberg.[7–9]

[43] T. Shimanouchi, "Tables of Molecular Vibration Frequencies," Parts 1–8, and future parts of this regularly appearing series. Natl. Bur. Stand., Natl. Stand. Ref. Data Ser., US Govt. Printing Office, Washington, D.C., 1967–1974.

[44] K. Nakamoto, "Infrared Spectra of Inorganic and Coordination Compounds, 2nd ed. Wiley, New York, 1970.

[45] S. Pinchas and I. Laulicht, "Infrared Spectra of Labelled Compounds." Academic Press, New York, 1971.

[46] T. J. Manuccia, in "Lasers in Chemistry" (M. A. West, ed.), p. 210. Elsevier, Amsterdam, 1977.

[47] G. M. Carter, Appl. Phys. Lett. 32, 810 (1978).

as direct absorption or acousto-optic (spectrophone) spectroscopy, and not by absolute frequency measurements.

Even less quidance is usually available in estimating the strength of the vibrational–rotational transitions of interest. Group theory can readily determine if a given transition is symmetry forbidden,[3] but cannot predict how strong it will be when it is symmetry allowed.

Simple harmonic oscillator (s.h.o) selection rules state that for allowed transitions, in one and only one mode can the number of quanta change by plus or minus one unit. These rules can be useful in telling if a transition will be strongly allowed, but are virtually useless if the transition falls into the technically interesting category of being symmetry allowed but s.h.o forbidden. This category includes difference, combination, and overtone bands. These three traditional designations of spectroscopic bands are characterized by different number of quanta changing in each mode of the molecule during, say, the act of optical absorption. In the first case, the number of quanta in one mode increases while the number in another mode decreases. For example, in CO_2, the normal 10 μm lasting transition $(001) \rightarrow (100)$ involves a change of two quanta and would be considered a difference band designated by $\nu_3 - \nu_1$ or $\nu_1 \rightarrow \nu_3$. For a combination band, the number of quanta in two or more modes must all increase or decrease, as in the CO_2 transition $(000) \rightarrow (011)$. An overtone band is one in which the number of quanta in one mode changes by more than one unit, e.g., $(000) \rightarrow (002)$ or $2\nu_3$.

The reason that these bands are observed (i.e., have nonzero oscillator strength) is that mechanical anharmonicity and other interactions present in real molecules mix the s.h.o. basis wavefunction and allow s.h.o. forbidden transitions to "borrow" strength from s.h.o. allowed transitions. Such behavior is exceedingly common and has wide-ranging repercussions in fields ranging from laser development and laser-induced chemistry[48] to the theory of radiationless transitions.[49] As was calculated in the classic paper by Statz *et al.*[50] for carbon dioxide, wavefunction mixing accounts for both the usable cross section on the s.h.o. forbidden 9 and 10 μm lasing bands as well as being intimately related to the repulsion between the (100) and (02° 0) lower laser levels, the so-called Fermi-resonance effect.[51] Without wavefunction mixing, the recently discov-

[48] D. Frankel and T. J. Manuccia, *Chem. Phys. Lett.* **54**, 451 (1978).

[49] J. Jortner and S. Mukamel, *in* "Molecular Energy Transfer" (R. Levine and J. Jortner, eds.), p. 178. Wiley, New York, 1976.

[50] H. Statz, C. L. Tang, and G. F. Koster, *J. Appl. Phys.* **37**, 4278 (1966).

[51] M. D. Harmony, "Introduction to Molecular Energies and Spectra," p. 257. Holt, New York, 1972.

ered $(100) \rightarrow (01^10)$ cascade lasing band at 14 μm,[52-54] the $(00n) \rightarrow$ $(1, 0, n - 1)$ and $(00n) \rightarrow (0, 2, n - 1)$ sequence bands at 10 and 9 μm,[55,56] and the $(101) \rightarrow (011)$ difference bands at 17 μm[57] would all be inactive. Figure 1 is a partial energy level diagram of CO_2 showing the traditional lasing bands as well as some of these less familiar lasing bands.

While theoretical progress on estimating the strengths of difference and combination bands is being made by Overend[58] and co-workers and others, the safest assumption is that if two states are of the same symmetry and are within 100 cm^{-1} of each other or show a perturbation from an equispaced harmonic oscillator model of the normal mode involved, then the chances are excellent that s.h.o. selection rules will not be appropriate.

Such behavior can be seen in many other molecules besides $^{12}CO_2$. The situation in the $^{13}CO_2$, N_2O, and CS_2 lasers is almost exactly analogous to CO_2, except that the accidental coincidence between the (100) and (020) states that exists in CO_2 is much further off-resonance in these other molecules and so the corresponding two states are much less admixed. The net result is that while everpresent mechanical anharmonicities still allow a reasonable transition moment on the $(001) \rightarrow (100)$ band in $^{13}CO_2$, N_2O, and CS_2, the $(001) \rightarrow (020)$ band, which involves a change of a total of three quantum numbers, is reduced in intensity in $^{13}CO_2$,[59] is very weak in N_2O[60-63] and has never been observed in CS_2.

The removal of s.h.o. selection rules in molecules becomes crucial as the generation of new IR lasers assumes increasing importance as a means to provide low-gain (i.e., scalable) sources at new IR wavelengths for photochemical and laser isotope separation applications. It is trivial to go to a compendium of vibrational frequencies such as Shimanouchi[43] and find numerous species that either show absorption at the desired optical pumping wavelength or a near resonance with a sensitizer mole-

[52] T. J. Manuccia, J. Stregack, N. Harris, and B. Wexler, *Appl. Phys. Lett.* **29**, 360 (1976).

[53] R. Osgood, *Appl. Phys. Lett.* **28**, 342 (1976).

[54] W. H. Kasner and L. D. Pleasance, *Appl. Phys. Lett.* **31**, 82 (1977).

[55] J. Reid and K. Siemsen, *J. Appl. Phys.* **48**, 2712 (1977).

[56] B. J. Feldman, R. A. Fisher, C. R. Pollock, S. W. Simons, and R. G. Tercovich, *Opt. Lett.* **2**, 166 (1978).

[57] M. I. Buchwald, C. R. James, H. R. Fetlerman, and H. R. Schlossberg, *Appl. Phys. Lett.* **29**, 300 (1976).

[58] J. Overend, *J. Chem. Phys.* **64**, 2878 (1976).

[59] F. O'Neill and W. Whitney, *Appl. Phys. Lett.* **31**, 270 (1977).

[60] N. Djeu and G. Wolga, *IEEE J. Quantum Electron.* **qe-5**, 50 (1969).

[61] G. Mullaney, H. Ahlström, and W. Christiansen, *IEEE J. Quantum Electron.* **qe-7**, 551 (1971).

[62] A. Demin, E. Kudryavtsev, Yu. A. Kulagin, and N. Sobolev, *Sov. J. Quantum Electron. (Engl. Transl.)* **4**, 1393 (1975).

[63] T. Chang and O. Wood, *Appl. Phys. Lett.* **24**, 182 (1974).

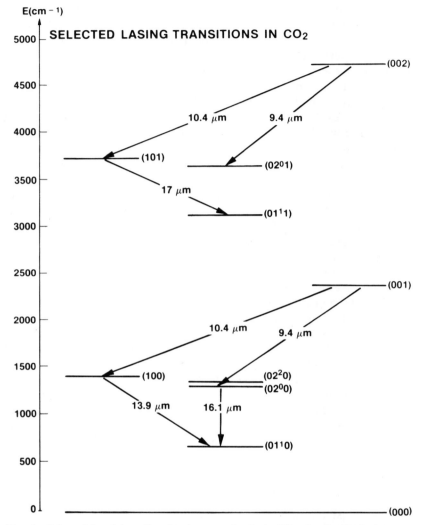

FIG. 1. Selected low-lying vibrational energy levels in CO_2 showing both the conventional 9 and 10 μm lasing transitions as well as some of the more recently discovered lasing bands.

cule such as CO, N_2, or D_2, and which have an energy level of the appropriate symmetry the appropriate spacing (say 16 μm) below the pumped level. Under the carefully controlled conditions of direct or sensitized optical excitation[64] or in an energy transfer mixing gas-dynamic laser (GDL), an inversion between the two levels can usually be attained at temperatures sufficiently low to remove the thermal population of the lower level. Because the two laser levels are often not connected by an s.h.o. allowed transition, there is always doubt as to whether or not there will be usable gain. It has not been until the past few years that experience has taught that quite often there will be a usable transition dipole, particularly if the total number of vibrational quanta that change during the transition does not exceed two or three.[65]

Many of the experiments cited above support this trend, as does the operation of sequence lasers in N_2O,[66] a very strong C_2H_2 $\nu_2 \rightarrow \nu_3$ 8 μm GDL,[67,68] and the CO-sensitized optical pumping experiments of Deutsch and Kildal[69] including C_2H_2 on the GDL transition, OCS ($\nu_3 \rightarrow \nu_1$) at 8.3 μm, and SiH_4 ($\nu_1 \rightarrow \nu_4$) and ($\nu_3 \rightarrow \nu_4$) at 7.99 μm. A good number of the optically pumped far-infrared (FIR) lasers listed in Rosenbluh et al.[70] are assumed to operate on difference bands, with the most blatent break from s.h.o. selection rules apparently occurring with the 190 μm (220) $\rightarrow$ (090) lasing band of DCN.[71] Optical pumping at 10 μm on s.h.o. forbidden combination bands has been recently used to advantage by Tiee and Wittig[72] to produce relatively high power mid-IR lasers in CF_4, NOCl, and CF_3I. Overtone emission has been used both as a diagnostic[73] as well as to produce a 2.8 μm CO $\Delta\nu = 2$ laser.[74]

7.1.2.2. **Collisional Processes.** After a pair of energy levels in a molecule has been singled out as having suitable spectroscopic properties for a hypothetical laser, the questions that must immediately be answered include: How do you pump the upper laser level (i.e., create molecules in this state)? Once a molecule is put into this state, how long will it stay there in the operating environment of the laser used for proof-of-principle experiments? How long will the molecule stay excited in the operating

[64] T. Deutsch and H. Kildal, *Chem. Phys. Lett.* **40**, 484 (1976).

[65] C. R. Jones, *Laser Focus* **14**, 68 (1978).

[66] K. Siemsen and J. Reid, *Opt. Commun.* **20**, 284 (1977).

[67] J. A. Stregack, B. L. Wexler, and G. A. Hart, *Appl. Phys. Lett.* **28**, 137 (1976).

[68] B. L. Wexler, unpublished data.

[69] T. Deutsch and H. Kildal, in "Tunable Lasers and Applications" (A. Mooradian, T. Jaeger, and P. Stokseth, eds.), p. 367. Springer, Berlin (1976).

[70] M. Rosenbluh, R. J. Temkin, and K. J. Button, *Appl. Opt.* **15**, 2635 (1976).

[71] J. Bruheau, P. Bellard, and D. Veron, *Opt. Commun.* **24**, 259 (1978).

[72] J. Tiee and C. Wittig, *J. Appl. Phys.* **49**, 61 (1978).

[73] R. Sileo and T. Cool, *J. Chem. Phys.* **65**, 117 (1976).

[74] R. C. Bergman and J. Rich, *Appl. Phys. Lett.* **31**, 597 (1977).

environment of a scaled-up version of the laser? Can molecules be removed from the lower state of the lasing transition and returned to the pool of molecules available for excitation, or will the lower state simply fill up and the laser be self-terminating? Finally, will the bandwidth of the transition be so large that the gain will be reduced to an unusable value?

The answers to all of these questions lie in the rates of various collisional processes in the gas. During collisions between molecules, energy transfer within or between degrees of freedom occurs on almost every collision. A fast-moving species may strike a slower one in an elastic collision and both will acquire new velocities. Such momentum-changing collisions are occasionally referred to as velocity relaxation and would result in changes in the width of spectral features due either to the Doppler effect or as a result of changing of the phase of each molecular oscillator.

If internal excitation of the collision partners undergoes a change in kind or degree, at least some of the energy usually winds up in the translational degrees of freedom. The collision would hence be inelastic, and the process is usually designated by which degrees of freedom lost and which degrees of freedom received the energy. Processes such as VT, VR, VV, RT, and EV relaxation are all important in molecular laser physics. As seen, abbreviations for the translation, rotation, vibration, and electronic degrees of freedom are commonly employed. Such notation is unfortunately not without ambiguity. Often, the available energy will be distributed to more than one degree of freedom. For example, deactivation of one vibrational mode of a polyatomic molecule to a lower-lying mode by a structureless collider can be referred to as either a VV or VT process to focus attention as desired at either vibration or translation as the accepting degree of freedom. With the inclusion of certain propensity rules, the physics of such a collision will be very similar to a case in which no vibration remained in the molecule, i.e., a pure VT event, and might be referred to in this way. On the other hand, because of subsequent use of the vibrationally excited species that remains (say to make a laser) it may be desirable to stress the VV and not the VT aspects of the event.

The kinetic theory of gases as discussed in standard texts[75-78] gives formulas relating the cross section for a binary collision process between molecules A and B to the average number of times per second, Q, the

[75] W. Kauzmann, "Kinetic Theory of Gases." Benjamin, New York, 1966.
[76] S. Benson, "The Foundations of Chemical Kinetics." McGraw-Hill, New York, 1960.
[77] J. O. Hirschfelder, C. F. Curtiss, and R. Bird, "Molecular Theory of Gases and Liquids." Wiley, New York, 1954.
[78] S. Chapman and T. Cowling, "The Mathematical Theory of Non-Uniform Gases." Cambridge Univ. Press, London and New York, 1952.

process under consideration occurs in a bulk sample of the gas under specified conditions, in units of events-cm^{-3} sec^{-1}. The cross section σ is defined as the number of times per second a transition is made from one region of phase space (reagents) to another region of phase space (products) at unit flux (cm^{-2} sec^{-1}) and specified relative initial velocities. The transition from σ to Q is then made by assuming an unperturbed Maxwell–Boltzmann distribution of velocities for A and B and, if necessary, an equation of state to predict the densities ρ (cm^{-3}) of A and B from pressure, temperature, and composition data.

In laser physics, with the possible exception of computer modeling, the quantity Q is only infrequently encountered as such. Rather, by dividing Q by either ρ_A or ρ_B, a pseudo-first-order molecular rate R in units of sec^{-1} is obtained. This is most similar to experimental qualities such as fluorescence decay times, and has the loose interpretation of being the number of times per second the process happens to a given molecule of A (if Q was divided by ρ_A) or to a molecule of B (if Q was divided by ρ_B). Note that the two rates R_A and R_B can be very different if, say, A is dilute in B. In this case, a given interaction between A and B will happen frequently to A, but only rarely to B.

This normalization process is usually carried one step further and the measured rate is divided by the density of the colliding species to get a rate constant k. The standard units for this quantity are sec^{-1} cm^3, but are often given by experimentalists as sec^{-1} $torr^{-1}$ or sec^{-1} atm^{-1}, or as the reciprocal, the so-called $\rho\tau$ of the process, most commonly in units of μsec-atm. Thus, $Q = k\rho_A\rho_B$. For comparison with chemical processes, the densities can be expressed in units of moles per liter and the rate constants expressed accordingly.

If desired for conceptual purposes, the rate constant for the process under consideration can be divided by the rate constant for some simple, fast reference process such as that for momentum transfer (i.e., collisions that determine bulk viscosity) and the ratio of the two rates taken to be the probability P of the event per gas-kinetic collision, or its reciprocal Z, the number of collisions between the event under consideration.

To allow rapid order-of-magnitude comparisons between the various conventions, if at 300 K the gas-kinetic reference rate for CO_2 is taken as 1.0×10^7 sec^{-1} $torr^{-1}$; then the same quantity in other units is 7.6×10^9 sec^{-1} atm^{-1}; 1.87×10^{14} cm^3 $mole^{-1}$ sec^{-1}; 1.87×10^{11} liter $mole^{-1}$ sec^{-1}; 1.87×10^8 m^3 $mole^{-1}$ sec^{-1}; 3.105×10^{-10} cm^3 $molecule^{-1}$ sec^{-1}; and 0.1315 nsec-atm. In this chapter we shall attempt to consistently use units of sec^{-1} $torr^{-1}$ and the collision number Z in keeping with their relatively widespread usage in the energy transfer literature.

Until now, the exact nature of the binary collision process was left unspecified except terming it as a transition from one portion of phase space to another.   This was done both for generality and because there are so many closely related cross sections, differing only in what parts of phase space are included as products, that confusion can readily occur.

For example, several values can often be found when looking up a molecular diameter (cross section).   The reason for this is that, indeed, the molecule does appear differently when studied by different techniques.   Measurements of viscosity, thermal conductivity, and diffusion all sample different regions of phase space, corresponding to different boundaries between reagents and products for effective momentum, energy, and particle transfer, respectively.   The second area susceptible to misinterpretation is the distinction between various state-to-state cross sections and total relaxation rates.   To be meaningful, any rate or cross section must be accompanied either implicitly or explicitly by a description of what final velocities, angles, quantum numbers were summed over, and what initial variables, if any, were averaged over.   It is only when such assumptions are left to be derived from context that misunderstanding can develop.

For example, the "rotational relaxation time" typically quoted for CO_2 is of order 100 nsec at 1 torr.[79,80]   This time is the time for the rotational distribution function to recover from a short, small perturbation.   On this basis, one would predict that at 10^3 torr, slightly above 1 atm, all of the rotational states could contribute to lasing for pulse lengths as short as ~ 0.5 nsec.   This is not experimentally observed because this conclusion is based on an inappropriate cross section.   The time of interest for the extraction of lasing energy is actually how long it takes to fill or drain an entire rotational manifold through one rotational state.   This time is considerably longer than the first time mentioned, because in the former case, essentially only one rotational state had to be filled or depleted by the rest of the rotational manifold.   The factor relating the two times is an "average" number of populated rotational states involved and is approximately equal to the value of the rotational partition function.

Of the various relaxation processes that can affect molecular laser operation, those involving transfer of the least amount of energy or the smallest quanta happen most readily.   The trend is that velocity relaxation is the most rapid (1 collision) followed by RT (1–10 collisions) and finally VT processes (10 to over 10^6 collisions), in order of decreasing rapidity.   Transfer of electronic energy can range from extremely rapid

[79] P. K. Cheo and R. L. Abrams, *Appl. Phys. Lett.* **14**, 47 (1969).

[80] R. R. Jacobs, S. J. Thomas, and K. J. Pettipiece, *IEEE J. Quantum Electron.* **qe-10**, 480 (1974).

(one or two orders of magnitude faster than gas kinetic) to completely metastable, depending on the individual species, states, and resonances involved.

7.1.2.2.1. VELOCITY AND ROTATIONAL RELAXATION. Velocity relaxation, sometimes known as spectral cross relaxation, has some importance in low-pressure, relatively low-power lasers where the Doppler contribution to the linewidth exceeds that component of the width, which is the result of pressure, lifetime, and power broadening. When this is the case, the medium is said to be inhomogeneously broadened and a monochromatic traveling wave passing through the medium interacts only with those molecules moving with velocities that Doppler shift their resonances into coincidence with the incident wave. If it were not for velocity-changing collisions, other molecules moving with different velocities could not interact with the radiation field. This would result in a decrease in the saturation flux. Since only those collisions that change the component of velocity v_z parallel to the propagation vector $\mathbf{k}$ of the incident wave are effective in this process, the rate of velocity relaxation can in principle be somewhat different than the rates of relaxation of either translational energies or momentia[78] in the gas. However, because of the averaging effects of the other two components of velocity, the differences between these rates is expected to be minimal. One-dimensional velocity relaxation is described analytically by a scattering kernel $R(v_z \rightarrow v_z')$, where the prime denotes quantities after collision.[81-84]

The effects of the finite rotational relaxation rate on molecular lasers are substantially more profound than those of the finite velocity relaxation rate even though the two processes proceed at similar rates. This is because, except in extreme cases, ever-present homogeneous line broadening mechanisms allow the field to interact to some extent with all of the velocity classes. When confronted by finite rotational relaxation rates, there simply are no mechanisms that allow a monochromatic field to have immediate access to the populations of other rotational states. Collisions are required. If the rate of rotational energy transfer is slow on a time scale of a pulsed laser or is the slowest step in the excitation–deexcitation cycle of a cw laser, the lasing output will be reduced.[85,86]

[81] J. Keilson and J. Sotrer, Q. Appl. Math. 10, 243 (1952).
[82] C. Freed and H. Haus, IEEE J. Quantum Electron. p. 56 9, 219 (1973).
[83] P. Berman, J. Levy, and R. Brewer, Phys. Rev. A 11, 1668 (1975).
[84] J. H. Henningsen and H. Jensen, IEEE J. Quantum Electron. p. 56 11, p. 248 (1975).
[85] G. T. Schappert, Appl. Phys. Lett. 23, 319 (1973).
[86] T. F. Stratton, in "Higher Power Gas Lasers—1975" (E. R. Pike, ed.), Conf. Ser. No. 29, p. 284. Inst. Phys., London.

With the exception of molecules with low moments of inertia such as H_2, D_2, CH_4, and the hydrogen halides, measurements of the rotational relaxation time for a variety of diatomics and triatomics cluster in the range of 1–10 collisions.[87,88] Anomalously long times (50–200 collisions) have been reported for H_2 and D_2 and are undoubtedly due to the spacing between rotational levels being comparable to kT. In the case of HF, the opposite behavior is seen, and rotational relaxation rates somewhat faster than the gas-kinetic rate have been reported.[89,90] In this case, very long range dipolar interactions together with the occurrence of hydrogen bonding make for very efficient transfer. Rotational relaxation of SF_6 is also about five times faster than gas kinetic,[91] probably due to the very closely spaced rotational states of this species. With the exception of H_2 and similar species, no strong temperature dependences are usually encountered. There is, however, a strong dependence of the state-to-state rate on the rotational quantum number J when the spacing between rotational states at high J approaches and exceeds kT. Polanyi and Woodall[92] have hypothesized that $k(J, J') \propto e(-C|\Delta E_{J,J'}|)$.

7.1.2.2.2. VIBRATIONAL RELAXATION. Of the various collisional processes, vibrational relaxation has the most profound influence on molecular IR lasers because of its direct control of excited-state populations. While it is usually only of secondary importance whether the translational and rotational distribution functions are equilibrated, the very existence of almost all molecular lasers depends on the existence of a nonequilibrium situation among the vibrational levels, or at least a large difference in effective temperature between the vibrational and other degrees of freedom. Because of the wide range of rate constants encountered for relaxation processes grouped under this heading and the theoretical difficulty in predicting individual rates, those rates that directly influence the inversion will usually be the most important unknown confronting the experimentalist.

Simple VT deactivation of the upper laser level at a constant pumping rate determines the maximum number density (cm^{-3}) in this level, and hence whether or not there exists an inversion between this level and a lower-lying level that generally contains substantial thermal population. For example, in a quasi-CW CO_2 laser, if the pumping rate into the upper

[87] C. Bamford and C. Tipper, eds., "Comprehensive Chemical Kinetics," Vol. 3. Elsevier, Amsterdam, 1969.

[88] B. Stevens, "Collisional Activation in Gases." Pergamon, Oxford, 1967.

[89] N. C. Lang, J. C. Polanyi, and J. Wanner, *Chem. Phys.* **4**, 219 (1972).

[90] J. J. Hinchen, *Appl. Phys. Lett.* **27**, 672 (1975).

[91] P. F. Moulton, D. M. Larsen, J. N. Walpole, and A. Mooradian, *Opt. Lett.* **1**, 51 (1977).

[92] J. Polanyi and K. Woodall, *J. Chem. Phys.* **56**, 1563 (1972).

(001) level is assumed to be known and does not vary too rapidly in time, then the population in this level (neglecting thermal population) cannot exceed the value given by obtaining the steady-state solution of

$$\frac{d}{dt}[(001)] = -\sum_i k_{\text{VT},i}[(001)][M_i] + R, \qquad (7.1.1)$$

where the M_i include all species that deactivate (001) at rates $k_{\text{VT},i}$, and R (cm^{-3} sec^{-1}) is the pumping rate.

The steady-state solution is simply

$$[(001)]_{\text{max}} = \frac{R}{\sum_i k_{\text{VT},i}[M_i]}. \qquad (7.1.2)$$

If excitation is applied in the form of a step-function at $t = 0$, the population of this state will exponentially approach $[(001)]_{\text{max}}$:

$$[(001)]_t = [(001)]_{\text{max}}[1 - \exp(-t/\tau_{\text{VT}})], \qquad (7.1.3)$$

where τ_{VT} is given by

$$\tau_{\text{VT}}^{-1} = \sum_i k_{\text{VT},i}[M_i]. \qquad (7.1.4)$$

If the excitation is pulsed with a duration considerably less than τ_{VT}, then

$$[(001)]_t \cong [(001)]_{t=0^+}e^{-t/\tau} \qquad (7.1.5)$$

where τ_{VT} is as given above.

Solutions to this and other highly simplified models allow rapid, order-of-magnitude comparisons between various effects. For example, a comparison of the photon build-up time with τ_{VT} allows one to see if energy extraction in high-pressure lasers is being limited by simple VT removal of the upper state.

Occasionally, the analysis of upper-state decay is complicated by some unique feature of the system under consideration. For example, in the 16 μm CO_2 laser[52] and in optically pumped lasers where a combination band is excited,[65] the most rapid process that removes upper-state population is a near-resonant VV process that produces one or two molecules in the lower laser state for each removed from the upper state. In the collision

$$(020) + (000) \longrightarrow 2(010) \qquad (7.1.6)$$

less than 50 cm^{-1} of energy is exchanged between translational and vibrational degrees of freedom and the process proceeds rapidly with a rate constant of about 10 gas kinetic collisions, or about 10^6 sec^{-1} torr^{-1}. In this process, for every molecule removed from (020), two appear in (010) and consequently, the inversion ($N_{upper}-N_{lower}$), which is the actual quantity of interest, decreases by three molecules.

Collisional excitation of the upper laser level is a common technique for providing efficient pumping of lasers. Typically, an easily pumped sensitizer species such as N_2, CO, or D_2 is either premixed with the lasing species and the mixture subjected to electron impact excitation or optical pumping, or else the sensitizer is excited separately and then rapidly mixed with the lasing species in a fast-flow mixing nozzle. The second approach does not subject the lasing species (which may be chemically reactive or unstable) to the harsh conditions that accompany electron impact excitation.

The advantages of the mixing scheme can be seen upon comparison of the 8 μm C_2H_2 laser operating in a mixing gas-dynamic laser[67] with premixed operation in an electron-beam sustainer device.[93] In the first case, the power output was strong and stable, almost equal to that of CO_2 used in the same device, whereas in the second configuration, lasing was considerably weaker than CO_2 and was accompanied by shot-to-shot performance degradation and the appearance of polymeric decomposition products.

Taking the typical CO_2-N_2-He mixture as an example of collisional excitation, pumping of the (001) state of CO_2 occurs by near-resonant VV exchange with the first excited vibrational state of nitrogen:

$$CO_2(000) + N_2(v = 1) \underset{k_r}{\overset{k_f}{\rightleftharpoons}} CO_2(001) + N_2(v = 0) - 17 \quad cm^{-1} \quad (7.1.7)$$

the rate constant for near-resonant collisions of this kind is usually in the range of 5–20 collisions and usually shows a weak negative temperature dependence.

If other sensitizer–acceptor systems such as $CO-CS_2$, $CO-C_2H_2$, and $CO-N_2O$[67] are examined, both laser experiments as well as energy transfer measurements show that the rate of transfer drops substantially if the energy mismatch is larger than kT (i.e., ~ 210 cm^{-1} at 300 K), and for a given mismatch is slower in the endothermic direction. This occurs because the ratio of these rates is a pseudo-equilibrium constant $e^{-\Delta E/kT}$. For an endothermic energy mismatch of a few kT, the rate constant can decrease by two or more orders of magnitude and the transfer will typi-

[93] L. Nelson, C. Fisher, S. Hoverson, S. Byron, F. O'Neill, and W. T. Whitney, *Appl. Phys. Lett.* **30**, 192 (1977).

cally show a strong positive temperature dependence. As the energy mismatch further increases, the collision takes on more character of a pure VT event and the temperature dependence of the rate can approach the celebrated Landau–Teller[88] form

$$k \sim T^{-1/6} \exp \left\{ \frac{A}{kT} - \frac{B}{(kT)^{1/3}} \right\}. \tag{7.1.8}$$

Vibrational relaxation of the lower laser level is generally beneficial to the lasing process. It forces the population in this level to approach the thermal equilibrium value, thereby increasing the unsaturated gain and hence power output of the laser. It also increases the saturation parameter of the medium by releasing molecules that have already participated in the lasing process to continue the excitation–lasing–deexcitation cycle. This also results in increased output.

In the absence of a saturating optical flux in the medium, if the lower-state relaxation rate is decreased, the population in the lower state will almost inevitably increase and this may eliminate any possibility for a population inversion. This occurs either because of limited selectivity of the excitation process or because one of the products of relaxation of the upper lasing level is the lower level. The second case occurs generally in all molecular lasers while the first is common in all but optically pumped systems.

In an electrical discharge in pure CO_2 in the absence of stimulated emission, the following simple model could be solved in the steady-state approximation (i.e., all time derivatives equal zero) to estimate [Eq. (7.1.13)] the population inversion:

$$e^- + (000) \xrightarrow{k_1} (001), \tag{7.1.9}$$

$$(001) + M \xrightarrow{k_{VT,u}} (000) + 2350 \quad cm^{-1}, \tag{7.1.10}$$

$$e^- + (000) \xrightarrow{k_2} (100), \tag{7.1.11}$$

$$(100) + M \xrightarrow{k_{VT,l}} (000) + 1388 \quad cm^{-1}, \tag{7.1.12}$$

$$\Delta N = [001] - [100] = \left[\frac{k_1[e^-]}{k_{VT,u}[M]} - \frac{k_2[e^-]}{k_{VT,l}[M]} \right] [(000)]. \tag{7.1.13}$$

While this model does show the effects of limited excitation selectivity, its quantitative predictions will be seriously in error due to the elimination of the (02⁰0), (02⁰0), and (01¹0) states from the model, as well as the neglect of routes of deactivation of (001) that pump the other two modes.

In particular, for every quantum lost from the asymmetric stretch, the sequence of events

$$(001) + (000) \longrightarrow (110) + (000) + 274 \quad cm^{-1}, \qquad (7.1.14)$$

$$(110) + (000) \longrightarrow (100) + (010) + 21 \quad cm^{-1}, \qquad (7.1.15)$$

places two quanta in the modes to which the lower state is coupled.

The state (100) does not relax directly to the ground state, but rather is rapidly coupled to adjacent and lower states by near-resonant processes such as

$$(100) + (000) \longrightarrow (02^2 0) + (000) + 50 \quad cm^{-1} \qquad (7.1.16)$$

$$(100) + (000) \longrightarrow (02^0 0) + (000) + 100 \quad cm^{-1} \qquad (7.1.17)$$

$$(100) + (000) \longrightarrow 2(01^1 0) + 50 \quad cm^{-1} \qquad (7.1.18)$$

$$(02^2 0) + (000) \longrightarrow 2(01^1 0) + 1 \quad cm^{-1} \qquad (7.1.19)$$

$$(02^0 0) + (000) \longrightarrow 2(01^1 0) - 50 \quad cm^{-1} \qquad (7.1.20)$$

$$(100) + (100) \longrightarrow (200) + (000) + 28 \quad cm^{-1} \qquad (7.1.21)$$

The overall removal of vibrational energy from the symmetric and bending modes is accomplished slowly, long after states in these modes have equilibrated with each other, by the process

$$(01^1 0) + (000) \longrightarrow 2(000) + 667 \quad cm^{-1} \qquad (7.1.22)$$

Thus, the model outlined in Eqs. (7.1.9)–(7.1.12) would have to be expanded at least by inclusion of Eqs. (7.1.14) and (7.1.15) and the following processes:

$$(100) \underset{}{\overset{k_{eq}}{\rightleftharpoons}} 2(01^1 0), \qquad (7.1.23)$$

$$(01^1 0) + (000) \xrightarrow{k_{VT,010}} 2(000), \qquad (7.1.24)$$

to even begin to approximate reality. In the case of a multicomponent gas mixture the kinetic modeling problem becomes even more severe, and rarely if ever in the development of a new laser will the energy transfer pathways be as well established as in CO_2.

These complexities illustrated by CO_2 do serve to point out the importance of three general phenomena in vibrational energy transfer: (a) the propensity of collisions to occur rapidly when the vibrational energy that is exchanged with the translational energy is less than kT; (b) the ability of certain modes of molecules to relax together, at a different rate than other modes relax at; and (c) the general phenomena in molecules of relaxation through the lowest vibrational state. This last effect is due to the rapid decrease in average spacing between energy levels once above the lowest-lying excited state and the propensity rule stated in (a), above.

During stimulated emission, the effects of slow relaxation of the lower laser level manifest themselves in the pulsed case in the form of a self-termination, while in the cw case in the form of a decreased saturation flux (the optical power density needed to reduce the gain on the transition of interest to one-half its small signal value).

Complete lower-state bottlenecking will limit energy extraction from a pulsed device to less than

$$E_{max} = \left[\frac{N_u^{(0)} - N_l(0)}{2} \right] h\nu, \qquad (7.1.25)$$

where the $N^{(0)}$ are the number of molecules in the upper and lower state, before lasing. The statement is presented as an inequality because (a) not all available optical energy can be extracted from a resonator, and (b) lasing will cease when gain equals loss, long before the inversion itself disappears.

During cw or quasi-cw, long-pulse operation, the possible existence of lower-state bottlenecking should be examined as a possible efficiency-reducing factor. If sufficient data are available, the net stimulated emission rate (as obtained, for example, by dividing the net optical power being generated by the photon energy) can be compared theoretically to the rates of lower-state relaxation and upper-state pumping.

Experimentally, it is often quite convenient to investigate this possibility by the examination of changes in N_u induced by lasing. Infrared fluorescence provides a commonly used, excellent probe. Given that resonator losses are known to be much less than the unsaturated gain of the system, a significant reduction in upper state population N_u upon lasing can only occur if the pumping process is the rate-limiting step. If little change is seen in upper-state population when the medium is saturated, the bottleneck occurs in the lower state and must be dealt with accordingly. While rotational bottlenecking (as discussed in the previous section) is usually of considerably less importance in the optimization of molecular lasers than vibrational bottlenecking, both processes should always be considered operative unless proven otherwise.

As the required level of detail and accuracy increases, it becomes obvious that there exist neither adequate predictive theories nor a sufficiently broad base of experimental measurements on which to base estimates of vibrational relaxation rates for species not already in common use as laser media. It is rare to obtain state-to-state rates at all and even more uncommon to have these available at the $\pm 20\%$ level of accuracy, often the minimum acceptable in laser development applications. Even for a molecule as thoroughly studied as CO_2, measurements of the rates of processes such as Eqs. (7.1.15)–(7.1.20) do not agree to better than a factor of two.[94,95] Similarly, experimental determination of the state-to-state VV and VT rates coupling the upper vibrational states of CO are subject to considerable technical and interpretational problems.[96] The difficulty often lies in the extraction of a large number of state-to-state rate constants, all possibly similar in magnitude, from very limited experimental data. The number of rate constants can be as large as $N(N-1)/2$, where N is the number of energy levels involved. Recent experimental data usually consist of fluorescence or double-resonance signals taken as a function of pressure, temperature, and gas composition. Only in rare cases is it possible to take data at many combinations of excitation and observation wavelengths.

Older experimental data taken by shock tube, spectrophone, ultrasonic dispersion, or similar classical methods provide a wealth of information,[87,88,97,98] and indeed should be examined, but such data are rarely detailed or accurate enough to provide guidance in the conception and development of new lasers.

The modern era of vibrational energy transfer research began with the use of lasers as selective excitation sources and is typified by the pioneering work done in the laboratories of George Flynn, C. B. Moore, and I. W. Smith, and their co-workers and students. Since this field spans the range of molecular complexity from hydrogen to N-carbon organics and in temperature from cryogenic to dissociation, only the most cursory summary of the results of this enormous body of research can be presented here. Only those trends and generalities applicable to laser development have been mentioned. For the rest, the reviews are mandatory.[99–108] Unfortunately, even the specialized reviews have not cov-

[94] O. P. Judd, "On the Fermi Resonance Levels in CO_2," Los Alamos Laboratory Internal Report LA-5892-MS. 1975.

[95] R. R. Jacobs, K. J. Pettipiece, and S. J. Thomas, *Phys. Rev. A* **11**, p. 54 (1975).

[96] H. T. Powell, *J. Chem. Phys.* **63**, 2635 (1975).

[97] T. L. Cottrell and J. C. McCoubrey, "Molecular Energy Transfer in Gases." Butterworth, London, 1961.

[98] G. M. Burnett and A. M. North, "Transfer and Storage of Energy by Molecules," Vol. 2. Wiley, New York, 1969.

[99] G. W. Flynn, in "Chemical and Biochemical Applications of Lasers" (C. B. Moore, ed.), Vol. 1, p. 163. Academic Press, New York, 1974.

[100] I. W. M. Smith, in "Molecular Energy Transfer" (R. Levine and J. Jortner, eds.), p. 85. Wiley, New York, 1976.

[101] E. E. Nikitin, "Theory of Elementary Atomic and Molecular Processes in Gases." Oxford Univ. Press (Clarendon), London and New York, 1974.

ered all work in this area and thus a thorough search of the original literature is also required if a particular rate is sought.

While order-of-magnitude estimates such as illustrated in this section are admittedly crude, they are often invaluable in defining the performance limits of newly discovered or proposed lasers. Often the uncertainties in and lack of basic data available on such systems makes more detailed analysis superfluous. On the other hand, if sufficient data are available, more sophisticated analyses such as the three- and four-level models discussed in Yariv[1] can be brought to bear. If the system matures to the point where detailed performance prediction is needed, the only recourse is a complete numerical model, which in addition to vibrational and rotational energy transfer and stimulated emission must include features such as plasma simulation, ion and neutral chemistry, gas transport, and diffusion, as needed. Examples of such large-scale simulation can be found for large CO_2 and other systems[109] and large HF/DF chemical lasers.[110]

7.1.3. Device Configurations

To achieve the seemingly easy goal of populating one molecular energy level while depopulating a lower level, a myriad of device configurations have been developed. These range from room-sized to hand held, can be single shot pulsed, high-repetition-rate pulsed, or cw. They can utilize static gas samples or gas flowing at supersonic velocities, and can be pumped by other lasers, lamps, heat (the thermally driven gas-dynamic laser), self-sustained glow discharges, electron beams, proton beams, or even nuclear reactors. This section will examine two methods that are commonly used in the development of new IR lasers. Because of its relative simplicity, optical pumping will be briefly discussed, while electron impact excitation will receive somewhat more attention. The other excitation sources mentioned above are rarely used in the development of new lasers and will not be discussed here.

[102] R. D. Levine and R. B. Bernstein, "Molecular Reaction Dynamics." Oxford Univ. Press (Clarendon), London and New York, 1974.

[103] E. Weitz and G. Flynn, *Annu. Rev. Phys. Chem.* **25,** 275 (1974).

[104] C. B. Moore, *Adv. Chem. Phys.* **23,** 41 (1973).

[105] R. L. Taylor and S. Bitterman, *Rev. Mod. Phys.* **41,** 26 (1969).

[106] D. Rapp and T. Kassal, *Chem. Rev.* **69,** 61 (1969).

[107] S. Ormonde, *Rev. Mod. Phys.* **47,** 193 (1975).

[108] B. L. Earl, L. A. Gamss, and A. M. Ronn, *Acc. Chem. Res.* **11,** 183 (1978).

[109] K. Smith and R. M. Thomson, "Computer Modeling of Gas Lasers." Plenum, New York, 1978.

[110] J. J. T. Hough and R. L. Kerber, *Appl. Opt.* **14,** 2960 (1975).

7.1.3.1. Optically Pumped IR Lasers. Optical pumping radiation can excite either the lasing molecule itself, a sensitizer molecule that absorbs IR photons[69] and then transfers its vibrational energy to the lasing species, or a sensitizer molecule that yields electronically excited fragments that can EV transfer[111-113] to the lasing species. Optical pumping is generally regarded as a "clean" pumping method because few other levels are populated by the excitation process and an inversion is usually rather straightforward to achieve merely by lowering the temperature of the medium. This removes population from lower-lying, thermally populated levels. When developing new optically pumped lasers, the principal areas of concern tend to center on those basic scientific issues such as spectroscopy and collisional processes, which were discussed in the previous section, rather than on the more technical and engineering issues normally associated with electron impact pumped lasers.

For example, the need for exact spectral overlap between the pumping source and the lasing medium is of central concern in the development of new optically pumped lasers. This requirement can be considerably relaxed if sensitized or EV excitation is employed. Because these latter approaches rely on near-resonant (within a few kT) collisional energy transfer, they are somewhat less selective in pumping a particular energy level than direct pumping. This concern increases in importance as the frequency of the pumped transition decreases into the far infrared (FIR) and the lower lasing level approaches the donor level and can thus be simultaneously excited with the upper laser level. On the other hand, sensitized excitation is a particularly general approach. For every pump/sensitizer combination, one has great freedom in the choice of acceptor/lasing species. For example, inactive vibrational modes can be excited using this approach because of the lack of strict optical selection rules in VV energy transfer. Relaxation of the spectral coincidence requirement in direct optical pumping can also be had by going to pump fluxes sufficiently large so that the Rabi frequency

$$\omega_R = \mu E/\hbar \qquad (7.1.26)$$

approaches the frequency mismatch between source and absorber.[114,115] In this limit, incoherent power broadening[4] or coherent Rabi sidebands[6] allow the upper state to be pumped at reasonable rates. This technique has recently been employed in the production of a powerful NH_3 laser

[111] A. B. Petersen, C. Wittig, and S. Leone, *Appl. Phys. Lett.* **27**, 305 (1975).

[112] A. B. Petersen, C. Wittig, and S. Leone, *J. Appl. Phys.* **47**, 1051 (1976).

[113] A. B. Petersen, L. Braverman, and C. Wittig, *J. Appl. Phys.* **48**, 230 (1977).

[114] H. Fetterman, H. Schlossberg, and J. Waldman, *Opt. Commun.* **6**, 156 (1972).

[115] T. Y. Chang and J. D. McGee, *Appl. Phys. Lett.* **29**, 725 (1976).

at 12 μm.[116-118] The distinction between direct linear absorption and Raman pumping becomes less distinct the further off resonance the pump source is.[115] High-incident fluxes can also be used to stimulate two-photon nonlinear absorption, thereby further increasing the probability of a spectral coincidence between a selected absorber molecule and a (presumably) efficient, line-tunable laser source. Lasing of CH_3F at 9.75 μm and a variety of NH_3 lines from 6 to 35 μm have been obtained in this way.[119,120]

Optically pumped gas lasers usually consist simply of a metal or glass cell surrounded by the appropriate optical components required to let the pump radiation efficiently interact with the medium while providing feedback (i.e., mirrors) at the laser frequency. Aside from the optics, the only design features that generally demand particular attention are: (a) ensuring a sufficiently large pumping flux to allow the gain coefficient to exceed distributed losses such as absorption by other transitions in the same molecule; (b) providing a sufficiently long path length to overcome mirror and diffractive losses; and (c) having a pumped volume that coincides with the mode volume of the optical resonator and that is large enough to supply the desired output energy. To achieve these ends, both longitudinal and transverse pumping have been employed. On occasion the lasing medium will be contained in a highly reflective optical light pipe or surrounded by flash lamps to ensure large pumping fluxes.

The gas-handling system associated with optically pumped devices is usually straightforward as only low flow rates are normally required for demonstration of lasing, low-repetition-rate experiments. In such experiments, the average power input to the medium is small and hence bulk temperature rise is effectively limited by direct heat conduction to the walls of the cell. Often, static gas fills can be employed because of negligible photolysis and decomposition of the lasing medium by the pump radiation.

Because of the simplicity of the experimental apparatus and the relative absence of complications in the initial excitation step, optical pumping is one of the first methods that should be considered for the production of new IR and especially FIR[121] wavelengths using gas phase molecular

[116] B. Walker, G. W. Chantry, and D. Moss, *Opt. Commun.* **20**, 8 (1977).

[117] R. G. Harrison and F. Al-Wathan, *Opt. Commun.* **20**, 225 (1977).

[118] E. Danielewicz, E. Malk, and P. D. Coleman, *Appl. Phys. Lett.* **29**, 557 (1976).

[119] D. Prosnitz, R. R. Jacobs, W. Bischel, and C. K. Rhodes, *Appl. Phys. Lett.* **32**, 221 (1978).

[120] R. R. Jacobs, D. Prosnitz, W. Bischel, and C. K. Rhodes, *Appl. Phys. Lett.* **29**, 710 (1976).

[121] P. Coleman, *J. Opt. Soc. Am.* **67**, 894 (1977).

laser technology. Optical pumping can even be employed at very high pressures (e.g., 42 atm) to obtain tunability where other excitation methods would be difficult if not impossible to implement. Often, the proof-of-principle experiments, estimates of line strengths, and initial kinetic data (quenching rates, etc.) can be obtained in an optical pumping configuration. In the near and mid-IR, larger-scale experiments aiming at high average power or good wallplug efficiency will generally employ some other pumping technique, if possible, because of limitations inherent to optical pumping.

Because of the selective nature of the optical pumping process, all of the energy input to the gain medium is funneled through a small number of states. This can result in vibrational, rotational, or velocity class bottlenecking due to finite collisional rates or saturation of the absorbing transition.[122-125] Because the maximum efficiency of any optically pumped laser is one photon out of every two absorbed, the maximum power or energy conversion efficiency will only be half of the ratio of photon energies. In practice, the actual efficiency will be considerably less than this because of process including relaxation in the long pulse and cw modes, finite photon build-up time in the short-pulse mode, and ever-present optical losses, which waste energy as well as terminating lasing before all the energy in the inversion can be extracted.

7.1.3.2. Electron Impact Excitation. Compared with optical pumping, the physics as well as the experimental apparatus for electron impact excitation of molecular IR lasers is usually more complex. The goal of this extra effort is generally efficiency and scalability. Considerable progress has been made toward these ends in the past several years. This is reflected by advances in high-power gas-dynamic lasers (thermal and electrical) as well as other fast-flow configurations, large electron-beam/sustainer devices, and numerical models capable of accuracy sufficient for the prediction of engineering performance data. The literature describing these developments is already well-reviewed, is voluminous, and cannot be covered within the stated emphasis of this part, i.e., the development of lasers operating at new wavelengths. For an overview of those aspects of electrically excited molecular IR lasers pertaining to efficient operation and moderate- to high-output powers, the reader is urged to consult five texts[11,12,14,16,109] mentioned earlier. These contain excellent descriptions and drawings of many commonly used devices such as the various longitudinal and transverse discharge geometries as

---

[122] C. O. Weiss, *IEEE J. Quantum Electron.* **qe-12**, 580 (1976).

[123] J. Henningsen and H. Jensen, *IEEE J. Quantum Electron.* **qe-11**, 248 (1975).

[124] G. Koepf and K. Smith, *IEEE J. Quantum Electron.* **qe-14**, 333 (1978).

[125] T. DeTemple and E. Danielewicz, *IEEE J. Quantum Electron.* **qe-12**, 40 (1976).

well as fast-flow configurations. Every attempt will be made to avoid duplication of this material herein.

Standing in contrast to this rapidly moving program is the rather low-key effort to produce new IR molecular lasers. The quest for brute-force power seems to have dominated the field for the past few years. With the exception of optically pumped sources, progress in new sources in the past 5 years has been slow compared with developmental work on existing transitions or compared with the level of activity in this area in the late 1960s or early 1970s.

Since 1974, the major developments in new electrically excited IR lasers have included: (a) the use of high pressures and isotopic substitution to broaden and shift emission lines to the point where gapless tunability over relatively wide ranges (i.e., 9–11 μm) can be routinely obtained; (b) the use of sensitized excitation in a fast-flow mixing configuration, the electric discharge gas-dynamic laser (EDGDL) of the Naval Research Laboratory; and (c) the development of lasers operating on new transitions in CO_2, N_2O, and CO, particularly the efficient 16 μm band of CO_2.

A significant negative result of this period has been the failure to achieve lasing from D_2–HCl mixtures with either electron-beam/sustainer pumping or in the EDGDL.[126,127] An electrically excited HCl laser would have offered operation in the same atmospheric transmission window that the chemical DF laser operates in without the necessity for carrying highly reactive fluorine containing fuels or changing the entire gas fill between shots.

7.1.3.2.1. HIGH-PRESSURE TUNABLE LASERS. For pressures over a few torr, the linewidth of vibrational–rotational transitions is determined by pressure broadening. Each line becomes Lorentzian in shape and the width is proportional to the rate of phase-interrupting collisions.[128] Since virtually all classes of collisions including VV, VT, RT, velocity changing, and M_z changing can induce a phase change in the molecular oscillator, the rate of phase-changing collisions will generally be larger than the sum of the rates of all of these processes. This will generally place it in the range of one to several times the gas kinetic collision frequency. For the 9 and 10 μm CO_2 bands, the pressure-broadening coefficient γ has been reported[129–131] to be of order 7 MHz/torr. Thus, at a pressure

[126] L. Y. Nelson, S. R. Byron and A. L. Pindroh, "Electrically Excited D_2-DF Transfer Laser," Final Report on Contract DAAH01-74-C-0618. U.S. Army Missile Command, Mathematical Sciences, Northwest, Bellevue, Washington, 1975.

[127] B. L. Wexler, private communication.

[128] R. G. Breene, Jr., "The Shift and Shape of Spectral Lines." Pergamon, Oxford, 1961.

[129] E. Gerry and D. Leonard, Appl. Phys. Lett. 8, 227 (1966).

[130] E. Murray, C. Kruger, and M. Mitchner, Appl. Phys. Lett. 24, 180 (1974).

[131] A. Biryukov, A. Yu. Volkow, E. Kudryavtsev, and R. Serikov, Sov. J. Quantum Electron. (Engl. Transl.) 6, 946 (1976).

of 5 atm, the widths of individual transitions will be about 1 cm^{-1}, comparable to the spacing between adjacent rotational lines, and one can expect continuous tunability, a major extension of the capabilities of most molecular lasers, which are only discretely (line) tunable. Because the pressure-broadening coefficients for different gases are similar in magnitude, similar behavior can be predicted for the N_2O and CS_2 lasers as well.

Continuous tunability was achieved[132–134] for all three gases using electron-beam sustainer excitation. In one device, using isotopically substituted variants of the lasing species, i.e., $^{13}CO_2$ and $^{13}CS_2$, nearly continuous tunability was achieved from 9.2 to 12.3 μm with one isotope and one chemical species lasing at a time. Because of arcing and mechanical strength limitations of the foil, windows, and other structural components, operation of such lasers near atmospheric pressure is highly desirable. Based on the results of Whitney and O'Neill, it seems clear that this could be done by utilizing mixtures of the isotopically substituted molecules to decrease the average spacing between adjacent rotational lines.

To achieve multiatmosphere operation in electric discharge lasers, special techniques must be employed to prevent the formation of arcs in the lasing medium. Arcs draw energy away from the rest of the discharge, destroy the optical homogeneity of the medium by virtue of large-index gradients, and destroy vibrational population inversions by virtue of their extremely high temperatures. They arise because of some spatial–temporal instability in the desired diffuse glow discharge. Instabilities in the electron density can arise from a large number of positive-feedback mechanisms such as temperature and electric field dependences of recombination, attachment, and ionization rates. Quantitative analysis of discharge stability is difficult and is complicated by the disproportionate effect minor species such as impurities and ion clusters play in determining the electron distribution function.[135] To overcome these instabilities, the most successful technique has been that of preionization or externally controlled ionization.

Providing a spatially uniform background ionization increases the energy threshold for many instabilities since useful amounts of energy can

[132] N. W. Harris, F. O'Neill, and W. T. Whitney, *Opt. Commun.* **10,** 57 (1976).

[133] F. O'Neill and W. T. Whitney, *Appl. Phys. Lett.* **28,** 539 (1976).

[134] V. N. Bagratashvils, I. Knyazev, V. S. Letokhov, and V. Lobko, *Sov. J. Quantum Electron (Engl. Transl.)* **6,** 541 (1976).

[135] K. Smith and R. M. Thomson, "Computer Modeling of Gas Lasers," Chapter 7. Plenum, New York, 1978.

be deposited in the medium at electric fields sufficiently low so that spatial variations in electron multiplication (i.e., ionization hot spots) can be dissipated by mechanisms such as diffusion and recombination before instability occurs.   While directed primarily at high-power applications, and extensively described in the appropriate literature,[16] this relatively new technique of externally sustained discharges has considerable potential for the production of new lasers.   In particular, external control of ionization allows more energy to be deposited and higher reagent pressures to be tolerated than in conventional self-sustained discharges.   This is particularly important where low gain is anticipated or when the lasing species is electron attaching and hence the discharge is more susceptible to instabilities.

O'Neill and Whitney[133] provided background ionization for the one microsecond duration of their pulse by means of a cold cathode, field emission electron gun.   A drawing of their apparatus is shown in Fig. 2. The 250 keV electrons from the high-vacuum chamber of the gun pass through a thin metal foil and enter the laser chamber.   In collisions with the neutrals, these fast electrons lose energy by ionizing a small fraction of the neutrals.   The secondary electrons so produced are generated with

TUNABLE HIGH-PRESSURE GAS LASER

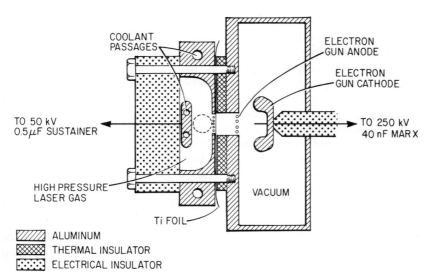

FIG. 2. The electron beam sustainer apparatus of O'Neill and Whitney[133] used to produce continuously tunable pulses from multiatmosphere gas mixtures.

near-thermal energies and are accelerated by the presence of a moderate (several kV/cm) electric field of the sustainer electrode to energies of order 1 eV, optimal for the excitation of the vibrational degrees of freedom of the medium, but below the energies required for a self-sustaining glow.

7.1.3.2.2. THE ELECTRIC DISCHARGE GAS-DYNAMIC LASER. The electric discharge gas-dynamic laser (EDGDL) was developed[67,136] specifically to enlarge the number of species that could be made to lase by electric discharge pumping. Previous to this device, lasers had been limited to molecules that could have their vibrational degrees of freedom either excited directly by electron impact or could be pumped by energy transfer from electrically excited N_2. With this device, D_2, CO, or N_2 can be used to provide a pool of vibrational energy available to resonantly transfer to the lasing species. By removing the lasing species from the discharge region and rapidly injecting it downstream, optimization of the vibrational excitation of the diatomic, donor species can be made essentially independently of conditions in the lasing region. In particular, rapid VT deactivation of the donor by the lasing species in the high-temperature environment of the discharge zone can be avoided, as can dissociation and other chemical reactions of the lasing molecule. Discharge stability problems are reduced by eliminating electron attaching gases from the discharge.

In the EDGDL (see Fig. 3), after passing through the discharge zone (the plenum in thermal GDL jargon), the primary gas flow passes through a band of converging–diverging nozzles to accelerate and cool the flow by converting the random thermal energy of the gas into directed kinetic energy.[137] The lasing species is injected into the supersonic flow field by small holes in each nozzle blade. Injection at this point can have several advantages for potential laser species: (a) promote anharmonic (Trainor) pumping of diatomics; (b) depopulate laser levels within a few kT_{room} of the ground state; (c) improve the mixing because of lower pressures; (d) decrease the rate of vibrational relaxation by virtue of the lower pressure and temperature; and (e) increase gain by narrowing linewidths and the rotational distribution function.

With this device, a number of unusual energy transfer cw lasers have been reported: D_2-CO_2, D_2-N_2O, $CO-CS_2$, $CO-C_2H_2$, $CO-N_2O$, and the 16 μm $N_2(D_2)-CO_2$ system, which will be discussed in the next section. During surveys of potential lasing species by monitoring the spontaneous infrared sidelight from the downstream region, several molecules

[136] J. A. Stregack, B. L. Wexler, and G. A. Hart, *Appl. Phys. Lett.* **27**, 670 (1975).
[137] W. Vincenti and C. H. Kruger, "Introduction to Physical Gas Dynamics." Wiley, New York, 1965.

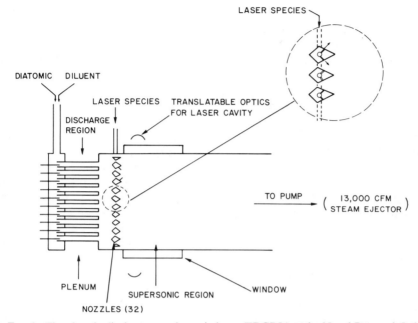

FIG. 3. The electric discharge gas-dynamic laser (EDGDL) at the Naval Research Laboratory. A detailed description of the individual components in the device can be found in Stragack *et al.*[67,136]

were found to emit strongly, suggesting that lasing would be possible with these species as well. These included the methyl halides, HCN, and HCl. This observation points out the potential and general applicability of this technique for the production of new lasers.

7.1.4. New Transitions

While it is rare to discover new and useful lasing transitions in seemingly well-studied systems, the examples which were briefly mentioned in Section 7.1.2. prove that success in this area is not impossible. In the short space of 2 or 3 years, at least six new transitions (not counting rotational components) were discovered in that most studied of all IR lasing molecules, CO_2. Two of these, the so-called sequence bands[55] had output powers comparable to the power of the usual 10.6 μm lasing transition in the same device, while the other two bands[52] had powers within a factor of four of that value. The remaining transitions[57] were optically pumped and will not be discussed further in this section.

The discovery[52] and evolution[54,138] of the electric discharge 16 μm CO_2 laser provides some insight into one strategy that can occasionally be employed in the development of a laser based on a new transition in a well-studied molecule. Because the lower states of the usual 9 and 10 μm lasing transitions in CO_2 (see Fig. 1) have strong optical transitions [(100)–(01^{1}0) and (02^{0}0–(01^{1}0)] terminating on a state some 600 cm^{-1} above the ground state, these transitions were prime candidates for new lasers. A population inversion on these could occur only when half the population in (001) exceeds the thermal population of (01^{1}0). Because the translational temperature in the low-pressure section of the NRL EDGDL is of order 100 K, and the effective vibrational temperature of the (00n) mode was known by independent measurement in the same device to exceed 1500 K, it was possible to predict that an inversion was possible on the desired transitions. The only obstacle that could prevent this was the fast relaxation processes known to equilibrate the symmetric and bending modes. To ensure that the population of (001) would be placed in the desired state before significant relaxation could occur, it was decided to adopt a "fail-safe" approach in which a fast-rising, externally generated saturating pulse would be used to instantaneously populate (100) or (02^{0}0), as desired. This combination of optical pumping and gas dynamic technology was successful and 14 and 16 μm lasing pulses were observed. From this point, the fail-safe approach was gradually discarded, first by letting the device generate its own 9 or 10 μm saturating pulse and then lasing in cascade on the other bands. After this was demonstrated, the rise time of the 9 μm pulse was lengthened until the device was operating cw, not pulsed. In another laboratory, operation of this laser was even further simplified by replacing gas dynamic cooling by a simple wall-cooled, slow-flow approach. The laser was now virtually indistinguishable from a typical 9 or 10 μm device.

The key element in the scientific development of the new laser transition was the incredible wealth of information already known about the system. This made it virtually impossible for the system not to have lased with the externally generated saturating pulse. It is certain that similar strategies can be applied to other potential new sources. This is certainly the case in the recently discovered overtone laser in CO.[74] In C_2H_2, there is virtually an identical situation to the 16 μm CO_2 laser in that there exists another 16 μm laser possibility arising from a known lasing transition. In this case, the oscillator strength of the proposed lasing transition is not as well known as in the case of the CO_2 transitions, and either independent measurements of its strength would have

[138] B. L. Wexler, T. J. Manuccia, and R. W. Waynant, *Appl. Phys. Lett.* **31**, 730 (1977).

to be made or else the actual medium should be probed to determine the gain length required for efficient operation.

Hopefully, this chapter has provided the reader with some of the flavor of the field and with a brief exposure to a useful number of fundamental scientific principles and useful devices in the area of new molecular IR laser development. Advances will likely come from the incorporation of seemingly unrelated concepts brought together via the driving force of application payoffs. Unique devices such as the electrically pumped, ultracompact waveguide laser[139] seem not to be yet fully applied to new mid-IR laser development and might provide one of the keys necessary for further progress.

---

[139] J. Degnan, *Appl. Phys.* **11**, 1 (1976).

7.2. Rare Gas Halide Lasers*

7.2.1. Introduction

Rare gas halide lasers generate short pulses of high-power ultraviolet radiation. These lasers are pumped electrically by electron beam, electron beam-sustainer, and discharge excitation of rare gas–halogen mixtures at atmospheric pressures. The excitation makes an upper laser level molecule, usually known as an excimer, by a bimolecular collision process. The excimer disappears by radiation to ground state. The bound-free transition makes high-power pulsed operation possible.

In this review all three electrical pumping techniques will be discussed, with the emphasis on discharge devices because of their low capital cost and portability. These features make a discharge device a practical laboratory tool. A discharge device can be used to make the principal rare gas halide lasers—ArF (193 nm), KrF (249 nm), and XeF (351, 353 nm), with peak powers of 10^5–10^7 W for 10–50 nsec. In addition, the same pumping apparatus can be used to make N_2 C–B (337 nm) and N_2^+ B–X (428 nm) lasers if desired. The discharge should uniformly excite a long thin volume in a pulse 10–50 nsec long. This requires transverse excitation with respect to the long dimension, which is the optical axis. A single pair of electrodes is used for this purpose. The discharge is started by preionization. In general, preionization provides initiatory electrons, which result in gas breakdown in the presence of the high electrical field between the long thin electrodes. The preionization technique is needed to uniformly start the discharge. Also the discharge starts at a much lower voltage across the electrodes. The preionizer generates UV light from a two-dimensional array of high-voltage sparks. The UV light photoionizes the laser gases and thereby creates the initiatory electrons. These electrons change the gas from an excellent insulator to a weak conductor. The initiatory electrons are accelerated by the electrical field and additional ionization results. For a stable discharge the electron concentration continues to grow until the production and loss processes balance each other.

---

* Chapter 7.2 is by S. K. Searles.

METHODS OF EXPERIMENTAL PHYSICS, VOL. 15B

7.2.2. Mechanism

In a rare gas halide laser, 1-atm mixtures of He with Xe, Kr, or Ar are used with F_2 or NF_3. The F_2 or NF_3 causes some initial steps to occur in addition to those already discussed for the general case. These steps and other detailed steps are listed in Table I. The electronegative F_2 rapidly converts the initiatory electrons to F^- by dissociative attachment, step (7), Table I. F^- is accelerated by the electrical field of ~ 15 kV/cm. Free electrons are finally created when an F^- strikes He with sufficient energy to detach the electron. These electrons gain energy from the electrical field, step (1), and lose energy through inelastic collisions with the rare gas–halogen mixture. A fraction of the electrons achieve a high enough energy to ionize the rare gas, step (2), and thereby increase the current across the discharge electrodes. This current amplification occurs rapidly. At the same time, energetic electrons excite the rare gas atoms to their lowest-lying electronic levels, step (3). These levels are $^3P_{0,2}$ and $^1P_1$, $^3P_1$, which are effectively metastable due to radiation trapping. The excited species R* reacts with F_2 or NF_3 via step (4) to produce the desired rare gas halide whose concentration must be $\sim 10^{15}/cm^3$ to achieve a gain coefficient of 0.1/cm. This bimolecular reaction takes place on every collision between R* and F_2. RF* disappears by emission, step (5), over a band several nanometers wide, by quenching, step (6), or by stimulated emission. These three processes compete with each other. The dominant process depends on the quenching gas pressure and on the intracavity laser flux.

TABLE I. Kinetic Mechanism[a]

	elec.
(1)	$e \xrightarrow{\text{field}} e'$
(2)	$e' + R \rightarrow R^+ + e + e$
	ΔE = 12.1 eV Xe, 14.0 eV Kr, 15.7 eV Ar
(3)	$e' + R \rightarrow R^* + e$
	ΔE = 8.4 eV Xe, 10.0 eV Kr, 11.6 eV Ar
(4)	$R^* + F_2$ (or NF_3) $\rightarrow RF^* + F$
(5)	$RF^* \rightarrow h\nu_{(uv)} + R + F$
	$\tau \approx$ 3–15 nsec
(6)	$RF^* + F_2 \rightarrow$ products
(7)	$e + F_2 \rightarrow F^- + F$
(8)	$e' + R^* \rightarrow R^+ + e + e$
	ΔE = 3.7 eV Xe, 4.0 eV Kr, 4.1 eV Ar

[a] R is a rare gas atom, Xe, Kr or Ar; R^+ is an ion; R* is an electronically excited state; e is a low-energy electron; e' is an energetic electron.

Steps (7) and (8) control the stability of the discharge. Once the concentration of R* reaches a high level, step (8) becomes important and releases electrons, which are partially removed from the discharge by (7). When the rate of step (8) exceeds the rate of step (7), the discharge becomes unstable and an arc forms. As a result a rare gas halide laser discharge is short-lived.

There are a number of underlying causes of the glow to arc transition. These causes distinguish the rare gas halide system from other systems. First of all, it is necessary to operate at a high E/N (ratio of the electric field to the number density) to obtain 8–10 eV electrons to excite Xe and Kr atoms to states that lead to XeF* and KrF* to cause laser emission. Xe* and Kr*, the lowest-lying electronically excited states, are within only ~4 eV of the ionization continuum. The cross section for exciting Xe* and Kr* by electrons is thought to be very large. It is impossible to make the needed 8–10 eV energy electrons without making 4 eV electrons. Thus two-step ionization is intrinsic to rare gas halide discharges and leads to instability as well as reduced electrical efficiency. The next section describes the experimental apparatus needed to make the high E/N discharge and to effect laser emission.

7.2.3. Apparatus

The electrical power supply is an important part of the apparatus. A supply with 10 J stored energy can make 100 mJ laser pulses. The capacitance is 30 nF for an electrode gap of 2 cm operated at 15 kV/cm-atm. The inductance can be determined from the equation $f = 1/2\pi(LC)^{1/2}$, where f is the inverse of the discharge duration. For the example cited above, $L = 2$ nH for a 50 nsec discharge. Care must be taken in order to obtain an inductance this low. Figure 1 shows a schematic drawing of the complete power supply. This particular power supply is a single-stage LC generator. Figure 2 indicates that the generator erects to only 25 kV before a large fraction of the stored energy is dissipated in the discharge. Furthermore, the ballasted voltage decreases at a time when the discharge impedance is rapidly diminishing due to two-step ionization. As a result, the discharge is stable on a transient basis. Figure 1 also shows the preionizer.

The discharge electrodes are usually made from aluminum and sandblasted to reduce reflections leading to optical parasitic phenomena. The electrodes must have a smooth continuously curved profile to prevent electrical field enhancement at the edges. A Rogowski profile is commonly used. The electrodes should be sized according to the laser output that is required. An active volume of 100 cm³ can generate about 0.1 J of

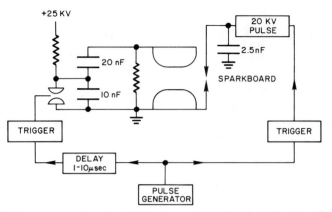

FIG. 1. Discharge-pumped rare gas halide laser schematic. Preionization of the laser gas mixture is provided by the spark-board photoionization source located approximately 4 cm from the laser axis. Energy storage in the main discharge capacitors is ~10 J. Overall efficiency of the laser is somewhat greater than 1%.

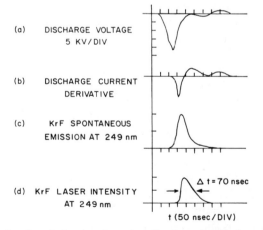

FIG. 2. Electrical and optical pulse characteristics from the discharge-pumped rare gas halide laser. (a) Discharge voltage. Gas breakdown occurs at approximately 10 kV/cm followed by rapid dumping of the stored electrical energy. (b) Time derivation of the discharge current. Current pulse width is approximately 50 nsec. Peak current approximately 10^4 A or 10^3 A/cm². (c) Spontaneous emission from KrF. Spontaneous emission follows discharge current very closely indicating rapid formation of KrF*. (d) KrF laser pulse. Laser initiation is delayed from onset of spontaneous emission by time required for the buildup of oscillation.

optical energy.   The electrodes and preionizer can be conveniently housed in a plexiglas container.   This container is connected to a gas-handling system.   The handling system introduces the required gases into a mixing tank.   Gas from the mixing tank flows through the laser chamber at about 1 liter-atm/min.   He with Xe, Kr, or Ar and with F_2 or NF_3 is needed. F_2 should be purchased as a mixture of He and F_2.   The mixture is safer to handle than pure F_2.   The mixture should be allowed to slowly flow through the entire system at a low pressure to passivate the metal surfaces.   It is more practical to run the dilute F_2 flow directly through the mechanical pump and change the oil periodically rather than trap out the F_2.   The pump exhaust must be vented outside the laboratory. NF_3 is an alternative to F_2.   It is noncorrosive but toxic. SF_6 has been tried but results to date have been negative.

Laser optics consist of Brewster angle windows and dielectrically coated mirrors.   The windows should be UV grade fused silica.   A separate set of mirrors are necessary for each rare gas halide laser.

7.2.4. Results

The performance of the UV preionized discharge lasers reported by Burnham and Djeu[1] can be summarized as follows.   At 1500 torr total pressure, the output of each of the rare gas halide lasers was optimum. For the XeF laser 0.3% NF_3 and 1% Xe gave 65 mJ energy per pulse from a cavity with a total reflector and a reflector with 70% T.   Substitution of F_2 for NF_3 decreased the output by 20%.   Laser lines on the bound–bound transition were at 351.1 and 353.2 nm.   Weaker transitions have also been reported at 348 nm.[2]   The 351 nm line has been mode-locked by Wittig and co-workers.[3]   They observed pulse durations of ≤ 2 nsec from a Blumlein drive device with acousto-optic gain modulation.

On the bound–free KrF transition laser emission[1] occurred at 248.4 nm with $\Delta\lambda = 0.3$ nm.   130 mJ was obtained from a 0.3% F_2, 15% Kr mixture in a cavity with 50% T.   Replacement of NF_3 for F_2 diminished the output by nearly a factor of three.   In a low-loss cavity a transition was also observed at 249.1 nm.   Burnham and Djeu[1] present a detailed survey of the effect of gas composition on laser performance of both XeF and KrF.   Wang and his associates[4] have utilized a Blumlein-type fast dis-

[1] R. Burnham and N. Djeu, *Appl. Phys. Lett.* **29**, 707 (1976).

[2] J. Tellinghuisen, G. C. Tisone, J. M. Hoffman, and A. K. Hays, *J. Chem Phys.* **64**, 479 (1976).

[3] C. P. Christensen, L. W. Braverman, W. H. Steier, and C. Wittig, *Appl. Phys. Lett.* **29**, 424 (1976).

[4] C. P. Wang, H. Mirels, D. G. Sutton, and S. N. Suchard, *Appl. Phys. Lett.* **28**, 326 (1976).

charge to achieve pulse repetition rates of 20 Hz. Sixty millijoules was obtained from ArF on its bound–free transition at 193.3 nm. The output spectrum was structured as a result of O_2 Shuman–Runge absorption within the laser cavity.

A fast-discharge device[4] used to make rare gas halide lasers can also be used to make other lasers with only minor modifications. Bigio and Begley[5] reported oscillation on five atomic fluorine lines near 710.0 nm from NF_3 and He mixtures. Peak power was 70 kW. Laser operation was also observed from excited Ne, only in fast discharge, 100 kW in 3 nsec at 540.1 nm, and N_2, a second positive band transition at 237.1 nm. One hundred eighty kilowatts predominantly at 427.8 nm was obtained by Landenslager *et al.*[6] with a preionized transverse discharge filled with He and N_2 and operated on the B–X transition of $N_2{}^+$. Thus a fast-discharge device can be used to make rare gas halide lasers efficiently and to make other lasers as well.

For some applications long laser pulses are necessary. These pulses can only be obtained from electron beam (e-beam) controlled discharges. Because the size and cost of an e-beam sustainer system is substantial, it is not a common device and little time will be spent discussing it. The device consists of two separate components—a pair of discharge electrodes and power supply and a cold cathode gun and its 200–300 keV power supply. The e-beam is generated by field emission. The beam enters the laser cavity through a thin metal window, usually a 25-μm-thick Ti foil. The 200–300 keV high voltage is required for the beam to pass through the foil. The beam ionizes the laser gases. The discharge electrical field accelerates the electrons resulting from ionization and excites rare gas atoms to their metastable levels. The beam remains on as long as the discharge is on. The other steps are the same as outlined in Table I for the preionized discharge. The important difference between the two kinds of discharge is that the processes of ionization and excitation are separate in the e-beam-controlled discharge. Separation of the two processes made the e-beam sustainer device a practical tool for CO_2 laser amplifiers.[7] In a CO_2 laser, 1–2 eV electrons cause excitation, 15 eV electrons cause one-step ionization, and 4 eV electrons cause two-step ionization at high current densities. Therefore, while the e-beam sustainer was quite practical for CO_2, it is unlikely to be as effective for rare gas halides. The energies needed for ionization and excitation lie too close together.

[5] I. J. Bigio and R. F. Begley, *Appl. Phys. Lett.* **28**, 263 (1976).

[6] J. B. Laudenslager, T. J. Pacala, and C. Wittig, *Appl. Phys. Lett.* **29**, 580 (1976).

[7] M. C. Richardson, K. Leopold, and A. J. Alcock, *IEEE J. Quantum Electron.* **qe-9**, 934 (1973).

Researchers at AVCO[8-10] have published a number of papers treating the theory of the controlled discharge. They have also reported laser emission from XeF in atmospheric pressure mixtures of 0.1% NF_3, 0.4% Xe, and 99.5% Ar.[11] The mean discharge current and voltage were 75 A/cm² and 11 kV/cm, respectively. The high e-beam current was 12 A/cm². In 0.1% F_2, 2% Kr, 97.9% Ar mixtures laser emission was observed on KrF.[12] Laser pulse energies were low compared to current unpublished work.

Short-pulse e-beams provide a third way of pumping rare gas halide lasers. In contrast to the e-beams used to sustain a controlled discharge,

TABLE II.   E-Beam Pumped Rare Gas Halides

Molecule	nm	Performance	Reference
XeF	351, 353	1.0 J, 0.2 × 10⁸ W	a
KrF	248.4, 249.1	108 J, 1.9 × 10⁹ W	b
		5.0% efficiency	c
ArF	193.3	92 J, 1.6 × 10⁹ W	b
XeCl	307.9, 308.1	1.5 × 10⁴ W	a
KrCl	222	10⁴ W	d,e
ArCl	175.0		f
XeBr	282.8	10³ W	a,g
KrBr	203ⁱ		h

ᵃ G. C. Tisone, A. K. Hays, and J. M. Hoffman, *in* "Electronic Transition Lasers" (J. I. Steinfeld, ed.), p. 191. MIT Press, Cambridge, Massachusetts, 1976.

ᵇ J. M. Hoffman, A. K. Hays, and G. C. Tisone, *Appl. Phys. Lett.* **28**, 538 (1976).

ᶜ M. L. Bhaumik, R. S. Bradford, Jr., and E. R. Ault, *Appl. Phys. Lett.* **28**, 23 (1976); R. S. Bradford, Jr., W. B. Lacina, E. R. Ault, and M. L. Bhaumik, *Opt. Commun.* **18**, 210 (1976).

ᵈ J. G. Eden and S. K. Searles, *Appl. Phys. Lett.* **29**, 350 (1976).

ᵉ J. R. Murray and H. T. Powell, *Appl. Phys. Lett.* **29**, 252 (1976).

ᶠ R. Waynant, *Appl. Phys. Lett.* **30**, 234 (1977). NOTE: A 1.6 m fast Blumlein discharge was used in this work rather than an e-beam.

ᵍ S. K. Searles and G. A. Hart, *Appl. Phys. Lett.* **27**, 243 (1975).

ʰ J. J. Ewing and C. A. Brau, *Phys. Rev. A* **12**, 129 (1975).

ⁱ Calculated, not yet observed.

[8] J. P. Daugherty, J. A. Mangano, and J. H. Jacobs, *Appl. Phys. Lett.* **28**, 581 (1976).

[9] J. H. Jacob and J. A. Mangano, *Appl. Phys. Lett.* **28**, 724 (1976).

[10] J. Hsia, *Appl. Phys. Lett.* **30**, 101 (1977).

[11] J. A. Mangano, J. H. Jacob, and J. B. Podge, *Appl. Phys. Lett.* **29**, 426 (1976).

[12] J. A. Mangano and J. H. Jacob, *Appl. Phys. Lett.* **27**, 495 (1975).

the short-pulse e-beam operates at orders of magnitude higher current densities for a duration of 10–50 nsec. The ability of a short-pulse e-beam machine to deliver very high pulse energies makes this device a good tool to discover new lasers. In fact, nearly all of the rare gas halide lasers were discovered by e-beam pumping. However, the size and expense of an e-beam machine limits the usefulness of this device. Table II lists some laser results obtained by e-beam pumping. The peak laser power of 1.9 GW obtained at Sandia with a 2 MeV, 6 kJ gun is a record for the UV and visible wavelengths.

Of all the rare gas halide lasers only the fluorides look interesting for an application where electrical efficiency is important. The KrF laser has been used to pump p-terphenyl and generate tunable laser emission from 335 to 346 nm.[13] The KrF laser itself has been tuned over 20 nm.[14] These lasers also can be used for spectroscopic studies in the UV or can be used for kinetic studies because the pulse duration is very short.

[13] D. G. Sutton and G. A. Capelle, *Appl. Phys. Lett.* **29**, 563 (1976).
[14] R. Burnham, private communication (1976).

8. CHEMICALLY PUMPED LASERS*

8.1. Introduction

During the past decade powerful devices have been developed that directly convert the energy release of chemical reactions into coherent electromagnetic radiation. The operation of all of these chemical lasers depends on the conversion of the energy release of chemical reactions into the specific (nonthermal) excitation of product energy states. The most successful chemical lasers to date operate at wavelengths in the near infrared (2.7–10.6 μm) part of the spectrum and are based on disequilibrium among the vibrational states of reaction product molecules. Chemical lasers operating on rotational and electronic transitions also exist, but have not yet been developed to the extent of the vibrational chemical lasers.

Continuous-wave HF (2.7 μm) and DF (3.8 μm) chemical lasers capable of power levels above 100 kW can now be constructed. Such devices operate at chemical efficiencies of a few percent and require little, if any, nonchemical energy for their operation. Pulsed HF chemical lasers with gigawatt peak powers and energies of several kilojoules have also been built.[1]

An extensive search is in progress for chemical reactions that efficiently form electronically excited metal oxide and metal halide reaction products. So far no one has succeeded in achieving chemical laser action between electronically excited states of diatomic reaction products, but a cw chemically pumped atomic iodine laser has been developed that is the first example of a purely chemical electronic transition laser.[2] This device operates at a 1.315 μm wavelength and shows much promise for development as a useful high-power output device.

Recently a class of excimer lasers has been discovered that are pumped by chemical reactions between halogen molecules and excited metastable

[1] R. A. Gerber, E. L. Patterson, L. S. Blair, and N. R. Greiner, *Appl. Phys. Lett.* **25**, 281 (1974).

[2] W. E. McDermott, N. R. Pchelkin, D. J. Benard, and R. R. Bousek, *Appl. Phys. Lett.* **32**, 469 (1978).

* Part 8 is by Terrill A. Cool.

95

rare gas atoms.[3–14] While these new excimer systems do not qualify as "purely chemical" devices, since they are dependent upon electrical excitation of the metastable atoms, they demonstrate that very high laser pulse energies at ultraviolet wavelengths can be achieved through chemical reaction. Only a little imagination is required to foresee the eventual development of purely chemically pumped excimer lasers at visible and ultraviolet wavelengths.

Several reviews[15–20] and a "chemical laser handbook"[21] are available that describe the historical development of the chemical laser field and provide specific information on virtually every type of chemical laser that had been developed by the end of 1974. "The Handbook of Chemical Lasers"[21] should be consulted for general discussions of the optical, kinetic, and gas dynamic aspects of chemical laser operation; this reference also contains an essentially complete bibliography of the chemical laser literature prior to 1974. The present review is intended to provide a brief overview of some of the more important developments in the chemical laser field between 1974 and the end of 1977. No attempt is made, however, to provide a complete bibliography of the chemical laser literature during this period.

Chapter 8.2 of the review deals with general kinetic aspects of chemical reactions that are useful in the discovery and development of chemical

[3] S. K. Searles and G. A. Hart, *Appl. Phys. Lett.* **27**, 243 (1975).

[4] J. J. Ewing and C. A. Brau, *Appl. Phys. Lett.* **27**, 350 (1975).

[5] C. A. Brau and J. J. Ewing, *J. Chem. Phys.* **63**, 4640 (1975).

[6] C. A. Brau and J. J. Ewing, *Appl. Phys. Lett.* **27**, 435 (1975).

[7] E. R. Ault, R. S. Bradford, Jr., and M. L. Baumik, *Appl. Phys. Lett.* **27**, 413 (1975).

[8] M. L. Baumik, R. S. Bradford, Jr., and E. R. Ault, *Appl. Phys. Lett.* **28**, 23 (1976).

[9] J. M. Hoffman, A. K. Hays, and G. C. Tisone, *Appl. Phys. Lett.* **28**, 538 (1976).

[10] R. Burnham, D. Harris, and N. Djeu, *Appl. Phys. Lett.* **28**, 86 (1976).

[11] D. G. Sutton, S. N. Suchard, O. L. Gibb, and C. P. Wang, *Appl. Phys. Lett.* **28**, 522 (1976).

[12] J. R. Murray and H. T. Powell, *Appl. Phys. Lett.* **29**, 252 (1976).

[13] R. Burnham and N. Djeu, *Appl. Phys. Lett.* **29**, 707 (1976).

[14] J. G. Eden and S. K. Searles, *Appl. Phys. Lett.* **29**, 350 (1976).

[15] A. N. Chester, *in* "High Power Gas Lasers—1975" (E. R. Pike, ed.), Conf. Ser. No. 29. Inst. Phys., London, 1976.

[16] N. G. Basov, V. I. Igoshin, J. I. Markin, and A. N. Oraevskii, *Kvantovaya Electron. (Moscow)* **2**, 3 (1971).

[17] D. H. Dawson and G. H. Kimbell, *Adv. Electron. Electron Phys.* **31**, 1 (1972).

[18] K. L. Kompa, *Fortschr. Chem. Forsch.* **37**, 1 (1973).

[19] C. E. Wiswall, D. P. Ames, and J. T. Menne, *IEEE J. Quantum Electron.* **qe-9**, 181 (1973).

[20] B. R. Bronfin, *Symp. (Int.) Combust. [Proc.]* **15**, 935 (1975).

[21] R. W. F. Gross and J. F. Bott, eds., "Handbook of Chemical Lasers." Wiley, New York, 1976.

lasers.   Chapter 8.3 gives a very brief discussion of supersonic chemical lasers, as exemplified by the HF and CO chemical lasers, which have received more extensive developmental efforts than any other chemical laser at present.   Chapter 8.4 discusses several purely chemical lasers that are currently under development and have unique features that complement the HF and DF devices.   Chapter 8.5 contains a brief discussion of rotational chemical lasers and problems in rotational energy transfer that are of current interest.   Finally Chapter 8.6 outlines problems in the search for new chemical lasers, with particular emphasis on electronic transition lasers capable of operation at visible wavelengths.

8.2. Disequilibrium in Reaction Product Energy States

The search for new chemical lasers has focused attention on the need for a better understanding of disequilibrium in chemical reactions.   In recent years the energy disposal among the quantum states of the products of many elementary reactions has been systematically studied.   In many cases it has been possible to account theoretically for the observed distributions of vibrational and rotational energy in reaction products with detailed calculations of the dynamics of reaction.   Much less success has been so far achieved in the prediction and analysis of the partitioning of electronic energy in reaction products.

Thermodynamic analysis, based on concepts of information theory, has been recently applied by Levine and co-workers[22-26] to the problem of the characterization of disequilibrium in simple chemical reactions.   This work has led to efficient methods for characterizing large quantities of experimental data.   More importantly it has provided a conceptual framework that relates the influence of dynamical constraints to observable departures from statistically favored product energy distributions.   Use of the principle of maximum entropy linked with dynamical constraints can provide the means for actual predictions of product energy distributions in cases where experimental data are lacking.

[22] R. D. Levine, in "The New World of Quantum Chemistry" (B. Pullman and R. Parr, eds.), pp. 103.   Reidel Publ., Dordrecht, Netherlands, 1976.

[23] R. D. Levine and R. B. Bernstein, in "Dynamics of Molecular Collisions" (W. H. Miller, ed.), Part B, pp. 323–364.   Plenum, New York, 1976.

[24] R. D. Levine and A. Ben-Shaul, in "Chemical and Biochemical Applications of Lasers" (C. B. Moore, ed.), Vol. 2, pp. 145–197.   Academic Press, New York, 1977.

[25] I. Procaccia and R. D. Levine, J. Chem. Phys. 63, 4261 (1975).

[26] A. Ben-Shaul, R. D. Levine, and R. B. Bernstein, J. Chem. Phys. 57, 5427 (1972).

In the absence of dynamical constraints all energetically allowed product states would be formed with equal *a priori* probabilities. A statistical distribution would result in which the fraction of products observed to be in a particular group of states would be proportional to the density of quantum states associated with that group. Such statistical distributions do not result in population inversions and cannot therefore lead to chemical laser action.

Chemical lasers are based on reactions that exhibit a pronounced specificity in the disposal of energy among reaction products. The experimental determination of the partitioning of vibrational, rotational, and translational energy among the products of reactions has been crucial to the development of vibrational and rotational chemical lasers. Much of the pioneering work in this field was done by J. C. Polanyi and co-workers with the "arrested relaxation" technique in which infrared chemiluminescence is monitored under single-collision conditions to determine the nascent energy distributions in reaction products.[27-30] More recently Pimentel and co-workers[31-34] have used chemical laser sources to obtain specific rate constant data formerly available only from the infrared chemiluminescence experiments.

A large number of simple atom exchange reactions of the type $A + BC \rightarrow AB\dagger + C$ ($AB\dagger$ denotes a vibrationally excited AB molecule) have been found to give rise to nascent vibrational distributions in the AB product molecule that differ in a predictable way from the statistical, or prior, distribution corresponding to the assumption of equal *a priori* probabilities for production of energetically allowed product states. The product vibrational distributions for these reactions are of the form[22-26]

$$P(v) = P^0(v) \exp(-\lambda_0 - \lambda_v f_v), \qquad (8.2.1)$$

where v is the vibrational quantum number, f_v is the fraction of the total energy in vibration, and $P^0(v)$ is the prior probability corresponding to the molecule being formed in vibrational state v:

$$P^0(v) = G(v; E)/ \sum_v G(v; E), \qquad (8.2.2)$$

[27] J. C. Polanyi, *J. Chem. Phys.* **34**, 347 (1961).
[28] J. C. Polanyi, *Appl. Opt., Suppl.* **2**, 109 (1965).
[29] P. D. Pacey and J. C. Polanyi, *J. Appl. Opt.* **10**, 1725 (1971).
[30] K. G. Anlauf, P. J. Kuntz, D. H. Maylotte, P. D. Pacey, and J. C. Polanyi, *Discuss. Faraday Soc.* **44**, 183 (1967).
[31] J. H. Parker and G. C. Pimentel, *J. Chem. Phys.* **51**, 91 (1969).
[32] R. D. Coombe and G. C. Pimentel, *J. Chem. Phys.* **59**, 251 (1973).
[33] M. J. Molina and G. C. Pimentel, *IEEE J. Quantum Electron.* **qe-9**, 64 (1973).
[34] M. J. Berry, *Annu. Rev. Phys. Chem.* **26**, 259 (1975).

where $G(v; E)$ is the number of quantum states having total energy E and vibrational quantum number v. The normalization condition $\Sigma_v P(v) = 1$ determines the parameter λ_0. The parameter λ_v is a constant for a given reaction, and may be either positive or negative. Negative values of λ_v can give inverted vibrational populations.

The probability distribution (8.2.1) follows theoretically from the maximization of the entropy subject to the constraint that the average energy in product vibration is fixed.[23,35] A dynamical explanation for the existence of this constraint has been given by Hofacker and Levine.[35,36] In many of the exothermic reactions of chemical laser interest, reaction conditions are dominated by attractive portions of the potential energy surface with high curvature along the reaction coordinate ("early downhill" surfaces).[36,37] In this type of reaction kinetic energy of the system can be efficiently converted to vibrational motion of the products.[37,38] The average fraction of energy appearing in vibration is fixed by the shape of a small portion of the potential energy surface of high curvature along the reaction coordinate.[35,36]

Figures 1–4 show the distribution of vibrational energy among product states observed for several reactions of importance in chemical lasers. The prior distributions, based on the assumption of equal *a priori* probabilities for all energetically allowed states (no dynamical constraints), are also indicated in Figs. 1–4. A useful way to characterize distributions of the type shown in Figs. 1–4 is in terms of the surprisal function defined as[22,26]

$$I(v) = -\ln[P(v)/P^0(v)] = \lambda_0 + \lambda_v f_v, \qquad (8.2.3)$$

which in terms of information theory gives a quantitative measure of the deviation between the actual distribution $P(v)$ and the prior statistical distribution $P^0(v)$. *The linear dependence of the surprisal on the fraction of energy in vibration, f_v, which is caused by the dynamical constraint on the vibrational energy release, is a feature found to exist in most of the chemical reactions which have been found to lead to vibrational chemical lasers.* This is a very important result, which provides a basis for the prediction of rate constants in the absence of experimental data. Specification of the surprisal, or equivalently the parameter λ_v, is sufficient to characterize the entire distribution and provides a compact way to represent

[35] G. L. Hofacker and R. D. Levine, *Chem. Phys. Lett.* **9**, 617 (1971); **15**, 165 (1972).

[36] A. Ben-Shaul and G. L. Hofacker, *in* "Handbook of Chemical Lasers" (R. W. Gross and J. F. Bott, eds.). Wiley, New York, 1976.

[37] M. H. Mok and J. C. Polanyi, *J. Chem. Phys.* **51**, 1451 (1969).

[38] K. G. Anlauf, D. S. Horne, R. G. Macdonald, J. C. Polanyi, and K. B. Woodall, *J. Chem. Phys.* **57**, 1561 (1972).

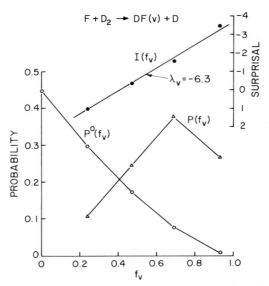

FIG. 1. Vibrational energy disposal in the $F + D_2 \rightarrow DF(v) + D$ reaction. The observed distribution $P(f_v)$ [J. C. Polanyi and K. B. Woodall, *J. Chem. Phys.* **57**, 1574 (1972)] and prior distribution $P^0(f_v)$ give a surprisal function $I(f_v) = -\ln[P(f_v)/P^0(f_v)]$, which varies linearly with f_v.

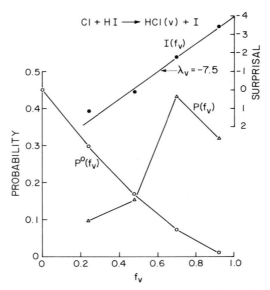

FIG. 2. Vibrational energy disposal in the $Cl + HI \rightarrow HCl(v) + I$ reaction. The observed distribution $P(f_v)$ is given by D. H. Maylotte, J. C. Polanyi, and K. B. Woodall, *J. Chem. Phys.* **57**, 1547 (1972).

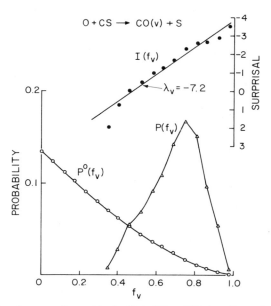

FIG. 3. Vibrational energy disposal in the O + CS → CO(v) + S reaction. The observed distribution $P(f_v)$ is given by O. Hancock, C. Morely, and I. W. M. Smith, *Chem. Phys. Lett.* **12,** 193 (1971).

experimental data. The dynamical description of the system is contained in the specification of the magnitude of the parameter λ_v, which can be calculated, by classical trajectory analysis, for example, or determined by experiment.

Surprisals can, of course, also be defined for distributions of electronic or rotational energies. At present, relatively little is known concerning electronic and rotational surprisals because of the lack of experimental data and because of computational difficulties in the case of electronic surprisals. Nonadiabatic transitions between electronic potential surfaces occur in general within well-defined regions of the potential energy surfaces.[39] For this reason one might expect dynamical constraints to exist that limit the phase space available to reaction products and provide specificity in the disposal of electronic excitation. On the other hand, if many such crossings can occur because of the complexities associated with many overlapping surfaces, then large regions of phase space become accessible and statistical distributions of product energies result. Surprisal analyses have been performed for the reactions of La, Y, and Sc

[39] J. C. Tully and R. K. Preston, *J. Chem. Phys.* **55,** 562 (1971).

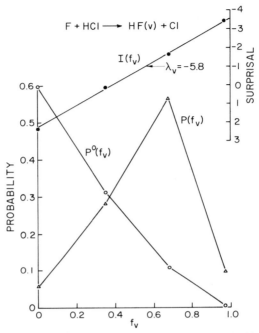

FIG. 4. Vibrational energy disposal in the F + HCl → HF(v) + Cl reaction. The observed distribution $P(f_v)$ is given by N. Jonathan, C. M. Melliar-Smith, S. Okuda, D. H. Slater, and D. Timlin, *Mol. Phys.* **22**, 561 (1971).

with several oxidizers.[24] The distributions of electronic energy in these cases were found to closely match the prior distribution and population inversions were not observed. On the other hand, reactions are known for which very nonstatistical distributions result and for which chemical laser action may be possible.[40]

8.3. Supersonic Chemical Lasers

Figure 5 illustrates the supersonic chemical laser concept, which has proved to be very successful for high-power cw operation of several molecular systems. Production of atoms and reactive radicals is accomplished by thermal dissociation of suitable parent compounds in a high-

[40] J. L. Gole, *Electron. Transition Lasers* [*Proc. Summer Colloq.*], *3rd, 1976* Vol. II, p. 136 (1977).

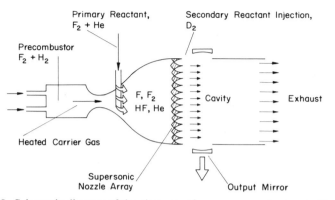

FIG. 5. Schematic diagram of the elements of a cw supersonic chemical laser.

temperature precombustor. The energy source for the precombustor is provided by a suitable self-sustaining chemical reaction such as the reaction of F_2/H_2 mixtures. The high-temperature and high-pressure partially dissociated gases leaving the precombustor provide a reservoir flow that feeds an array of supersonic nozzles designed to rapidly cool and expand the flow to low pressures suitable for rapid mixing with secondary reactant gases. The mixing is accomplished with use of injectors designed to produce complete diffusive mixing between primary and secondary reactants in the shortest possible time. The supersonic nozzles provide a convenient means to tailor the temperature and pressure of the reaction zone for optimal diffusive mixing and chemical reaction conditions. The high velocities associated with the supersonic flow provide a high mass flow and lead to high power outputs with very compact flow dimensions. Laser power is extracted from an optical cavity with an axis located transverse to the flow direction as illustrated in Fig. 5. In most devices the static pressure in the optical cavity is only a few torr. Flow pumping requirements are considerably reduced if a suitably designed supersonic diffuser is employed for exhaust gas pressure recovery.

Table I indicates several different chemical laser systems that have been designed for operation as supersonic chemical lasers. Extensive engineering and development efforts have been directed to the fluid-dynamical aspects of the chemically reacting, diffusively mixed flows occurring in supersonic chemical lasers.[21] Figure 6 illustrates a nozzle and mixing geometry commonly employed in early designs of such devices.[41]

[41] D. J. Spencer, H. Mirels, and T. A. Jacobs, *Appl. Phys. Lett.* **16**, 384 (1970).

TABLE I. Examples of Supersonic Purely Chemical Lasers

Precombustor reactants	Primary reactants	Additive gases	Secondary reactants	Laser species	Wavelength (μm)
F_2/H_2	F_2	He	D_2	DF	3.6–4.1
$NF_3/CH_4/H_2$	CS_2	N_2, N_2O	O_2	CO	4.9–5.7
$CH_4/CO/O_2$	F_2	He	D_2	CO_2	9.5–10.6
F_2/H_2	F_2	He	H_2, N_2F_4	$NF(b^1\Sigma^+)$	0.53^a
ClO_2/NO^b	$ClO_2{}^b$	He	HI	HCl	3.5–3.9

^{a} Assumes laser operation on the $b^1\Sigma^+ \rightarrow X^3\Sigma^-$ band is possible.
^{b} This laser does not employ a precombustor for thermal dissociation. Chlorine atoms are formed by the branched-chain reactions (8.4.5)–(8.4.7).

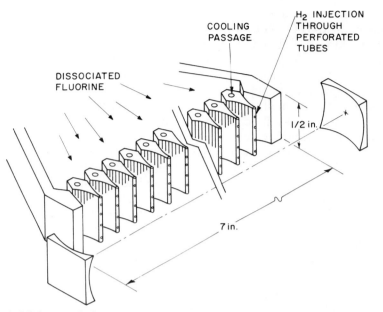

FIG. 6. Mixing nozzle for a cw supersonic HF chemical laser [D. J. Spencer, H. Mirels, and T. A. Jacobs, *Appl. Phys. Lett.* **16**, 385 (1970)].

8.3.1. The HF (DF) and CO Chemical Lasers

There are, at present, two major types of chemical lasers capable of the continuous efficient conversion of chemical energy into laser radiation. These are the HF (DF) lasers based on the reaction

$$F + H_2 \longrightarrow HF(v) + H, \qquad v = 0, 1, 2, 3, \qquad (8.3.1)$$

and CO chemical lasers pumped by the reaction

$$O + CS \longrightarrow CO(v) + S, \qquad v = 0, 1, 2, \ldots, 15. \qquad (8.3.2)$$

Both of these reactions result in the conversion of about two-thirds of the reaction exothermicity into vibrational excitation of the products. Both reactions are of the simple atom exchange type $A + BC \rightarrow AB\dagger + C$ and require the initial generation of atoms, and in the CO case, the generation of CS radicals. These reactions are quite rapid at room temperature and have activation energies of about 2 kcal/mole.

A major kinetic difference between these two laser systems arises from differences in the vibrational deactivation ($V \rightarrow R, T$) rates for collisions among product molecules. Collisions between HF molecules result in deactivation with a probability of about 10^{-2} at room temperature,[42] whereas for CO–CO collisions the deactivation probability is only about 10^{-9} at 300 K.[43,44] For CO molecules the V–V exchange processes are typically more than 10^6 times faster than V → R, T deactivation.[45–47] Under usual operating conditions V → V processes significantly alter the nascent vibrational distributions produced by chemical reaction for both HF and CO laser molecules. In the HF (DF) case the relatively large influence of V → R, T deactivation tends to dominate all other kinetic factors. This occurs to such an extent that generally only partial population inversions result in cw HF (DF) chemical lasers.[48] Most CO chemical lasers, however, operate with complete population inversions. In both laser devices the overall rate of chemical reaction is limited by the diffusive mixing rate. This limitation is quite severe in the HF (DF) lasers, where the deactivation rates are quite large. Much effort has gone into the design of gas dynamic mixing injectors to provide the greatest possible reduction in the scale size of unmixed portions of the gas flows so that diffusive mixing and reaction times can be minimized for HF (DF) lasers.[49]

Figure 7 shows the relative rate constants for reaction (8.3.1) into specific vibrational states. The distribution is strongly peaked at $v = 2$ for HF and at $v = 3$ for DF. Many HF (DF) lasers operate with the chain reaction

$$F + H_2 \longrightarrow HF(v) + H, \qquad v = 0, 1, 2, 3, \qquad \Delta H = -32 \quad \text{kcal/mole}, \quad (8.3.1)$$

$$H + F_2 \longrightarrow HF(v) + F, \qquad v = 0, 1, 2, \ldots, 9, \qquad \Delta H = -98 \quad \text{kcal/mole}, \quad (8.3.3)$$

Specific rate constants for vibrational excitation of the HF product of reaction (8.3.3) are given in Fig. 8. This reaction tends to favor higher vibrational states than does reaction (8.3.1). The influence of reaction (8.3.3) is quite evident in pulsed HF(DF) lasers based on the chain re-

[42] N. Cohen and J. F. Bott, *in* "Handbook of Chemical Lasers" (R. F. Gross and J. F. Bott, eds.), Chapter 2. Wiley, New York, 1976.

[43] R. C. Millikan, *J. Chem. Phys.* **38**, 2855 (1963).

[44] M. G. Ferguson and A. W. Reed, *Trans. Faraday Soc.* **61**, 1559 (1965).

[45] I. W. M. Smith and C. Wittig, *Trans. Faraday Soc.* **69**, 939 (1973).

[46] H. T. Powell, *J. Chem. Phys.* **59**, 4937 (1973).

[47] J. C. Stephenson and E. R. Mosburg, Jr., *J. Chem. Phys.* **60**, 3562 (1974).

[48] D. I. Rosen, R. N. Sileo, and T. A. Cool, *IEEE J. Quantum Electron.* **qe-9**, 163 (1973).

[49] R. W. F. Gross and D. J. Spencer, *in* "Handbook of Chemical Lasers" (R. W. F. Gross and J. F. Bott, eds.), Chapter 4. Wiley, New York, 1976; G. Grohs and G. Emanuel, *ibid.*, Chapter 5.

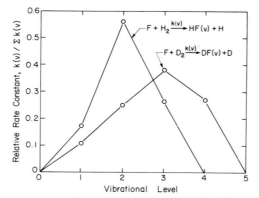

FIG. 7. Relative rate constants for reaction into specific vibrational states of HF and DF [J. C. Polanyi and K. B. Woodall, *J. Chem. Phys.* **57**, 1574 (1972)].

action[50]; in this case laser action is observed from several vibrational levels accessible only to reaction (8.3.3). For cw devices, however, the influence of slow mixing and rapid deactivation rates prevents laser operation from the highest vibrational levels, and most of the laser output comes from the $1 \rightarrow 0$ and $2 \rightarrow 1$ vibrational bands of HF and the $1 \rightarrow 0$, $2 \rightarrow 1$, and $3 \rightarrow 2$ bands of DF. Laser wavelengths for HF cover the range from about 2.6 to 3.0 μm; for DF from about 3.6 to 4.0 μm.

Figure 9 shows relative rate constant data for the CO laser pumping reaction (8.3.2). Here again the reaction is fairly specific and tends to favor vibrational levels near $v = 13$. Under usual operating conditions CO chemical laser output spectra include vibrational bands from $2 \rightarrow 1$

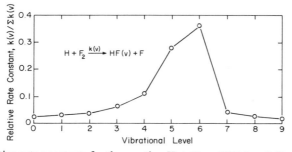

FIG. 8. Relative rate constants for the reaction $H + F_2 \rightarrow HF(v) + F$ [J. C. Polanyi and J. J. Sloan, *J. Chem. Phys.* **57**, 4988 (1972)].

[50] S. N. Suchard and J. R. Airey, *in* "Handbook of Chemical Lasers" (R. W. F. Gross and J. F. Bott, eds.), Chapter 6. Wiley, New York, 1976.

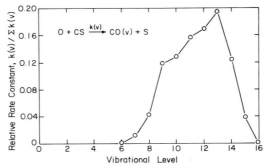

FIG. 9. Relative rate constants for reaction into specific vibrational states of CO [G. Hancock, C. Morely, and I. W. M. Smith, *Chem. Phys. Lett.* **12**, 193 (1971)].

through $13 \to 12$. Laser wavelengths cover the range from about 4.9 to 5.6 μm. Operation on the $1 \to 0$ band is ordinarily prevented by the absorption from ground-state CO molecules.

The design of present large-scale HF (DF) and CO chemical lasers exemplifies the use of gas dynamic techniques in the achievement of a favorable kinetic environment.[49,51] The supersonic flow concept of Fig. 5 has proven its worth in the design of high-power devices. The DF chemical laser, for example, may be run with an F_2/H_2 precombustor designed to operate at temperatures above 1200 K to permit virtually complete dissociation of the excess F_2. The high-pressure and high-temperature combustion products and dissociated F_2 are expanded to pressures of a few torr through the supersonic nozzle array. The expanded primary flow is then mixed with a secondary D_2 flow and chemical reaction occurs within the supersonic mixing zone immediately downstream from the nozzles and D_2 injectors.

It has been found useful to dissociate nearly all of the F_2 so that operation is based primarily on reaction (8.3.1) alone, rather than on the chain reactions (8.3.1) and (8.3.3).[52] Such lasers can operate at high chemical efficiencies at multikilowatt power levels. As much as 15% of the reaction exothermicity of reaction (8.3.1) is converted to useful laser output in such lasers.[52]

The kinetics of the chemical reaction and energy transfer processes in HF (DF) have been extensively studied and reviewed.[42] Elaborate computational programs have been developed to permit optimization of

[51] B. R. Bronfin and W. Q. Jeffers, *in* "Handbook of Chemical Lasers" (R. W. F. Gross and J. F. Bott, eds.), Chapter 11. Wiley, New York, 1976.

[52] G. Grohs and G. Emanuel, *in* "Handbook of Chemical Lasers" (R. W. F. Gross and J. F. Bott, eds.), Chapter 5. Wiley, New York, 1976.

high-power lasers.[51,53,54] Present models include the essential aspects of reaction and deactivation kinetics, diffusive mixing, and the interaction with the cavity radiation field. Generally good agreement between computed performance predictions and experimental results is obtained.

Continuous-wave CO chemical lasers based on reaction (8.3.2) have been operated by mixing O atoms, produced by either an electrical discharge or thermal dissociation, with CS_2 in the presence of large amounts of helium or argon diluent.[55-61] The reaction kinetics of this system is complex[62,63]; several key reactions are given in Table II. In this type of CO laser the CS radicals are created by the primary reaction (1a) of Table II.

TABLE II. Key Reactions of the $O/O_2/CS/CS_2$ System[a]

Primary reactions		
1a.	$O + CS_2$	$\rightarrow CS + SO$
1b.		$\rightarrow OCS + S$
1c.		$\rightarrow CO + S_2$
2.	$O + CS$	$\rightarrow CO\dagger + S$
3.	$S + O_2$	$\rightarrow SO + O$
Termination reactions		
4.	$O + OCS$	$\rightarrow CO + SO$
5.	$S + CS_2$	$\rightarrow CS + S_2$
6.	$SO + O + M$	$\rightarrow SO_2 + M$
Branching reactions		
7.	$SO + O_2$	$\rightarrow SO_2 + O$
8.	$CS + SO$	$\rightarrow OCS + S$
9.	$CS + O_2$	$\rightarrow OCS + O$
10.	$SO + SO$	$\rightarrow SO_2 + O$

[a] W. Q. Jeffers, H. Y. Ageno, and C. E. Wiswall, *J. Appl. Phys.* **47**, 2509 (1976).

[53] G. Emanuel, in "Handbook of Chemical Lasers" (R. W. F. Gross and J. F. Bott, eds.), Chapter 8. Wiley, New York, 1976.
[54] H. V. Lilenfeld and W. Q. Jeffers, *J. Appl. Phys.* **47**, 2520 (1976).
[55] C. Wittig, J. C. Hassler, and P. D. Coleman, *Appl. Phys. Lett.* **16**, 117 (1970); *Nature (London)* **226**, 845 (1970).
[56] R. D. Suart, P. H. Dawson, and G. H. Kimbell, *J. Appl. Phys.* **43**, 1022 (1972).
[57] R. D. Suart, S. J. Arnold, and G. H. Kimbell, *Chem. Phys. Lett.* **7**, 337 (1970).
[58] K. D. Foster, *J. Chem. Phys.* **57**, 2451 (1972).
[59] C. J. Ultee and P. A. Bonczyk, *IEEE J. Quantum Electron.* **qe-10**, 105 (1974).
[60] W. Q. Jeffers and C. E. Wiswall, *Appl. Phys. Lett.* **23**, 626 (1973).
[61] L. R. Boedeker, J. A. Shirley, and B. R. Bronfin, *Appl. Phys. Lett.* **21**, 247 (1972).
[62] D. W. Howgate and T. A. Barr, Jr., *J. Chem. Phys.* **59**, 2815 (1973).
[63] S. H. Bauer and N. A. Nielson, Jr., unpublished; N. A. Nielson, Jr., Ph.D. Thesis, Cornell University, Ithaca, New York (1975).

Recently a purely chemical CO laser has been developed.[64,65] A schematic of the flow and reaction sequence is indicated in Fig. 10. The laser depends on reactions (2) and (3) of Table II. A thermochemical combustor operating with NF_3–CH_4–H_2–CS_2 mixtures produces HF, N_2, CS, S, and residual CS_2, which is expanded through a supersonic nozzle array and then mixed with the secondary reagent O_2, and additional gases N_2 and N_2O. The S atoms react with O_2 via reaction (3) of Table II to produce O atoms. The O atoms react via reaction (2) to form CO† for laser operation. Reactions (2) and (3) form an efficient chain for the formation of CO†. Since residual CS_2 competes with the CS for O atoms via reactions (1), the CS to CS_2 ratio and the CS to S ratio for a practical combustor must be near 5 and 20, respectively.[65]

The heat capacity of the N_2 diluent and the supersonic expansion serve to moderate the temperature in the reaction zone to about 400 K. The pressure in this zone is about 35–40 torr. The additive N_2O serves to enhance population inversion in CO by selective vibrational energy transfer processes.

The reaction kinetics and energy transfer processes present in CO chemical lasers are reasonably well understood at present.[51] A number of interesting kinetic effects occur in the CO system including: (1) V–V pumping of high vibrational levels at the expense of lower levels because of anharmonicity of the vibrational energy levels,[66] and, (2) power output

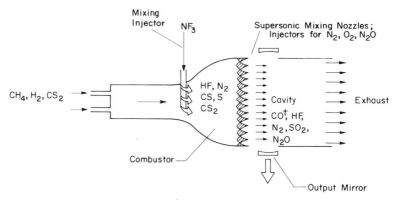

FIG. 10. A purely chemical CO laser [R. J. Richardson and V. R. Buonadonna, *J. Appl. Phys.* **48**, 2509 (1977); R. J. Richardson and C. E. Wiswall, *Int. Quantum Electron. Conf., 10th, 1978* Paper H. 7 (1978)].

[64] R. J. Richardson and V. R. Buonadonna, *J. Appl. Phys.* **48**, 2509 (1977).

[65] R. J. Richardson and C. E. Wiswall, *Int. Quantum Electron. Conf., 10th, 1978* Paper H. 7 (1978).

[66] J. T. Yardley, *Appl. Opt.* **10**, 1760 (1971).

enhancement effects caused by V → V transfer to various additive gases such as cold CO or N_2O.[60]

Just as in the case of HF (DF) lasers, detailed computer codes have been successfully used for laser performance prediction.

8.4. New Purely Chemical Lasers

Although present large-scale HF (DF) and CO chemical lasers are examples of the use of fairly sophisticated fluid dynamic, kinetic, and optical methods for efficient production of high laser powers, purely chemical lasers of great simplicity have been operated. The first examples of lasers capable of operation by purely chemical means were the HF, DF, $HF-CO_2$, and $DF-CO_2$ subsonic devices developed by Cool and Stephens.[67,68] Other extremely simple lasers were the O_2/CS_2 diffusion flame CO lasers developed by Pilloff, Searles, and Djeu,[69] and, more recently, the N_2O flame laser of Benard, Benson, and Walker.[70] In their simplest forms these devices operated by the simple mixing of reagents in a suitably designed optical cavity. Quite recently other examples of purely chemical lasers of great interest have been developed, including the first electronic transition chemical laser.[2]

8.4.1. A Purely Chemical Electronic Transition Laser

The first example of a cw chemical laser based on pumping of an electronic transition was recently reported by McDermott et al.[2] Laser output originates from the 1315 nm $^2P_{1/2}$ to $^2P_{3/2}$ transition of atomic iodine. The $^2P_{1/2}$ upper level was selectively pumped by energy transfer from chemically formed $O_2(^1\Delta)$ metastable molecules; see the energy level diagram of Fig. 11.

Although pulsed 1315 nm iodine lasers have been readily operated by simple photodissociation of several parent compounds,[71] the development of the cw chemically pumped iodine laser is considerably more complicated and has required a sustained effort by several investigators over the

[67] T. A. Cool and R. R. Stephens, J. Chem. Phys. 51, 5175 (1969).

[68] J. A. Shirley, R. N. Sileo, R. R. Stephens, and T. A. Cool, AIAA Aerosp. Sci. Meet., 9th, 1971 Paper 71-27 (1971).

[69] H. S. Pillof, S. K. Searles, and N. Djeu, Appl. Phys. Lett. 19, 9 (1971).

[70] D. J. Benard, R. C. Benson, and R. E. Walker, Appl. Phys. Lett. 23, 82 (1973).

[71] J. V. V. Kasper, J. H. Parker, and G. C. Pimentel, Phys. Rev. Lett. 14, 352 (1965); J. Chem. Phys. 43, 1827 (1965).

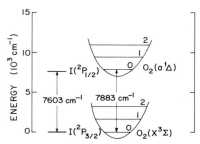

FIG. 11. Energy level diagram showing the near resonance in energy between the $I^*(^2P_{1/2})$ and $O_2(a^1\Delta)$ $v = 0$ states.

past three years.[72-74] The operation of this laser depends on the successful optimization of various steps in a complex reaction sequence. Key reactions in laser pumping include:

$$O_2(^1\Delta) + O_2(^1\Delta) \longrightarrow O_2(^1\Sigma) + O_2(^3\Sigma), \qquad (8.4.1)$$

$$O_2(^1\Sigma) + I_2 \longrightarrow O_2(^3\Sigma) + 2I(^2P_{3/2}), \qquad (8.4.2)$$

$$O_2(^1\Delta) + I(^2P_{3/2}) \rightleftharpoons O_2(^3\Sigma) + I^*(^2P_{1/2}), \qquad (8.4.3)$$

$$I^*(^2P_{1/2}) + O_2(^1\Delta) \longrightarrow O_2(^1\Sigma) + I(^2P_{3/2}). \qquad (8.4.4)$$

Two of the most difficult problems that were solved in laser development were, first, the development of a suitable chemical generator of $O_2(^1\Delta)$ molecules and, second, the establishment of the proper kinetic conditions for maximization of the concentration of upper state $I^*(^2P_{1/2})$ atoms. A crucial aspect of reactions (8.4.1)–(8.4.3) for the formation of $I^*(^2P_{1/2})$ is the virtually complete dissociation of I_2 ensured by reaction (8.9.2); I_2 is a very efficient quencher of $I^*(^2P_{1/2})$ atoms.[75] The chemical generation of $O_2(^1\Delta)$ molecules was based on the reaction of hydrogen peroxide (H_2O_2) with sodium hypochlorite (NaOCl).

A schematic diagram of the experimental apparatus employed by McDermott et al.,[2] is shown in Fig. 12. Chemically formed $O_2(^1\Delta)$ molecules mixed with ground-state $O_2(^3\Sigma)$ molecules were passed through a cold trap to remove residual Cl_2 left over from the chemical generator. The

[72] R. D. Franklin, A. K. MacKnight, and P. J. Modreski, Electron. Transition Lasers [Proc. Summer Colloq., 2nd, 1975 p. 119 (1976).

[73] W. E. McDermott, R. E. Lotz, and M. C. Delong, Electron. Transition Lasers [Proc. Summer Colloq.], 3rd, 1976 II, L. E. Wilson, S. N. Suchard, p. 302 (1977).

[74] A. T. Pritt, Jr., R. D. Coombe, D. Pilipovich, R. I. Wagner, D. Benard, and C. Dymek, Appl. Phys. Lett. 31, 745 (1977).

[75] D. H. Burde and R. A. McFarlane, J. Chem. Phys. 64, 1850 (1976).

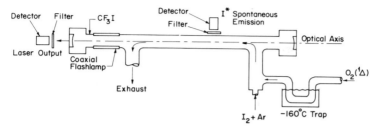

FIG. 12. Schematic diagram of the first purely chemical cw electronic transition chemical laser [W. E. McDermott, N. R. Pchelkin, D. J. Benard, and R. R. Bousek, *Appl. Phys. Lett.* **32**, 469 (1978)].

concentration of $O_2(^1\Delta)$ in $O_2(^3\Sigma)$ molecules exceeded 40% as determined by an electron paramagnetic spectrometer (EPR). The primary flow of molecular oxygen was then rapidly mixed with molecular iodine entrained in an argon stream, and the combined flows were introduced into a longitudinal optical cavity with a 70 cm active length. Optical alignment of the cavity was provided by a small (10-cm-long) flashlamp used for the generation of atomic iodine laser emission by the photodissociation of CF_3I. The difficult task of optimization of the many flow variables was facilitated with the use of two detectors used to monitor the $I^*(^2P_{1/2})$ to $I(^2P_{3/2})$ spontaneous emission, both along the laser optical axis and at right angles to it. These detectors provided a sensitive means for the detection of on-axis stimulated emission near laser threshold.

The device illustrated in Fig. 12 produced cw laser outputs of a few milliwatts. The use of a transverse optical axis and larger flow rates should permit scaled-up devices of modest size to operate at power levels of a few watts.

The operating characteristics of this first purely chemical electronic transition laser are similar in many respects to longitudinal-flow subsonic hydrogen halide chemical lasers[76] and $DF-CO_2$ transfer chemical lasers.[77] Typical operating pressures are near 1 torr. The long radiative lifetime of the $I^*(^2P_{1/2})$ to $I(^2P_{3/2})$ transition ensures that volume collisional processes dominate radiative processes, thereby permitting scaling to large dimensions. Collisional quenching is significant and places limitations on flow pressures and the characteristic time for gas mixing.

If the cw iodine chemical laser can be scaled to large sizes and operated with a reasonably high chemical efficiency, then the iodine chemical laser may offer advantages for many applications over the longer wavelength chemical lasers.

[76] T. A. Cool, R. R. Stephens, and J. A. Shirley, *J. Appl. Phys.* **41**, 4038 (1970).
[77] T. A. Cool, R. R. Stephens, and T. J. Falk, *Appl. Phys. Lett.* **15**, 318 (1969).

8.4.2. Purely Chemical HCl and HCl–CO₂ Lasers

Recently Arnold *et al.*[78] have developed a purely chemical HCl laser based on the reaction sequence

$$NO + ClO_2 \longrightarrow NO_2 + ClO, \qquad\qquad (8.4.5)$$

$$NO + ClO \longrightarrow NO_2 + Cl, \qquad\qquad (8.4.6)$$

$$Cl + ClO_2 \longrightarrow 2ClO, \qquad\qquad (8.4.7)$$

$$Cl + HI \longrightarrow HCl(v = n) + I, \qquad v = 0, 1, 2, 3, 4. \qquad (8.4.8)$$

Reactions (8.4.5)–(8.4.7) form a branched chain reaction for the chemical production of Cl atoms. The reaction of Cl atoms with HI molecules provides an efficient source of vibrationally excited HCl molecules (See Chapter 8.2); this reaction has been often used in both pulsed and cw chemical lasers.[49,50] Figure 13 schematically illustrates the apparatus employed by Arnold *et al.*[78] Although several modes of operation are possible for this device, depending on the sequence of gas mixing, the most successful method consists of the initial mixing of two parts NO with approximately one part ClO_2. This is followed by subsequent injection of HI for reaction with Cl atoms formed in the first stage.

A laser with a flow transverse to the optical axis produced 13 W from HCl at wavelengths from 3.6 to 4.0 μm. The flow cross section was 40 × 1 cm. The cavity pressure was 2.4 torr; the flow velocity was 220 m/sec. A chemical efficiency of 8%, based on the exothermicity of reaction (8.4.8) was obtained. This chemical efficiency is considerably higher

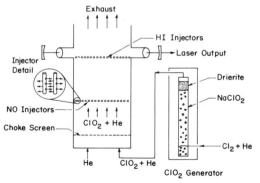

FIG. 13. A purely chemical HCl laser [S. J. Arnold, K. D. Foster, D. R. Snelling, and R. D. Suart, *Appl. Phys. Lett.* **30**, 637 (1977)].

[78] S. J. Arnold, K. D. Foster, D. R. Snelling, and R. D. Suart, *Appl. Phys. Lett.* **30**, 637 (1977).

than that obtained in comparable HF/DF devices.[49] The laser was also operated as an HCl–CO$_2$ transfer chemical laser at 10.6 μm.

8.4.3. A Candidate for a Purely Chemical Laser at Visible Wavelengths

Another example of a system of much promise for cw chemical laser operation at visible wavelengths is currently being extensively studied.[79,80] The kinetic mechanism consists of the key steps:

$$H(^2S) + NF_2(^2B_1) \longrightarrow HNF_2^*(^1A) \longrightarrow HF(X^1\Sigma^+) + NF(a^1\Delta), \quad (8.4.9)$$

$$HF(v \geqslant 2) + NF(a^1\Delta) \longrightarrow HF(v-2) + NF(b^1\Sigma^+), \quad (8.4.10)$$

$$NF(b^1\Sigma^+) \longrightarrow NF(X^3\Sigma^-) + h\nu(5288 \text{ Å}). \quad (8.4.11)$$

Reaction (8.4.9) is known to be an excellent source for the selective production of NF radicals in the metastable (a$^1\Delta$) state.[81] The radiative lifetime of NF(a$^1\Delta$) is very long, of the order of seconds,[81] and these metastables provide a convenient reservoir for chemically produced electronic energy. Chemiluminescence experiments have demonstrated that vibrationally excited HF (formed by chemical reaction) provides an effective means for the conversion of NF(a$^1\Delta$) molecules into NF(b$^1\Sigma^+$) molecules by process (8.4.10) without also introducing severe competitive quenching of the NF(a$^1\Delta$) molecules.[81,82] The radiative lifetime for the emission process (8.4.11) has been determined to be 1.5×10^{-2} sec,[82] which makes this transition a very favorable candidate for high-power cw laser operation in the important wavelength region near 5300 Å.

Two laboratories have assembled apparatus designed for the eventual demonstration of laser operation. Herbelin *et al.*[79] have employed a subsonic device, patterned after cw HF lasers, in which SF$_6$ is subjected to an electrical discharge for the production of F atoms. N$_2$F$_4$ and H$_2$ molecules are then added to the flow to produce NF$_2$ radicals, (by thermal dissociation of N$_2$F$_4$) and H atoms and vibrationally excited HF molecules (by the reaction F + H$_2$ → HF† + H). Herbelin *et al.*[79] report peak densities of NF(a$^1\Delta$) and NF(b$^1\Sigma^+$) molecules of 1.4×10^{15} and 1.4×10^{13} cm^{-3}, respectively.

Another group has adapted a cw supersonic HF laser for proposed NF

[79] J. M. Herbelin, D. J. Spencer, and M. A. Kwok, *J. Appl. Phys.* **48** 3050 (1977).

[80] D. J. Miller and J. A. Betts, *Int. Quantum Electron. Conf., 10th, 1978* Paper H. 8.

[81] J. M. Herbelin and N. Cohen, *Chem. Phys. Lett.* **20**, 603 (1973).

[82] M. A. Kwok, J. M. Herbelin, and N. Cohen, *Electron. Transition Lasers* [*Proc. Summer Colloq., 2nd, 1975* p. 8 (1976).

chemical laser operation.[80]   An H_2/F_2 combustor operating with an excess of F_2 is employed for production of F atoms by thermal dissociation of F_2.   This is done in a high-pressure plenum chamber located immediately upstream of a supersonic nozzle array. H_2 is rapidly mixed with the expanded flow downstream of the nozzles to produce H atoms and vibrationally excited HF by the F + H_2 reaction.   A second set of mixing nozzles is employed for injection of N_2F_4 into the supersonic flow consisting primarily of H atoms, vibrationally excited HF, and helium diluent.   The flow conditions are tailored so that the region at which N_2F_4 is injected is appropriate for rapid mixing and NF_2 radical production (T = 1000 K, p = 1.5 torr).   This system is an excellent example of the use of staged combustion and rapid mixing techniques to achieve the desired kinetic environment necessary for the proper sequencing of chemical reactions for laser operation.   Densities of $NF(a^1\Delta)$ radicals of 3 × 10^{14} cm^{-3} and $NF(b^1\Sigma^+)$ densities of 1.4 × 10^{13} cm^{-3} have been measured in this device.[80]   It is estimated that about 1% of the NF_2 radicals are converted to $NF(b^1\Sigma)$ molecules.   If further experiments with this device are successful, powerful purely chemical laser operation in the visible spectral region will be realized for the first time.

8.5. Rotational Chemical Lasers

Chemical laser action between rotational states of HCl, HBr, HF, OH, and OD has been reported at wavelengths in the 10–40 μm region.[83–85] All of these lasers are initiated by electrical discharges in premixed reagents.   The kinetics of these systems has not been carefully studied. Detailed reaction mechanisms are presently unknown; the roles of vibration to vibration (V–V) and vibration to rotation (V–R) energy transfer and radiative cascading are subject to considerable conjecture.   Rotational chemical lasers have not yet been of sufficient interest for applications to warrant extensive development; nevertheless, the kinetic aspects of rotational nonequilibrium in these devices are of much fundamental interest and many challenging questions remain unanswered at present.

Studies of the infrared chemiluminescence from reactions forming vibrationally excited hydrogen halides have shown that considerable departures from rotational equilibrium can result from chemical reaction.[86–89]

[83] T. F. Deutsch, *Appl. Phys. Lett.* **11**, 18 (1967).

[84] D. P. Akitt and J. J. Yardley, *IEEE J. Quantum Electron.* **qe-6**, 113 (1970).

[85] T. W. Ducas, L. D. Geoffrion, R. M. Osgood, Jr., and A. Javan, *Appl. Phys. Lett.* **21**, 42 (1972).

[86] J. C. Polanyi and K. B. Woodall, *J. Chem. Phys.* **56**, 1563 (1972).

Polanyi and Woodall have discussed bimodal rotational distributions in hydrogen halide reaction products.[86] One peak in the distribution corresponds to a portion of the molecules having thermalized rotational energies while a second peak at much higher rotational quantum numbers gives evidence of specificity in the partitioning of rotational energy by chemical reaction. They have shown that the bimodal nature of the distribution can be accounted for if the rotational relaxation rates for highly excited states are significantly less than those for lower states near the peak of the thermalized distribution. An "exponential gap" model has been found useful in describing rotational relaxation in these systems.[86]

Other evidence exists that also suggests that characterization of the rotational disequilibrium in chemical lasers depends significantly on relaxation processes. Krogh and Pimental have reported V–R laser transitions from rotational states excited considerably beyond those states initially formed by chemical reaction.[90] They studied the photolytically initiated reactions of ClF, ClF_3, and ClF_5 with H_2. These reactions produced chemical laser transitions from both vibrationally excited HF and HCl molecules. Laser transitions from very high rotational states, not directly pumped by chemical reaction, were observed. These laser transitions were characterized by late threshold times and extended durations following reaction initiation. This behavior is consistent with the assumption of efficient transfer of energy from vibration to rotation by (V–R) processes of the type

$$HF(v, J) + M \longrightarrow HF(v - 1, J + \Delta J) + M, \qquad (8.5.1)$$

where ΔJ can be as large as 13 or so. In addition, V–V pumping processes

$$HF(v = n) + HF(v = m) \longrightarrow HF(v = n - 1) + HF(v = m + 1), \qquad (8.5.2)$$

with $n > m$, analogous to processes of great importance in CO lasers, may be operative here. References to the importance of V–R processes in HF exist in the literature,[91] but definitive experiments and theoretical calculations are still lacking. It should also be noted that chemical lasers operating on pure rotational transitions typically operate at very high J

[87] D. H. Maylotte, J. C. Polanyi, and K. B. Woodall, *J. Chem. Phys.* **57**, 1547 (1972).
[88] K. G. Anlauf, D. S. Horne, R. B. Macdonald, J. C. Polanyi, and K. B. Woodall, *J. Chem. Phys.* **57**, 1562 (1972).
[89] J. C. Polanyi and K. B. Woodall, *J. Chem. Phys.* **57**, 1574 (1972).
[90] O. D. Krogh and G. C. Pimentel, *J. Chem. Phys.* **67**, 2993 (1977).
[91] R. L. Wilkins, *J. Chem. Phys.* **67**, 5838 (1977).

values in HF, HCl, HBr, OH, and OD.[83-85] These J values are considerably higher than those corresponding to direct chemical pumping by any known reaction mechanism. The populations of these high-J states are many orders of magnitude higher than populations corresponding to thermal equilibrium.

The use of chemical lasers in double-resonance experiments seems to offer the best means at present for the study of the rotational relaxation processes of interest. Although these techniques have not yet been used for V–R measurements, Hinchen and Hobbs have demonstrated their use in rotational relaxation measurements.[92] The apparatus used by Hinchen and Hobbs, shown schematically in Fig. 14, employed two HF chemical lasers. A pulsed HF laser operated on a single P branch transition of the $v(1 \rightarrow 0)$ band was used to permit selective excitation of a single rotational state in the $v = 1$ level of HF. A second HF probe laser was operated on the $v(2 \rightarrow 1)$ band. The cw laser was carefully tuned to the narrow-frequency range associated with the linewidth of the pulsed pump laser, which was localized near the center of the Doppler-broadened velocity distribution of absorbing molecules. The temporal variation in the absorption of the cw laser output was used as a probe for monitoring the populations of rotational states at or near the state populated by the pump laser.

The temporal variation in absorption of the $P_2(3)$ transition for the cw HF laser following excitation of the $J = 3$ state of the $v = 1$ level with the $P_1(4)$ transition of the pump laser exhibited a double exponential decay.

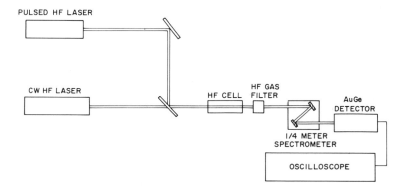

FIG. 14. Experimental apparatus employed by Hinchen and Hobbs [J. J. Hinchen and R. H. Hobbs, *J. Chem. Phys.* **65**, 2732 (1976)] for double-resonance measurements of rotational relaxation in HF. The pulsed HF laser (pump) operates on the $v(1 \rightarrow 0)$ P branch transitions and the cw probe laser on $v(2 \rightarrow 1)$ P branch transitions.

[92] J. J. Hinchen and R. H. Hobbs, *J. Chem. Phys.* **65**, 2732 (1976).

An initial rapid decay was observed on a time scale associated with the rise times of populations in other rotational levels (determined in separate experiments). A subsequent decay with a time constant about two orders of magnitude slower was also observed. This decay is believed to monitor momentum transfer collisions that result in a spreading of the initially sharply peaked velocity distribution into the thermal Gaussian distribution of velocities. The initial rotational relaxation corresponded to a rate constant greater than 2.2×10^{-9} cm^3/molecule/sec; the secondary momentum transfer rate constant was observed to be $3.1 \pm 0.6 \times 10^{-11}$ cm^3/molecule/sec.

In order to determine the relative rate constants for coupling of rotational energy to nearby rotational states, a careful series of experiments were performed in which the temporal variations in absorption of the cw laser was observed when the laser was tuned for absorption by levels differing by one to five rotational quanta from the excited level. Table III lists the results of these studies.

Examination was given of a kinetic model for rotational relaxation originally employed by Polanyi and Woodall in their treatment of $R \rightarrow T$ relaxation processes.[86] That is, the rate constant for a molecule in $v = 1$,

TABLE III. Rate Constants for Rotational Energy Transfer[a]

Pump J	Probe J	Transfer rate constant[b] (sec^{-1} torr^{-1})
2	3	22×10^6
2	4	6.3
2	5	1.6
2	6	0.42
2	7	0.083
3	4	6.0
3	5	1.4
3	6	0.42
3	7	0.12
3	8	0.02
4	5	3.4
4	6	0.76
4	7	0.21
4	8	0.017
5	6	1.5
5	7	0.23

[a] J. J. Hinchen and R. H. Hobbs, *J. Chem. Phys.* **65**, 2732 (1976).
[b] The transfer rate constant k_{ij} is the overall rate of transfer of molecules from the initially pumped level i to the probed level j. That is, $k_{ij} = \Sigma k_{ijkl}$, where the summation is over the unobserved rotational levels of the ground vibrational state collision partner.

$J = i$ for transfer to the $v = 1$, $J = j$ state in a collision with a ground-state partner that goes from $v = 0$, $J = k$ to $v = 0$, $J = l$ was represented with an exponential form based on the rotational energy discrepancy for the collision:

$$k_{ijkl} = k_0 \bar{n}_j \bar{n}_l \exp[-\alpha|(E_j^{v=1} - E_i^{v=1}) + (E_l^{v=0} - E_k^{v=0})|/kT], \quad (8.5.3)$$

where k_0 and α are adjustable parameters to be determined by experiment and $\bar{n}_j$ and $\bar{n}_l$ are the fractions of molecules in the $J = j$ and $J = l$ states given by the Boltzmann distribution function. Computer solutions of the rate equations that couple the various rotational states were compared with the experimental results of Table III to give values for the adjustable parameters:

$$k_0 = 6.2 \pm 1.2 \times 10^{-9} \quad \text{cm}^3/\text{molecule}/\text{sec}, \qquad \alpha \approx 1.$$

On this basis it was possible to calculate the fraction of molecules in a given J level that relax by rotational collisions with $\Delta J = 1, 2, 3$, etc. It was found that the losses from $J = 3$ occurred as 42.5% by $|\Delta J| = 1$: 31.5% by $|\Delta J| = 2$; 23.0% by $|\Delta J| = 3$; and $<3\%$ by $|\Delta J| > 3$. The importance of multiquantum rotational energy transfer processes is clearly demonstrated by these results.

8.6. Problems in the Search for New Chemical Lasers

The discovery and development of existing chemical lasers has required advancement in many areas, including the fluid dynamics of mixing, the design of high-Fresnel-number optical systems, kinetic diagnostic techniques, computer modeling of the optics and kinetics of transverse flow, and supersonic diffusion lasers,[21] but perhaps most crucial has been knowledge concerning the chemical reactions themselves. Fortunately a very wide class of exothermic chemical reactions produces the requisite disequilibrium for vibrational–rotational chemical lasers, as already discussed in Chapter 8.2. Characterization of disequilibrium among electronically excited states has proven to be more difficult. In contrast to the situation only five years ago, many reactions are now known that lead to efficient production of electronically excited reaction products. Unfortunately, however, nearly all of these reactions do not appear to exhibit specificity in the disposal of electronic energy. Even a high degree of specificity is not enough, by itself, to ensure the establishment of population inversions on appropriate laser transitions. It is useful to examine some of the kinetic constraints that define the characteristics of reaction systems suitable for electronic transition chemical laser operation at visible wavelengths.

8.6.1. Kinetic Considerations in Chemical Laser Operation at Short Wavelengths

Consider the reaction of a metal atom A with a molecule BC leading to the formation of the diatomic reaction product AB:

$$A + BC \longrightarrow AB(I, v, J) + C, \qquad (8.6.1)$$

where the internal states of the AB molecule are characterized by the electronic state I, the vibrational state v, and the rotational state J. The fraction, $\Phi_i(v, j)$ of the total reaction rate constant k that leads to the product state (I, v, j) is separable into respective fractions expressing the probability of formation of molecules in the vibrational state v and the rotational state J, where J is taken to be the rotational state of maximum population at a given rotational temperature T. Even in favorable cases values of $\Phi_i(v, J)$ will not exceed about 10^{-3} (see Fig. 15) for a diatomic molecule. Thus the dilution of the energy of reaction among many rovibronic product states is a serious difficulty. For this reason it is not surprising that the first example of a chemically pumped electronic transition laser (see Chapter 8.4) is based on selective energy transfer to a single atomic state rather than direct operation on a chemically pumped rovibronic transition.

The chemical pumping rate must greatly exceed the threshold fluorescence rate $\Delta N_c/\tau_s$ for a powerful chemical laser (see Fig. 16). For a Doppler-broadened line[93]

$$\Delta N_c/\tau_s = (4\pi^2/\lambda^3)(8kT/\pi m/c)^{1/2}/\tau_p, \qquad (8.6.2)$$

- Assume $\Phi_i(v, J) = \phi_e \, \phi_v \, (N_{vJ}/N_v)_{max}$
  where
  - ϕ_{e_i} = fraction of reactions leading to electronically excited state I = A, B, C, etc.
  - ϕ_{v_i} = fraction of molecules in state I excited to the vibrational state v
  - $(N_{vJ}/N_v)_{max} \approx \left(\dfrac{2hcB}{kT}\right)^{1/2} e^{-0.5} \approx 0.04$; $(N_{vJ}/N_v)_{max}$ is the maximum rotational state population for a given vibronic level with rotational temperature T.
  - In favorable cases $(\phi_{e_i})(\phi_{v_i})$ can be as large as .08.
  - Thus $\Phi_i(v, J) \lesssim 3 \times 10^{-3}$.

FIG. 15. Branching fractions.

---

[93] A. Yariv, "Quantum Electronics." Wiley, New York, 1967.

$$A + BC \longrightarrow AB(X,v,J) + C \qquad \text{ground state}$$
$$\longrightarrow AB(A,v,J) + C$$
$$\longrightarrow AB(B,v,J) + C$$
$$\cdot$$
$$\cdot \qquad\qquad\qquad \text{electronically excited states}$$
$$\cdot$$
$$\longrightarrow AB(?,v,J) + C$$

- Define $\Phi_i(v,J)$;

$$\frac{d[AB(I,v,J)]}{dt} = k\Phi_i(v,J)[A][BC]$$

- Then for a powerful laser with upper state $AB(I,v,J)$

$$\boxed{k\Phi_i(v,J)[A][BC] \gg \Delta N_c/\tau_s}$$

FIG. 16. Minimum chemical pumping rate.

where ΔN_c is the critical inversion density [$\Delta N_c = N_2 - N_1(g_2/g_1)$, where N_2 and N_1 are the densities of upper and lower laser states, respectively], τ_s the radiative lifetime for the laser transition of interest, and τ_p is the mean lifetime of a laser photon in the optical cavity. The fluorescence rate at threshold is proportional to λ^{-3}, which necessitates a larger chemical pumping rate at visible wavelengths than is necessary for near-infrared operation.

The condition that the chemical pumping rate to the upper laser state exceed the threshold fluorescence rate provides an estimate of the minimum required pump rate. As an example, if $\lambda = 5000$ Å, then $\Delta N_c/\tau_s$ ought to exceed 2×10^{16} sec^{-1}cm^{-3} for a reasonable laser gain (10^{-3}cm^{-1}). Taking $\Phi_i(v, J)$ to be 10^{-3} and $k = 10^{-11}$cm^3/sec (1/10 gas kinetic), then we have $[A][BC] \gg 2 \times 10^{30}$cm^{-6} for chemical laser operation from the state (I, v, J). Thus we conclude that an adequate pumping rate requires operation at pressures above about 0.1–1.0 torr. The rate of chemical reaction at pressures in this range and higher is limited by the rate of diffusive mixing of reagents for reactions fast enough to be of laser interest; this conclusion is well established by numerous studies of HF (DF) chemical laser performance.[49] The characteristic time for such mixing is expressible (see Fig. 17) in terms of a diffusion coefficient D and the characteristic dimension l of unmixed portions of the flow. At the present state of the art, achievable characteristic times for diffusive mixing in this pressure regime are of the order $\tau_m = 3 \times 10^{-22}$[M] sec, where [M] is the total gas density in molecules/cm^3. For a diffusion-limited reaction rate we have the relationship from Fig. 17 that $[A] \gg 3 \times 10^{-22}$[M]($\Delta N_c/\tau_s$)/$\Phi_i(v, J)$. Thus, with the use of the values given for ($\Delta N_c/\tau_s$) and $\Phi_i(v, J)$, we find for [M] $= 3 \times 10^{16}$cm^{-3} that $[A] \gg 2 \times 10^{14}$cm^{-3}. Since the required concentration of the atomic or

● If diffusion limited:

$$-\frac{d[A]}{dt} \simeq [A](D/l^2) = [A]/\tau_m,$$

● Compared with premixed case:

$$-\frac{d[A]}{dt} = k[A][BC] \gg (\Delta N_c/\tau_s)/\Phi_i(v,J),$$

● Thus in diffusion limited case

$$[A] \gg \tau_m(\Delta N_c/\tau_s)/\Phi_i(v,J)$$

FIG. 17. Minimum concentration requirement for mixing-limited chemical reaction.

free radical species [A] is large, efficient atom and free-radical production methods must be developed.

This example suggests that cw chemical laser operation at 5000 Å might be possible for reagent pressures of about 1 torr. However, at this pressure diffusive mixing times are not likely to be much shorter than 10^{-5} sec with existing fluid-mixing techniques. This sets a lower limit on collisional and radiative lifetimes of the upper state if a population inversion is to be maintained by selective pumping. Most fully allowed visible electronic transitions in molecules have radiative lifetimes of much less than 10^{-5} sec; moreover, at 1 torr the collision frequency per molecule is about 10^7/sec and thus collisional quenching probabilities larger than 10^{-2} are likely to prevent cw laser operation. These considerations are based only on laser pumping requirements. A population inversion cannot be sustained without adequate relaxation of the lower-state population.

The important features of a successful cw electronic transition chemical laser based on reactions of the type (8.6.1) capable of operation at visible wavelengths may be summarized[94]:

(1) The radiative lifetime of the upper laser state should be 10^{-5} sec or greater.

(2) The overall probability for collisional deactivation of the upper laser level by all species in the laser medium should not exceed 10^{-2}. Anticipated operating pressures are about 1 torr.

(3) Reagent densities should exceed about 10^{15}–10^{16} cm^{-3}. Efficient means for atom or free-radical production must be employed.

[94] A more extensive consideration of the prospects for electronic transition chemical lasers has been presented; see L. E. Wilson, S. Benson, T. Cool, A. Javan, A. Kuppermann, A. Schawlow, S. Suchard, and R. Zare, "New Gas Lasers Committee Report on Electronic Transition Chemically and Electronically Excited Lasers," Tech. Rep. AFWL-TR-73-60, U.S. Air Force Weapons Lab., Kirtland AFB, New Mexico, 1973.

(4) Rapid chemical reactions with a high degree of specificity are required. The fraction of product molecules in the upper rovibronic laser state should approach 10^{-3} or greater.

(5) A population inversion requires that the lower laser state population be minimized. This may require collisional relaxation or rapid convective removal of bound lower-state populations. Excimer systems with dissociative lower states may be attractive candidates. Radiative decay or diffusion processes cannot be relied upon for lower-state relaxation since lasers based on such relaxation mechanisms cannot be scaled up to large sizes.

A comparison of the characteristics of a hypothetical cw diatomic visible chemical laser with those of the DF chemical laser and a representative atomic gas discharge laser are given in Table IV. It is of interest to note the contrast between the DF chemical laser and the atomic gas discharge laser. While the DF laser is capable of operation at power densities greater than 10^7 times the critical fluorescence power density at laser threshold, the atomic gas discharge laser saturates at power levels only about an order of magnitude above the threshold fluorescence power density. The reason for this is that the saturation intensity for the DF laser is not determined by radiative relaxation processes but rather by col-

TABLE IV.   cw Chemical Laser Operation

	DF chemical laser	Atomic gas discharge laser	Diatomic visible chemical laser
Typical wavelength	$\lambda = 4 \ \mu m$	$\lambda = 5000 \ \text{Å}$	$\lambda = 5000 \ \text{Å}$
Laser transition radiative lifetime (sec)	$\tau_s = 10^{-2}$	$\tau_s = 3 \times 10^{-7}$	$\tau_s = 10^{-3}$ to 10^{-7}
Collisional lifetime (p in torr)(sec)	$\tau_d = 5 \times 10^{-4}/p$	$\tau_d = 10^{-5}/p$	$\tau_d = 10^{-5}/p$
Mixing time (sec)	$\tau_m = 3 \times 10^{-5}p$	Premixed	$\tau_m = 3 \times 10^{-5}p$
Mean photon lifetime ($\alpha/L = 2 \times 10^{-3} \text{cm}^{-1}$)(sec)	$\tau_p = 1.7 \times 10^{-8}$	$\tau_p = 1.7 \times 10^{-8}$	$\tau_p = 1.7 \times 10^{-8}$
Threshold fluorescence rate $\Delta N_c/\tau_s = (4\pi^2/\lambda^3)(\bar{v}/c)\tau_p^{-1}$ (cm^{-3} sec^{-1})	4×10^{12}	2×10^{16}	2×10^{16}
Threshold inversion density ΔN_c (cm^{-3})	4×10^{11}	6×10^9	2×10^9 to 2×10^{13}
Critical fluorescence Power density (W/cm^3)	2×10^{-6}	7×10^{-3}	7×10^{-3}
Output power densities (W/cm^3)	$>10^2$	≤ 1	(?)

lisional relaxation processes and flow convection. In contrast typical gas discharge lasers rely on the difference in radiative lifetimes of upper and lower states for population inversion and have relatively small saturation intensities.

Pulsed electronic transition chemical laser operation appears to be possible in a much wider class of reaction systems than those defined by the foregoing considerations. Reaction initiation in premixed reagents by pulsed techniques can provide population inversions on time scales short compared with fluid mixing times and relaxation times for lower laser states. Such techniques include photolysis, laser-induced dissociation, electrical discharge production of atoms or metastables, and shock-induced dissociation and pyrolysis. Typically the pulsed production of atom or free-radical densities exceeding about $10^{13} cm^{-3}$ should be sufficient for pulsed chemical laser operation. The reaction initiation process should occur on a time scale short compared to the time for combined collisional and radiative relaxation of the upper laser level. At the present writing the prospects for pulsed chemical lasers capable of excitation at UV or visible wavelengths appear to be limited by a lack of sufficient information concerning reaction systems capable of the requisite specificity of product formation. The situation could change rapidly as evidenced by the recent development of rare gas–halogen excimer lasers.

8.6.2. Rare Gas Monohalide Excimer Lasers

Very recently pulsed laser action in the ultraviolet from excimer states of several rare gas monohalides has been reported. Laser wavelengths from 1930 to 3530 Å are observed from $^2\Sigma_{1/2}^+ \to {}^2\Sigma_{1/2}^+$ transitions connecting the lowest excimer states to the dissociative ground state in ArF, KrF, XeF, KrCl, XeCl, and XeBr molecules[3,14]; see Fig. 18. Pumping of these lasers is thought to be provided by simple atom exchange reactions between rare gas metastables and halogen molecules, e.g.,

$$Kr(^3P_{0,2}) + F_2 \longrightarrow KrF^* + F. \qquad (8.6.3)$$

The first of these lasers was the XeBr laser reported by Searles and Hart,[3] following earlier flow tube studies of Velazco and Setser[95] and Golde and Thrush.[96] Although these lasers require an electrical source for initial reaction of the $^3P_{0,2}$ rare gas metastables, laser pumping is dependent on chemical reactions of the type (8.6.1), and these lasers might well be termed chemical lasers. It should be noted that rare gas metastables can

[95] J. E. Velazco and D. W. Setser, *J. Chem. Phys.* **62**, 1990 (1975).
[96] M. G. Golde and B. A. Thrush, *Chem. Phys. Lett.* **29**, 486 (1974).

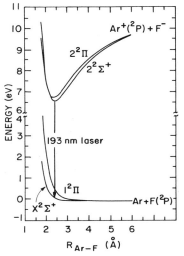

FIG. 18. Potential energy diagram for the ArF excimer laser [W. R. Wadt and P. J. Hay, *J. Chem. Phys.* **68**, 3850 (1978)].

also be produced by impact with high-energy protons[97] or fusion fragments. It is conceivable that excimer lasers could be nuclear powered.

Pulse energies exceeding 100 J for a KrF laser have been reported[9]; this value is over 10^4 times larger than pulse energies reported for previous lasers in the UV spectral region, such as the N_2 laser.

The new excimer lasers promise to be very convenient sources for laboratory experiments in photochemistry, isotope separation, and energy transfer. The ArF laser at 1930 Å, for example, can be operated in a small capacitive discharge configuration with pulse energies exceeding 50 mJ.[13]

8.6.3. Promising Reaction Systems

In the past four years several research groups have actively searched for chemical reaction systems capable of the rather stringent requirements for laser operation. Most emphasis has been placed upon atom exchange reactions of the type (8.6.1), which form electronically excited diatomic metal oxides and metal halides. Such reactions are highly exothermic and are capable of the excitation of many low-lying electronic states in the product molecules. Reactions forming polyatomic products are regarded as less favorable because a much greater number of product

[97] C. H. Chen, M. G. Payne, and J. P. Judish, *Int. Quantum Electron. Conf., 10th, 1978* Paper X.10.

states are accessible and the chances of selective excitation of a few favored states is accordingly less likely. The cw iodine chemical laser is an example of the successful utilization of chemical energy via selective energy transfer. The chemical energy release appears directly as the excitation energy of $O_2(^1\Delta)$ metastables, which are unsuitable for direct laser action; specificity is ensured by the selective energy transfer.

In the search for selective excitation of reaction products, consideration has been given to reactions that lead to products with electronic ground states that do not adiabatically correlate with reactant states. In a given reaction system, however, complications such as nonadiabatic transitions, energy barriers, and E → V transfer processes may be of sufficient importance that conventional adiabatic correlations may be of limited use.

Another way to characterize prospective diatomic reaction products has been considered by Sutton and Suchard.[98] In this approach diatomic reaction products are sought that have potential energy curves for upper and lower electronic states that are favorable for the establishment of population inversions. When Boltzmann-like vibrational distributions are present, molecules with large Franck–Condon factors for transitions between low vibrational levels of an upper electronic state with high vibrational levels of a lower electronic state are favored. An example of such a diatomic molecule with a large displacement in the equilibrium internuclear distances between upper and lower states is GeO, as illustrated in Fig. 19. Reactions of Ge and Si with several oxidizers have been of considerable recent interest.[99]

The electronic states of many reaction products consist of two spin state manifolds. Intercombination transitions in such products may have favorably long radiative lifetimes.

The photon yields for visible chemiluminescence have been determined for several reactions of the type $A + BC \rightarrow AB(I, v, j) + C$. The "photon yield" is ordinarily taken to be simply the rate of visible photon emission per unit volume divided by the rate of chemical reaction per unit volume. At low pressures, under single-collision conditions, such a measurement can be associated with the quantity $\Sigma\Phi_i(v, J)$, where the summation includes contributions from all electronic states I, which spontaneously radiate in the region of sensitivity of the photomultiplier used as a detector (usually 3000–9000 Å). At higher pressures such a simple interpretation is no longer possible because of the influence of secondary collisional processes.

[98] D. G. Sutton and S. N. Suchard, *Appl. Opt.* **14**, 1898 (1975).
[99] See P. M. Swearengen, S. J. Davis, and T. N. Niemczyk, *Chem. Phys. Lett.* **55**, 274 (1978), and references contained therein.

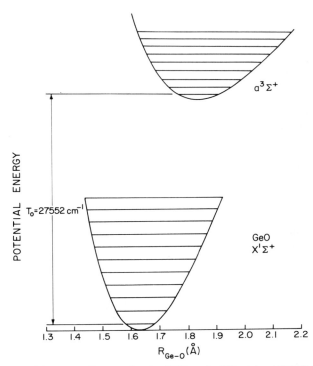

FIG. 19. Potential energy diagram for the $X^1\Sigma^+$ and $a^3\Sigma^+$ states of GeO [G. Hager, R. Harris, and S. G. Hadley, *J. Chem. Phys.* **63**, 2810 (1975)].

Tables V and VI summarize some recent photon yield measurements for several reactions that produce diatomic metal oxide or metal halide reaction products. A hundred or so additional reactions have been screened as possible chemical laser candidates and found to give relatively low yields compared with the best reactions of Tables V and VI.[100–148] Most of the reactions for which data are available, both at high

[100] C. Ottinger and R. N. Zare, *Chem. Phys. Lett.* **5**, 243 (1970).

[101] C. D. Jonah and R. N. Zare, *Chem. Phys. Lett.* **9**, 65 (1971).

[102] C. D. Jonah, R. N. Zare, and C. Ottinger, *J. Chem. Phys.* **56**, 263 (1972).

[103] J. L. Gole and R. N. Zare, *J. Chem. Phys.* **57**, 5331 (1972).

[104] R. H. Obernauf, C. J. Hsu, and H. B. Palmer, *Chem. Phys. Lett.* **17**, 455 (1972).

[105] R. H. Obernauf, C. J. Hsu, and H. B. Palmer, *J. Chem. Phys.* **57**, 5607 (1972); **58**, 4693 (1973).

[106] M. Menzinger and D. J. Wren, *Chem. Phys. Lett.* **18**, 431 (1973).

[107] D. J. Wren and M. Menzinger, *Chem. Phys. Lett.* **20**, 471 (1973).

[108] G. A. Capelle, R. S. Bradford, and H. P. Broida, *Chem. Phys. Lett.* **21**, 418 (1973).

[109] C. R. Jones and H. P. Broida, *J. Chem. Phys.* **59**, 6677 (1973).

[110] R. C. Oldenborg, J. L. Gole, and R. N. Zare, *J. Chem. Phys.* **60**, 4032 (1974).

[111] R. W. Field, C. R. Jones, and H. P. Broida, *J. Chem. Phys.* **60**, 4377 (1974); **62**, 2012 (1975).

[112] A. Schultz and R. N. Zare, *J. Chem. Phys.* **60**, 5120 (1974).

[113] R. W. Field, *J. Chem. Phys.* **60**, 2400 (1974).

[114] P. J. Dagdigian, H. W. Cruse, and R. N. Zare, *J. Chem. Phys.* **60**, 230 (1974).

[115] D. J. Eckström, S. A. Edelstein, and S. W. Benson, *J. Chem. Phys.* **60**, 2930 (1974).

[116] G. Black, M. Luria, D. J. Eckstrom, S. A. Edelstein, and S. W. Benson, *J. Chem. Phys.* **60**, 3709 (1974).

[117] G. Hager, L. E. Wilson, and S. G. Hadley, *Chem. Phys. Lett.* **27**, 439 (1974).

[118] C. J. Hsu, W. D. Krugh, and H. B. Palmer, *J. Chem. Phys.* **60**, 5118 (1974).

[119] C. R. Jones and H. P. Broida, *J. Chem. Phys.* **60**, 4369 (1974).

[120] D. J. Wren and M. Menzinger, *Chem. Phys. Lett.* **27**, 572 (1974).

[121] S. A. Edelstein, D. J. Eckstrom, B. E. Perry, and S. W. Benson, *J. Chem. Phys.* **61**, 4932 (1974).

[122] G. A. Capelle, C. R. Jones, J. Zorskie, and H. P. Broida, *J. Chem. Phys.* **61**, 4777 (1974).

[123] M. Menzinger, *Can. J. Chem.* **52**, 1688 (1974).

[124] P. J. Dagdigian, H. W. Cruse, and R. N. Zare, *J. Chem. Phys.* **62**, 1824 (1975).

[125] M. Menzinger, *Chem. Phys.* **5**, 350 (1974).

[126] J. B. West and H. P. Broida, *J. Chem. Phys.* **62**, 2566 (1975).

[127] R. S. Bradford, Jr., C. R. Jones, L. A. Southall, and H. P. Broida, *J. Chem. Phys.* **62**, 2060 (1975).

[128] G. A. Capelle, H. P. Broida, and R. W. Field, *J. Chem. Phys.* **62**, 3131 (1975).

[129] D. Husain and J. R. Wiesenfeld, *J. Chem. Phys.* **60**, 4377 (1974).

[130] S. Rosenwaks, R. E. Steele, and H. P. Broida, *J. Chem. Phys.* **63**, 1963 (1975); *Chem. Phys. Lett.* **38**, 121 (1976).

[131] D. J. Eckstrom, S. A. Edelstein, D. L. Huestis, B. E. Perry, and S. W. Benson, *J. Chem. Phys.* **63**, 3828 (1975).

[132] C. J. Hsu, W. D, Krugh, H. B. Palmer, R. H. Obernauf, and C. F. Aten, *J. Mol. Spectrosc.* **53**, 273 (1974).

[133] R. W. Field, G. A. Capelle, and C. R. Jones, *J. Mol. Spectrosc.* **54**, 156 (1975).

[134] C. R. Dickson and R. N. Zare, *Chem. Phys.* **7**, 361 (1975).

[135] G. Hager, R. Harris, and S. G. Hadley, *J. Chem. Phys.* **63**, 2810 (1975).

[136] G. A. Capelle and J. M. Brom, Jr., *J. Chem. Phys.* **63**, 5168 (1974).

[137] A. Fontijn, W. Felder, and J. J. Houghton, *Symp. (Int.) Combust. [Proc.]* **15**, 951 (1975).

[138] H. B. Palmer, W. D. Krugh, and C. J. Hsu, *Symp. (Int.) Combust. [Proc.]* **15**, 951 (1975).

[139] W. Felder and A. Fontijn, *Chem. Phys. Lett.* **34**, 398 (1975).

[140] R. C. Oldenborg, C. R. Dickson, and R. N. Zare, *J. Mol. Spectrosc.* **58**, 283 (1975).

[141] M. Luria, D. J. Eckstrom, and S. W. Benson, *J. Chem. Phys.* **64**, 3103 (1976); **65**, 1581 and 1595 (1976).

[142] M. Luria, D. J. Eckstrom, B. Perry, S. A. Edelstein, and S. W. Benson, *J. Chem. Phys.* **64**, 2247 (1976).

[143] C. L. Chalek and J. L. Gole, *J. Chem. Phys.* **65**, 2845 (1976).

[144] S. Rosenwaks, *J. Chem. Phys.* **65**, 3668 (1976).

[145] A. Yokozeki and M. Menzinger, *Chem. Phys.* **14**, 427 (1976).

[146] G. A. Capelle and C. Linton, *J. Chem. Phys.* **65**, 5361 (1976).

[147] J. I. Steinfeld, ed., "Electronic Transition Lasers." MIT Press, Cambridge, Massachusetts, 1976.

[148] F. Engelke, R. K. Sander, and R. N. Zare, *J. Chem. Phys.* **65**, 1146 (1976).

TABLE V. Summary of Some Recent (%) Photon Yield Measurements under Single-Collision Conditions ($\sim 10^{-4}$ torr)

Metal	N$_2$O	NO$_2$	O$_3$	F$_2$	Cl$_2$
			Oxidizers		
Ba	2.3[b]	0.2[b]	—	—	—
Eu	0.2[a]	0.007[a]	0.3[a]	0.9[a]	—
Sc	—	—	—	≥10[c]	—
Sm	0.3[a]	0.07[a]	0.9[a]	11.8[a,b]	—
Y	—	—	—	—	≥5[c]

[a] G. Hager, R. Harris, and S. G. Hadley, *J. Chem. Phys.* **63**, 2810 (1975).
[b] C. R. Dickson, S. M. George, and R. N. Zare, *J. Chem. Phys.* **67**, 1524 (1977).
[c] J. L. Gole, *Electron. Transition Lasers* [*Proc. Summer Colloq.*], *3rd, 1976* p. 136 (1977).

and low pressures, exhibit a low photon yield at low pressure ($\sim 10^{-3}$ torr) and a high yield at high pressures (~ 1 torr). In some cases the photon yields are remarkably high (e.g., SmF and SnO), exceeding 50% at high pressures. In general, these reactions do not produce a high degree of specificity of excitation either at low or high pressures. The chemiluminescence from Sm + O$_3$, illustrated in Fig. 20, indicates the typically broad featureless emission observed for many of the reactions inves-

TABLE VI. Summary of Some Recent (%) Photon Yield Measurements at High Pressures (~ 1 torr)

Metals	N$_2$O	NO$_2$	O$_3$	O$_2$	F$_2$	NF$_3$	Cl$_2$
				Oxidizers			
Ba	25[a]	—	36[a]	8[a]	4.4[a]	50[a]	0.6[c]
Ca	—	—	—	—	0.3[a]	—	—
Eu	20[a]	—	5[a]	2.5[a]	2.5[a]	—	—
Ge	0.15[b]	0.05[b]	—	0.2[b]	0.07[b]	—	10^{-3} [b]
Mg	—	—	—	—	0.02[c]	—	—
Sm	38[a]	—	8[a]	9[a]	64[a]	70[a]	—
Sn	6.7[d](50[e])	—	—	0.08[d]	0.3[d]	—	<10^{-2} [d]
Sr	—	—	—	—	0.1[c]	—	—

[a] D. J. Eckstrom, S. A. Edelstein, D. L. Huestis, B. E. Perry, and S. W. Benson, *J. Chem. Phys.* **63**, 3828 (1975).
[b] G. A. Capelle and J. M. Brom, Jr., *J. Chem. Phys.* **63**, 5168 (1974).
[c] D. J. Eckstrom, S. A. Edelstein, and S. W. Benson, *J. Chem. Phys.* **60**, 2930 (1974).
[d] G. A. Capelle and C. Linton, *J. Chem. Phys.* **65**, 5361 (1976).
[e] W. Felder and A. Fontijn, *Chem. Phys. Lett.* **34**, 398 (1975).

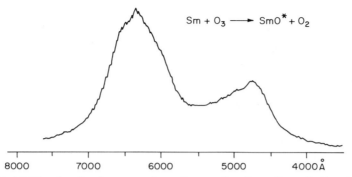

FIG. 20. Visible chemiluminescence from the reaction $Sm + O_3 \rightarrow SmO + O_2$ under single-collision conditions [C. R. Dickson and R. N. Zare, *Chem. Phys.* **7**, 361 (1975)].

tigated to date. These reactions include Sm, Eu, and Yb with the oxidizers Cl_2, F_2, N_2O, O_3, and NO_2.[134,145] Spectra observed at high pressures are very similar to the low-pressure spectra for these systems.

The most extensively studied reactions to date are the $Ba + O_3 \rightarrow BaO + O_2$ and $Ba + N_2O \rightarrow BaO + N_2$ reactions. These reactions give similar spectra. At high pressures the emission is dominated by $A^1\Sigma^+ \rightarrow X^1\Sigma^+$ bands, which are responsible for the high-photon yields of Table VI. Careful studies by several groups—Field *et al.*,[111] Palmer *et al.*,[132] and Eckstrom *et al.*[131]—have established that the $A^1\Sigma^+ \rightarrow X^1\Sigma^+$ emission results from energy transfer from one or more precursor states. The $BaO(a^3\pi)$ state and high vibrational levels of the $BaO(X^1\Sigma^+)$ state have been suggested as such possible primary reservoir states for the energy release of chemical reaction.

The several reactions of Tables V and VI that exhibit a strongly pressure-dependent photon yield apparently conform to a simple mechanistic explanation.[134] A large fraction of the energy release of reaction is divided in a nearly statistical fashion among the various accessible internal product states. These states include high vibrational states of the ground electronic state and somewhat lower vibrational states of several electronically excited states. Communication between these electronic and vibrational states is maintained by rapid collisional processes. At low pressures the primary contributions to the photon yield come from directly excited states with relatively short radiative lifetimes. At higher pressures electronic states of long radiative lifetimes and high vibrational levels of the ground state are coupled to one or more electronic states of short radiative lifetimes to provide an efficient path for radiation as a major energy loss mechanism.

Reactions of oxidizers with metal atoms excited to metastable states are also of interest and in some cases have exhibited very high photon yields. Examples are the reactions of the metastable $Ca(^3P)$, $Sr(^3P)$, and $Mg(^3P)$ atoms with N_2O.[149]

In contrast to all other reactions studied to date, several reactions forming ScF and YCl have a high degree of specificity and high photon yields under single-collision conditions. Chemiluminescent spectra observed by Gole et al.[150,151] for two of these reactions are shown in Figs. 21 and 22. The narrow-emission features near 3500 and 3950 Å for ScF* and YCl*, respectively, are thought to arise from transitions from a $^3\Sigma^+$ upper state to the $X^1\Sigma^+$ ground state. Although reliable radiative lifetimes are not yet available for most of the singlet and triplet states of interest in YCl and ScF, many of the molecular transitions are essentially analogous to two-electron transitions between atomic orbitals in Y or Sc and therefore radiative lifetimes of the order of 10^{-5} sec or longer are expected.

These considerations suggest that the reactions of Figs. 21 and 22 have the desired specificity and lead to reaction product states of long radiative lifetimes. A third desirable characteristic of these reactions, a high photon yield, has been confirmed in recent studies of Gole[151] and co-workers. Their measurements indicate a photon yield of 4.8%, under single-collision conditions, for the 3500 Å $^3\Sigma^+ \to X^1\Sigma^+$ band of ScF* and a comparable photon yield for the corresponding band of YCl* at 3950 Å. Even though at present nothing is known concerning the collisional quenching of ScF* and YCl*, these molecules appear to be favorable candidates for electronic transition chemical lasers.

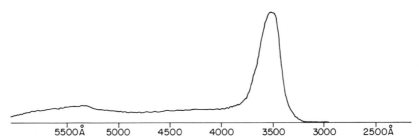

FIG. 21. Visible chemiluminescence from the reaction $Sc + F_2 \to ScF + F$ under single-collision conditions [C. Chalek, D. R. Preuss, and J. L. Gole, *Electron. Transition Lasers* [*Proc. Summer Colloq.*], 2nd, 1975 p. 50 (1976); J. L. Gole, *Electron Transition Lasers*, [*Proc. Summer Colloq.*], 3rd, 1976 p. 136 (1977)].

[149] P. J. Dagdigian, *Chem. Phys. Lett.* **55**, 239 (1978).
[150] C. Chalek, D. R. Preuss, and J. L. Gole, *Electron. Transition Lasers* [*Proc. Summer Colloq.*], 2nd, 1975 p. 50 (1976).
[151] J. L. Gole, *Electron. Transition Lasers* [*Proc. Summer Colloq.*], 3rd, 1976 p. 136 (1977).

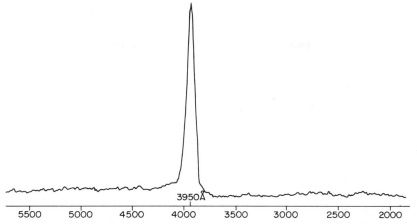

FIG. 22. Visible chemiluminescence from the reaction $Y + Cl_2 \rightarrow YCl + Cl$ under single-collision conditions [C. Chalek, D. R. Preuss, and J. L. Gole, *Electron. Transition Lasers* [*Proc. Summer Colloq.*], *2nd, 1975* p. 50 (1976); J. L. Gole, *Electron. Transition Lasers* [*Proc. Summer Colloq.*], *3rd, 1976* p. 136 (1977)].

Perhaps the most extensive studies of a candidate reaction for visible chemical laser operation have been those of the $Ba + N_2O \rightarrow BaO + N_2$ reaction performed by a group at the Stanford Research Institute.[152]   A resistively heated oven was used to vaporize Ba metal, which was fed to an array of small nozzles and jetted into a surrounding blanket of N_2O in the mixing-flame apparatus of Fig. 23.   The flame zone had a transverse extent of about 40 cm to provide a long optical path for laser cavity tests. The flame was run with a large excess of N_2O and a substantial argon carrier gas flow was employed for convection of Ba atoms from the oven. Most tests were carried out at a total pressure of 5–8 torr, where the photon yield for the $Ba + N_2O$ reaction is high.[119]

Spontaneous-emission measurements carried out under conditions of high and low Ba atom concentrations revealed a strong quenching of the electronic chemiluminescence of the BaO at high Ba concentrations.   At low Ba concentrations the predominant visible emission was from the $A^1\Sigma \rightarrow X^1\Sigma$ band of BaO; at high concentrations of about 10^{16}–$10^{17} cm^{-3}$, this emission was replaced by strong emission from electronically excited Ba atoms.

A partial potential energy diagram for BaO is shown in Fig. 24.   At low pressures the BaO product molecules are thought to be primarily formed

---

[152] D. J. Eckstrom, S. A. Edelstein, D. L. Huestis, B. E. Perry, and S. W. Benson, "Study of New Chemical Lasers," Tech. Rep., SRI MP74–40.   Standard Research Institute, 1974.

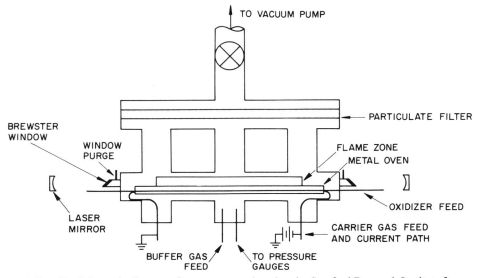

FIG. 23. Schematic diagram of apparatus employed at the Stanford Research Institute for laser studies of the Ba + N_2O → BaO + N_2 reaction (D. J. Eckstrom, S. A. Edelstein, D. L. Huestis, B. E. Perry, and S. W. Benson, Semiannual Technical Report No. 1, SRI No. MP74-40. Stanford Res. Inst., Menlo Park, California 1974).

in the nonradiating $a^3\Pi$ and highly vibrationally excited $X^1\Sigma^+$ states.[111,131,132] The hopes for laser operation in this system were based on the large photon yield observed for $A^1\Sigma^+ \to X^1\Sigma^+$ emission at high pressures, which results from collisional transfer from nonradiating excited states. Large Franck–Condon factors exist for transitions from the first two vibrational levels of the $A^1\Sigma^+$ state to levels near $v = 6$ in the $X^1\Sigma^+$ state.[119] Thus it was argued that if conditions could be found for which sufficiently small populations could be maintained near $v = 6$ in the $X^1\Sigma^+$ state, then at large Ba concentrations the overall chemical pumping rate of the $BaO(A^1\Sigma^+)$ state would be adequate to support laser oscillations on the $BaO(A^1\Sigma^+ \to X^1\Sigma^+)$ band.[153]

An extensive and careful series of tests of this system performed with the apparatus of Fig. 23 failed to demonstrate laser action or gain. Collisional quenching of the $BaO(A^1\Sigma^+)$ emission was cited as a severe difficulty at high pumping rates; at present it is unresolved whether or not population inversions on the $BaO(A^1\Sigma^+ \to X^1\Sigma^+)$ band exist at somewhat lower reagent pressures.

[153] C. R. Jones and H. P. Broida, *Laser Focus* **10**, 37 (1974).

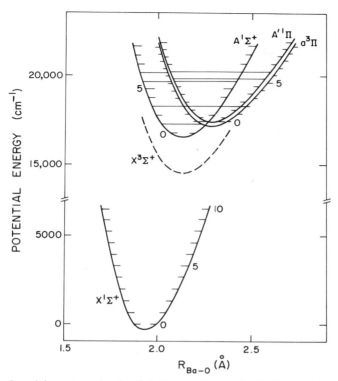

FIG. 24. Low-lying energy levels of BaO according to R. W. Field, C. R. Jones, and H. P. Broida, *J. Chem. Phys.* **60**, 4377 (1974); **62**, 2012 (1975).

8.6.4. Special Initiation Techniques

An important requirement, already mentioned, for either pulsed or cw electronic transition chemical laser operation is the achievement of large metal atom or free-radical concentrations. Moreover, these reagent concentrations must be made available in a manner that ensures that the chemical pumping rate of the upper laser level is large enough compared with the rates of the competing processes of collisional quenching and radiative decay that an upper-level population is created sufficient to exceed laser threshold requirements.

In pulsed systems, techniques are needed that produce metal atoms or free radicals in a time much shorter than that associated with deactivation processes. Typically this time scale will be of the order of 10^{-7}–10^{-8} sec for many systems of interest. A large variety of such techniques have been employed in the search for new laser systems; several are listed in

Table VII. Of particular interest are the processes of laser-induced pyrolysis and dissociation. Bauer and co-workers[154,155] have successfully
initiated a large number of chemical reactions by pyrolysis following
the heating of gas mixtures through CO_2 laser absorption by SF_6. This
technique permits the large deposition of energy on a microsecond time
scale. A more direct means for atom or radical production is through the
photolysis of a parent compound. A new technique of much promise is
photolysis by the multiple absorption of infrared photons.[156,157]

Experiments directed toward cw chemical laser operation require electrical, thermal, or preferably chemical sources of atoms or radicals. A
novel approach under study by Benson et al.[141,142] is the use of a ternary
flame consisting of various volatile polyhalides (e.g., SCl_2, PCl_3, $SnCl_4$,
CCl_4), an alkali metal, and various oxidizers. The essential feature of
these reaction systems is that gas phase reactions of the polyhalide molecules with alkali atoms can lead to the production of free metal atoms by
sequential stripping reactions of the type

$$MX_n + nNa \longrightarrow M + nNaX, \qquad (8.6.4)$$

TABLE VII. Techniques for Pulsed Production of Metal Atoms
and Free Radicals

1. Laser-induced pyrolysis
a. $h\nu + SF_6 \rightarrow SF_6\dagger$, absorption
b. $SF_6\dagger + A \rightarrow SF_6 + A$, collisional thermalization
(A is an arbitrary collision partner)
c. $MR \xrightarrow{\Delta} M + R$, pyrolysis
2. Dissociation by electrical discharge
a. Direct impact of high-energy electrons
b. Dissociative attachment processes
c. Photolysis by discharge-produced photons
3. Direct photolysis at UV wavelengths
4. Photolysis by multiple absorption of infrared photons
a. $h\nu + SF_6 \rightarrow SF_6\dagger$
b. $SF_6\dagger \rightarrow SF_5 + F$, etc.
5. Pyrolysis by shock heating
6. Laser-induced vaporization of metal films
7. Vaporization by exploded metal wires, films

[154] W. M. Shaub and S. H. Bauer, Int. J. Chem. Kinet. **7**, 509 (1975).

[155] S. H. Bauer, E. Bar-Ziv, and J. A. Haberman, IEEE J. Quantum Electron **qe-14**, 237 (1978).

[156] R. V. Ambartzumian, N. V. Chekalin, V. S. Letokov, and E. A. Ryabov, Chem. Phys. Lett. **36**, 301 (1975).

[157] J. L. Lyman and S. D. Rockwood, J. Appl. Phys. **47**, 595 (1976).

where MX_n denotes a polyhalide molecule of interest.   The presence of an oxidizer then can produce pumping reactions, e.g.,

$$M + N_2O \longrightarrow MO^* + N_2, \tag{8.6.5}$$

which may lead to laser action.

Shock-tube-driven supersonic mixing flames have recently been employed in experiments directed toward the discovery of new chemical laser systems.   Figure 25 shows a schematic diagram of the apparatus employed by Rice et al.[158] in their studies of quasi-continuous metal atom oxidation lasers.

A somewhat similar experimental apparatus shown schematically in Fig. 26 is currently in use in the author's laboratory for studies of the reactions

$$Sc + F_2 \longrightarrow ScF^* + F, \tag{8.6.6}$$

$$Y + Cl_2 \longrightarrow YCl^* + Cl. \tag{8.6.7}$$

A heated section of the shock tube permits vaporization of either $ScCl_3$ or YCl_3 in the presence of an argon diluent.   The heated gases are initially confined between thin metal diaphragms, which are ruptured by the passage of the shock wave.   Shock wave heating accomplishes dissociation of the $ScCl_3$ or YCl_3 before the gases are accelerated through a supersonic nozzle array.   This primary flow is then mixed with a secondary flow of either F_2 or Cl_2 through slots at the trailing edge of each nozzle blade.   Reactions (8.6.6) and (8.6.7) are initiated in the supersonic mixing zone immediately downstream of the nozzle array.   The pressures and temperatures in this zone are typically about 5–20 torr and 800–1200 K, respectively.

Another technique of interest has been developed by Friichtenicht and Tang[159] and applied by Zare et al.[160]   They have made use of a laser-induced vaporization technique for the rapid vaporization and dispersal of a metal into a surrounding oxidizing gas.   This method accomplishes the same objectives as the exploding wire method of Rice et al.,[158,161–165] but

[158] W. W. Rice, W. H. Beattie, R. C. Oldenborg, S. E. Johnson, and P. B. Scott, *Phys. Lett.* **28**, 444 (1976).

[159] J. F. Friichtenicht and S. P. Tang, *Electron. Transition Lasers* [*Proc. Summer Colloq.*], *2nd, 1975* p. 36 (1976).

[160] C. R. Dickson, H. U. Lee, R. C. Oldenborg, and R. N. Zare, *Electron. Transition Lasers* [*Proc. Summer Colloq.*], *2nd, 1975* p. 43 (1976).

[161] W. W. Rice and R. J. Jensen, *Appl. Phys. Lett.* **22**, 67 (1973).

[162] W. W. Rice and W. H. Beattie, *Chem. Phys. Lett.* **19**, 82 (1973).

[163] W. W. Rice, *IEEE J. Quantum Electron.* **qe-11**, 689 (1975).

[164] R. J. Jensen, in "Handbook of Chemical Lasers" (R. W. F. Gross and J. F. Bott, eds.), Chapter 13. Wiley, New York, 1976.

[165] J. G. DeKoker and W. W. Rice, *J. Appl. Phys.* **45**, 2770 (1974).

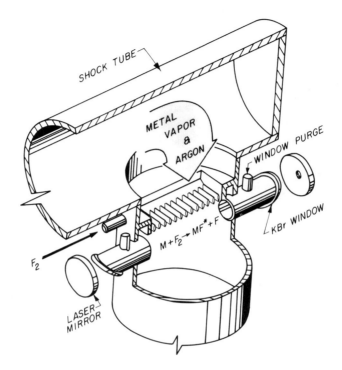

(a)

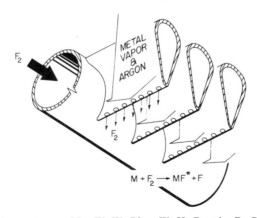

(b)

FIG. 25. (a, b) Apparatus used by W. W. Rice, W. H. Beattie, R. C. Oldenborg, S. E. Johnson, and P. B. Scott, *Phys. Lett.* **28,** 444 (1976) in the demonstration of chemical laser operation from the B + F$_2$ → BF + F and Al + F$_2$ → AlF + F reactions.

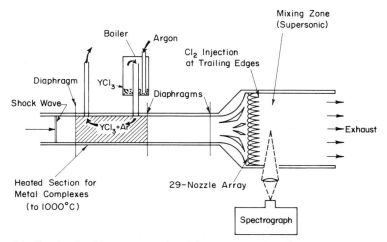

FIG. 26. Shock-tube-driven supersonic mixing apparatus for study of the reactions Sc + $F_2 \rightarrow$ ScF + F and Y + $Cl_2 \rightarrow$ YCl + Cl.

so far the laser "blow off" technique has not resulted in the discovery of new laser systems.

8.6.5. An Ideal Reaction Mechanism for an Efficient Visible Chemical Laser

The remarkably high photon yields observed for several of the reactions of Table VI hold much promise for efficient chemical laser operation. These reactions demonstrate that radiant energy can be efficiently obtained from chemical reactions even when the released energy initially resides among nonradiating states. What is needed is a kinetic mechanism that acts to ensure that this radiant energy is emitted from the upper state of an inverted population.

An ideal mechanism of this type is illustrated in Fig. 27. Given a chemically produced metastable atom or molecule A*, suppose that an association reaction A* + BC $\rightarrow$ (ABC)* can occur to form an electronically excited excimer molecule (ABC)*. If radiative transitions to a repulsive energy hypersurface account for the major source of decomposition of the excimer, then efficient laser action may be possible analogous to that of the rare gas halides already discussed.

An example of such a system that has been investigated recently[166] consists of the reaction steps

[166] T. A. Cool, D. G. Harris, and M. S. Chou, *Proc. Conf. Chem. Mol. Lasers, 5th, 1977* (1977).

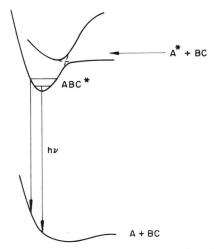

FIG. 27. Schematic illustration of potential energy curves for a hypothetical chemically pumped excimer laser.

$$P(^4S) + N_2O(X^1\Sigma^+) \longrightarrow PO(^4\Pi) + N_2(X^1\Sigma^+), \qquad (8.6.8)$$

$$PO(^4\Pi) + PO(X^2\Pi) \longrightarrow (PO)_2{}^*, \qquad (8.6.9)$$

$$(PO)_2{}^* \longrightarrow h\nu + PO(X^2\Pi) + PO(X^2\Pi). \qquad (8.6.10)$$

Reaction (8.6.8) is expected to lead to the efficient production of the metastable $PO(^4\Pi)$ state illustrated in Fig. 28. Studies of the chemiluminescence resulting from the discharge initiation of chemical reactions in mixtures of PH_3 and N_2O have revealed a strong continuous emission extending from about 3200 Å to beyond 2 μm, with a broad maximum near 7500 Å. This emission is believed to arise from the $(PO)_2{}^*$ excimer emission of process (8.6.10). To date attempts to achieve laser emission from this system have been inconclusive, although photon yields for wavelengths from 3200 Å to 2 μm have been found to exceed 10^{-3} under favorable conditions.

8.7. Conclusion

In the past few years the development of efficient large-scale chemical lasers at near infrared wavelengths has demonstrated the unique capabilities such devices offer compared with other types of lasers. It remains to be seen whether or not comparably useful chemical lasers can be developed at shorter wavelengths. The primary obstacle to the development

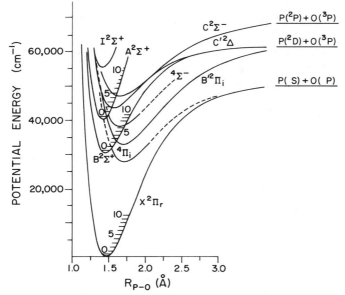

FIG. 28. Low-lying potential energy curves for PO, adapted from R. D. Verma, M. N. Dixit, S. S. Jois, S. Nagaraj, and S. R. Singhal, *Can. J. Phys.* **49**, 3180 (1971).

of electronic transition chemical lasers is the present incomplete knowledge of the reaction kinetics, radiative lifetimes, and quenching mechanisms for candidate chemical systems. An unfilled need exists for high-power cw and pulsed lasers at visible wavelengths; the potential advantages of chemically pumped visible wavelength lasers are great enough to warrant considerable additional fundamental research on candidate systems.

The recent discovery of the rare gas halide excimer lasers provides a good example of the capability of molecular lasers for high energy output at short wavelengths. Indeed the kinetic similarities of these devices to existing and proposed chemical lasers offer much encouragement for further chemical laser development.

Chemical lasers will continue to have an inherent fascination because of what they reveal concerning the partitioning of energy in reaction products. The information-theoretic approach to the thermodynamic analysis of disequilibrium in reaction products has contributed much valuable insight. If the predictive capabilities of the methods of Levine and Bernstein and co-workers[22-26] are fully developed, traditional approaches to molecular dynamics will be supplemented by powerful new concepts that may lead to revolutionary advances in this field. The development of chemical lasers has had a crucial influence on fundamental research on reaction dynamics, energy transfer processes, and the gas dynamics of reacting flows. The intellectual climate and knowledge gained from advances in these areas has contributed much to related research in laser-induced chemistry, laser isotope separation, and laser-induced nuclear fusion.

9. NONLINEAR OPTICAL DEVICES*

9.1. Introduction

In the microwave and radio frequency regions of the spectrum, frequency mixing has long been a useful technique. With the advent of lasers, similar mixing processes have become practical in the optical region of the spectrum. While these processes are indeed similar to those at lower frequencies, there is a significant difference: With the exception of the traveling-wave amplifier, the lower-frequency interactions are almost always localized, and propagation effects are not important. In the optical region, however, one uses the nonlinearity of the optical properties of crystals, and the mixing takes place in a volume large compared to the wavelength. Therefore, the interaction is a traveling-wave interaction, and phase relationships between the different waves are of paramount importance.

The most often used nonlinearity is the second-order dependence of the polarization of a material on the electric field: $P^{NL} = dEE$. If the incident electric fields are two optical waves with frequencies ω_1 and ω_2, the nonlinear polarization has components with frequencies $2\omega_1$, $2\omega_2$, $\omega_1 + \omega_2$, $\omega_1 - \omega_2$, and a zero frequency term. These components of the polarization wave of course radiate electromagnetic waves. Thus, using this nonlinearity one can, in principle, generate the second harmonic of each input, and the sum and difference frequencies.†

If no dispersion were present in the medium, all these frequencies would indeed be generated. However, in the actual case, where there is dispersion between the polarization wave and the electromagnetic waves that it radiates, only that frequency for which this dispersion has somehow been compensated is generated efficiently. For this frequency the interaction is said to be phase matched.

The output power of such a mixing interaction is proportional to the product of the input powers. Thus, even if one input signal is weak, one can still generate a relatively strong output by making sure that the other input signal is strong. This has interesting consequences. For example,

---

† Hereafter, these interactions are referred to as SHG, SFG, and DFG, respectively.

* Part 9 is by F. Zernike.

METHODS OF EXPERIMENTAL PHYSICS, VOL. 15B

it can be used to upconvert a weak infrared signal by mixing it with a strong visible signal, giving an output in the visible, which can be more easily detected than the original infrared signal. A more far-reaching consequence is in difference frequency generation. It can be shown (see Section 9.2.2) that in the interaction with the frequency relationship $\omega_1 - \omega_2 = \omega_3$, the highest-frequency photon, that at ω_1, splits into two photons, one at ω_2 and one at ω_3. Thus, both the signals at ω_2 and at ω_3 are amplified, but again, the signal at ω_3 is proportional to the product of the signals at ω_1 and ω_2, and so the interaction has gain. Moreover, the signal at ω_2 does not have to be supplied externally; even the noise at this frequency is amplified, and if the signal at ω_1 is powerful enough, this amplified noise can become large enough to be observed. If feedback is added, either at both ω_2 and ω_3, or only at ω_2, such a parametric amplifier can be made into an oscillator. This is the principle of the optical parametric oscillator (OPO).

This new field of nonlinear optics started in 1961 when Franken and his co-workers generated the second harmonic of a ruby laser in a quartz crystal.[1] In the years immediately following, the different possible interactions were explored in detail, and discrete frequencies were generated in almost all regions from the UV to the far IR. Much work was also done on parametric oscillators, which allow the generation of tunable frequencies starting from one input frequency. The more recent development of tunable dye lasers, however, has made possible the generation of pairs of tunable input frequencies over the entire visible range, thereby greatly simplifying the generation of useful amounts of tunable, narrowband frequencies anywhere in the spectral region from the near UV to the mid-IR (200 nm to 20 μm) and in the far-IR.

This part describes some of the experimental methods. Many of the equations are given without derivation. For a more theoretical background, the reader is referred to the literature.[2-2c]

9.2. General

9.2.1. The Nonlinear Polarization

The polarization of a material is a function of the applied electric field, $P = \alpha E$. However, as mentioned in the Introduction, the polarizability

---

[1] P. A. Franken, A. E. Hill, C. W. Peters, and G. Weinreich, *Phys. Rev. Lett.* **7**, 118 (1961).

[2] J. A. Armstrong, N. Bloembergen, J. Ducuing, and P. S. Pershan, *Phys. Rev.* **127**, 1918 (1962).

[2a] R. W. Minck, R. W. Terhune, and C. C. Wang, *Proc. IEEE* **54**, 1357 (1966).

[2b] F. Zernike and J. E. Midwinter, "Applied Nonlinear Optics." Wiley, New York, 1973.

[2c] N. Bloembergen, "Nonlinear Optics." Benjamin, New York, 1965.

of a material, α, is not independent of the applied electric field. Developing the polarizability into a power series one has

$$\alpha = \alpha_0(1 + a_1E + a_2E^2 + \cdots).$$

The coefficients a_1, a_2, etc. are small, and so this nonlinearity of the polarizability becomes significant only when E is large, comparable in magnitude to the interatomic fields.   This part considers only the effects due to the first nonlinear term, a_1. The nonlinear polarization caused by this term is proportional to the square of the field. If the material has a center of symmetry, the polarizability has to be the same for E and for $-E$, so that $a_1 = 0$. Consequently, this second-order nonlinear polarization occurs only in materials that lack a center of symmetry. However, only one (class 1 in the triclinic system) completely lacks any symmetry. In all the others there are always certain directions in which a particular symmetry occurs. For those directions, again, the nonlinear polarizability is identically zero.* In general a polarization in any direction P_i can be generated by any combination of the fields $E_j(\omega_1)$ and $E_k(\omega_2)$ or, in tensor notation,

$$P_i(\omega_1, \omega_2) = d_{ijk}E_j(\omega_1)E_k(\omega_2), \tag{9.2.1}$$

where both j and k run from 1 to 3, and summation over repeated indices is assumed. In Eq. (9.2.1), one clearly cannot distinguish between $E_j(\omega_1)E_k(\omega_2)$ and $E_k(\omega_2)E_j(\omega_1)$, so that

$$d_{ijk} \approx d_{ikj}. \tag{9.2.2}$$

Equation (9.2.2) is often used to contract the last two indices as follows: $d_{i1} = d_{i11}$, $d_{i2} = d_{i22}$, $d_{i3} = d_{i33}$, $d_{i4} = \frac{1}{2}(d_{i23} + d_{i32})$, $d_{i5} = \frac{1}{2}(d_{i13} + d_{i31})$, and $d_{i6} = \frac{1}{2}(d_{i12} + d_{i21})$. It is then pointed out that the nonlinear coefficients d_{ij} thus formed obey the same symmetry rules as the piezoelectric coefficients.[3] However, in the contracted piezoelectric tensor notation one has $d_{ik} = d_{ilm} + d_{iml}$ for $l = m$. Therefore, if in the piezoelectric case a coefficient in one of the last three columns of the contracted tensor is shown to be equal to two times a coefficient in one of the first three columns, this factor of two should be omitted in the nonlinear optics case.
    The polarization is now found from

---

[3] J. F. Nye, "Physical Properties of Crystals." Oxford Univ. Press (Clarendon), London and New York, 1960.

---

* Hereafter, whenever the terms polarizability and polarization are used, only the nonlinear parts are meant.

$$
\begin{vmatrix} P_1 \\ P_2 \\ P_3 \end{vmatrix} = \begin{vmatrix} d_{11}\,d_{12}\,d_{13}\,d_{14}\,d_{15}\,d_{16} \\ d_{21}\,d_{22}\,d_{23}\,d_{24}\,d_{25}\,d_{26} \\ d_{31}\,d_{32}\,d_{33}\,d_{34}\,d_{35}\,d_{36} \end{vmatrix} \begin{vmatrix} E_1(\omega_1)E_1(\omega_2) \\ E_2(\omega_1)E_2(\omega_2) \\ E_3(\omega_1)E_3(\omega_2) \\ E_2(\omega_1)E_3(\omega_2) + E_3(\omega_1)E_2(\omega_2) \\ E_1(\omega_1)E_3(\omega_2) + E_3(\omega_1)E_1(\omega_2) \\ E_1(\omega_1)E_2(\omega_2) + E_2(\omega_1)E_1(\omega_2) \end{vmatrix} \quad (9.2.3)
$$

For $\omega_1 = \omega_2$, i.e., for second harmonic generation, it can be seen that the factor of 2 has been moved from the matrix of the coefficients to the column matrix of the fields. For interactions with two distinguishable incident fields this cannot be done. For example, if one has two incident beams with frequencies ω_1 and ω_2, polarized respectively in the 1 and the 2 directions, then $E_1(\omega_1)E_2(\omega_2) + E_2(\omega_1)E_1(\omega_2) = E_1(\omega_1)E_2(\omega_2)$ since $E_2(\omega_1) = E_1(\omega_2) = 0$. Thus, in this case an automatic assumption of a factor of two would give the wrong result.

In addition to the tensor symmetry expressed in Eq. (9.2.2), Kleinman[4] has conjectured that all three indices can be freely permuted if the polarization is purely electronic in origin, and if the medium is lossless in a spectral region that contains all the frequencies involved in the interaction. This conjecture has been shown to be true in most cases of second harmonic generation. A few deviations have been observed.[5,6] It does not, of course, hold for far-IR difference frequency generation.[7]

In general, for collinear interactions, the beams travel through the crystal at an angle θ to the z axis (the direction labeled 3 in the tensor), and in a plane that makes an angle ϕ with the x axis. One can then calculate an effective nonlinearity d_{eff} for which these conditions have been taken into account. Table I shows d_{eff} for the 11 uniaxial crystal classes that have nonzero coefficients, for the case where Kleinman's symmetry does *not* hold. The equations for d_{eff} where Kleinman's symmetry does hold can easily be derived from Table I by realizing that, in that case, the interactions have the same d_{eff} for type I and type II phase matching. That is to say, d_{eff} for the interaction in which two o rays produce an e ray is the same as that for the interaction of an o ray and an e ray giving an o ray, and the same holds for interactions of two e rays and one o ray. Thus, if Kleinman's symmetry does hold, the expressions in both columns have to be the same. For example, for class 3 this means that, if Kleinmann's symmetry holds, $d_{31} = d_{15}$, and $d_{14} = 0$.

[4] D. A. Kleinman, *Phys. Rev.* **126**, 1977 (1962).
[5] M. Okada and S. Ieiri, *Phys. Lett. A* **34**, 63 (1971).
[6] S. Singh, W. A. Bonner, and L. G. van Uitert, *Phys. Lett. A* **38**, 407 (1972).
[7] F. Zernike and P. R. Berman, *Phys. Rev. Lett.* **15**, 999 (1965); erratum, **16**, 177 (1966).

TABLE I. Effective Coefficients for the 11 Uniaxial Crystal Classes That Have Nonzero Coefficients[a]

Class	Nonzero Coefficients	Type I	Type II
$\bar{4}2m$	$d_{14} = d_{25}, d_{36}$	$-d_{36}\sin\theta\sin 2\phi$ $-d_{14}\sin 2\theta\cos 2\phi$	$-d_{14}\sin\theta\sin 2\phi$ $-\frac{1}{2}(d_{14}+d_{36})\sin 2\theta\cos 2\phi$
$\bar{6}m2$	$d_{22} = -d_{21} = -d_{16}$	$-d_{22}\cos\phi\sin 3\phi$ $-d_{22}\cos^2\theta\cos 3\phi$	Same Same
$\bar{6}$	$d_{11} = -d_{12} = -d_{26}$ $d_{22} = -d_{21} = -d_{16}$	$(+d_{11}\cos 3\phi - d_{22}\sin 3\phi)\cos\theta$ $-(d_{11}\sin 3\phi + d_{22}\cos 3\phi)\cos^2\theta$	Same Same
32	$d_{11} = -d_{12} = -d_{26}$ $d_{14} = -d_{25}$	$d_{11}\cos\theta\cos 3\phi$ $-d_{11}\cos^2\theta\sin 3\phi + d_{14}\sin 2\theta$	Same $-d_{11}\cos^2\theta\sin 3\phi - d_{14}\sin\theta\cos\theta$
$\bar{4}$	$d_{14} = d_{25}, d_{15} = -d_{24}$ $d_{31} = -d_{32}, d_{36}$	$-(d_{36}\sin 2\phi + d_{31}\cos 2\phi)\sin\theta$ $(d_{15}\sin 2\phi - d_{14}\cos 2\phi)\sin 2\theta$	$-(d_{14}\sin 2\phi + d_{15}\cos 2\phi)\sin\theta$ $\frac{1}{2}[(d_{15}+d_{31})\sin 2\phi - (d_{14}+d_{36})\cos 2\phi]\sin 2\theta$
$4mm$ and $6mm$	$d_{15} = d_{24}$ $d_{31} = d_{32}, d_{33}$	$d_{31}\sin\theta$ 0	$d_{15}\sin\theta$ 0
6 and 4	$d_{14} = -d_{25}, d_{15} = -d_{24}$ $d_{31} = d_{32}, d_{33}$	$d_{31}\sin\theta$ $-d_{14}\sin 2\theta$	$d_{15}\sin\theta$ $d_{14}\sin\theta\cos\theta$
$3m$	$d_{15} = d_{24}, d_{31} = d_{32}$ $-d_{21} = d_{22} = -d_{16}, d_{33}$	$-d_{22}\cos\theta\sin 3\phi + d_{31}\sin\theta$ $-d_{22}\cos^2\theta\cos 3\phi$	$-d_{22}\cos\theta\sin 3\phi + d_{15}\sin\theta$ Same
3	$d_{11} = -d_{12} = -d_{26}, d_{31} = d_{32}$ $d_{22} = -d_{21} = -d_{16}, d_{15} = d_{24}$ $d_{14} = -d_{25}, d_{33}$	$(d_{11}\cos 3\phi - d_{22}\sin 3\phi)\cos\theta + d_{31}\sin\theta$ $(d_{11}\sin 3\phi + d_{22}\cos 3\phi)\cos^2\theta - d_{14}\sin 2\theta$	$(d_{11}\cos 3\phi - d_{22}\sin 3\phi)\cos\theta + d_{15}\sin\theta$ $(d_{11}\sin 3\phi + d_{22}\cos 3\phi)\cos^2\theta + d_{14}\sin\theta\cos\theta$

[a] For each class four equations are shown corresponding, in the first column, to two o rays producing an e ray and e rays producing an o ray; and in the second column to an o ray and an e ray producing an o ray, and an o ray and an e ray producing an e ray.

9.2.2. The Coupled Amplitude Equations

The frequency mixing discussed here is a three-wave interaction; the two input waves interact to generate an output, and the output in turn interacts with each of the inputs. However, the interaction does not reach an equilibrium, but rather there is a periodic energy exchange. The exact nature of this exchange depends on the phase relationships between the different waves. The two input waves generate a polarization wave and this "driven" wave radiates a "free" electromagnetic wave. For energy to be coupled from the driven to the free wave, the latter has to have a phase lead. For optimum coupling this lead has to be 90°. The free wave starts with the proper phase to satisfy this condition. If there is no dispersion, the phase lead is maintained and the free wave grows. Assume, for example, that the inputs have frequencies ω_1 and ω_2, and that the output for which there is no dispersion is at $\omega_3 = \omega_2 + \omega_1$. After some length then, one of the inputs, say the one at ω_2, goes to zero. At this point there are two waves, at ω_1 and ω_3, and these interact to regenerate a wave at ω_2, for which there is, of course, also no dispersion. Thus, the output is a periodic function of the interaction length: if all of the signal at ω_2 is converted to a signal at ω_3 in a length L, then the input conditions will be re-created after a total length $2L$.

A different situation obtains in SHG: when the entire input has been converted to the second harmonic, there is no wave left for this second harmonic to interact with to regenerate the fundamental. In other words, the subharmonic is not generated. In reality the subharmonic can be generated by interaction between the fundamental and the noise at the second harmonic frequency, but this interaction has a threshold.

If the interaction is not phase matched, i.e., if there is dispersion, the phase relationships between the three interacting waves change with the length, and complete conversion is never achieved. Instead, the wave at ω_3 increases only over a length L_{coh}, known as the coherence length. At that point, the phase difference between the driven wave and the free wave has changed by π rad. The polarization wave now lags the electromagnetic wave by 90°, and thus couples energy out of it. Consequently, the wave at ω_3 decreases again, and becomes zero after two coherence lengths.

The interaction can be compared to the energy exchange between two weakly coupled pendulums. If the pendulums have identical periods, all the energy is transferred from one to the other and back again in a given time. If the periods are unequal, only a part of the energy is transferred, and this transfer takes place in a shorter time.[8]

---

[8] W. H. Louisell, "Coupled Mode and Parametric Electronics." Wiley, New York, 1960.

To put this into proper equation form, consider the general interaction between three collinear waves propagating in the z direction, with electric fields of the form

$$E_i(\omega_i, z, t) = E_i(z) \exp[i(k_i z - \omega_i t)] + \text{c.c.},$$

where $k_i = \omega_i n(\omega_i)/c$, and $E_i(z)$ is a complex amplitude, which changes slowly with z in both phase and magnitude as a result of the interaction. By adding the nonlinear polarization as a source term to Maxwell's equations, one finds, taking into account that $k\, dE/dz \gg d^2E/dz^2$ because of the slowly varying property of $E_i(z)$,

$$dE_1/dz = iK_1 E_2^*(z) E_3(z) e^{i\,\Delta kz}, \tag{9.2.4}$$

$$dE_2/dz = iK_2 E_1^*(z) E_3(z) e^{i\,\Delta kz}, \tag{9.2.5}$$

$$dE_3/dz = iK_3 E_1(z) E_2(z) e^{i\,\Delta kz}. \tag{9.2.6}$$

Here, $\Delta k = k_1 + k_2 - k_3$,

$$K_i = 8\pi \omega_i^2 d/k_i c^2, \tag{9.2.7}$$

where d is the effective nonlinear coefficient. The coupled amplitude equations (9.2.4)–(9.2.6) pertain to the sum frequency interaction in which two waves $E_1(\omega_1, z, t)$ and $E_2(\omega_2, z, t)$ generate a third wave $E_3(\omega_3, z, t)$, with the frequency relation

$$\omega_1 + \omega_2 - \omega_3 = 0 \tag{9.2.8}$$

The power flow in the interaction can be found by multiplying each one of Eqs. (9.2.4), (9.2.5), and (9.2.6) by the complex conjugate of the amplitude in the left-hand side. This leads to

$$\frac{n_1 c}{\omega_1} \frac{d}{dz} (E_1 E_1^*) = \frac{n_2 c}{\omega_2} \frac{d}{dz} (E_2 E_2^*) = -\frac{n_3 c}{\omega_3} \frac{d}{dz} (E_3 E_e^*), \tag{9.2.9}$$

and, since the power flow per unit area in a material with refractive index n is

$$I = (cn/8\pi)EE^* \tag{9.2.10}$$

one arrives at the Manley–Rowe relation

$$-\frac{1}{\omega_1} \frac{dI_1}{dz} = -\frac{1}{\omega_2} \frac{dI_2}{dz} = \frac{1}{\omega_3} \frac{dI_3}{dz}. \tag{9.2.11}$$

The physical interpretation of Eq. (9.2.11) is that if the number of photons that passes through 1 cm² of the wavefront per second increases by a certain amount in the wave with frequency ω_3, the number of photons in the waves at ω_1 and ω_2 decreases by the same amount.

By substituting ω_3 for ω_1, $-\omega_1$ for ω_2, and ω_2 for ω_3, in Eqs. (9.2.4)–

(9.2.8) and observing that $E(-\omega_i) = E^*(\omega_i)$, it can easily be shown that the set of Eqs. (9.2.4)–(9.2.6) also represents the difference frequency generation for which

$$\omega_3 - \omega_1 - \omega_2 = 0. \qquad (9.2.12)$$

This is, of course, as it should be, since as mentioned before the interaction is a three-wave interaction: which way the power will actually flow depends on the initial conditions. For the interaction of Eq. (9.2.12), the Manley–Rowe relation means that for every photon per cm²/sec that is annihilated in the wave at ω_3, one photon is generated at ω_1 and one at ω_2. In other words, in difference frequency generation, in addition to the generation of the output, there is an *amplification* of the lowest-frequency *input*.

To obtain the equations for SHG, one cannot just substitute $\omega_2 = \omega_1$ in Eqs. (9.2.4)–(9.2.6). This is a simple consequence of the mathematical equalities $(a + b)^2 = a^2 + 2ab + b^2$ and $(a)^2 = a^2$. If there are two separate frequencies in the input, their cross product is taken twice; if there is only one, its square is taken once. Thus, for second-harmonic generation one has

$$dE_1/dz = iK_1E_1^*(z)E_2(z)e^{i\,\Delta kz}, \qquad (9.2.13)$$

$$dE_2/dz = \tfrac{1}{2}iK_2E_1^2(z)e^{i\,\Delta kz}, \qquad (9.2.14)$$

where K_i is as given in Eq. (9.2.7) and $\Delta k = 2k_1 - k_2$.

In the small-signal approximation one assumes that so little output energy has been generated that the change in the input energies can be neglected, i.e., $dE_1/dz = 0$ and $dE_2/dz = 0$. Equation (9.2.6) can then easily be integrated to give

$$E_3 = -\frac{16\pi^2}{n_3\lambda_3\,\Delta k}\,dE_1E_2(e^{i\,\Delta kL} - 1), \qquad (9.2.15)$$

where L is the interaction length, and λ_3 the wavelength corresponding to the frequency ω_3.

By combining Eqs. (9.2.15) and (9.2.10), one finds for the output power per unit area

$$I_3 = \frac{512\pi^5L^2d^2I_1I_2}{n_1n_2n_3\lambda_3^2c}\left(\frac{\sin x}{x}\right)^2, \qquad (9.2.16)$$

where $x = \Delta kL/2$.

For second-harmonic generation one finds similarly

$$I_2 = \frac{128\pi^5L^2d^2I_1^2}{n_1^2n_2\lambda_2^2c}\left(\frac{\sin x}{x}\right)^2. \qquad (9.2.17)$$

Equations (9.2.16) and (9.2.17) give the power per unit area, and, since the cgs system is used, the dimension is ergs/cm². In most practical cases one is concerned with total input and output powers in watts. To avoid confusion with various factors of 10^7, convenient equations, using hybrid dimensions, are

$$W_3 = \frac{52.2 L^2 d^2}{n_1 n_2 n_3 \lambda_3^2 A} W_1 W_2 \left(\frac{\sin x}{x}\right)^2 \qquad (9.2.18)$$

for SFG, and

$$W_2 = \frac{13.1 L^2 d^2}{n_1^2 n_2 \lambda_2^2 A} W_1^2 \left(\frac{\sin x}{x}\right)^2 \qquad (9.2.19)$$

for SHG. Here W_i is in watts, d is in cgs, L and λ are in centimeters, and the cross-sectional area A of the beam is in square centimeters.

Equations (9.2.18) and (9.2.19) are convenient to make preliminary estimates of the efficiency of a given interaction. It should be borne in mind, however, that they are small-signal approximations. As such they are reasonably accurate for efficiencies up to 10%. In addition, the interaction was assumed to take place between plane waves. As the efficiency of the interaction is dependent on the energy density, one normally employs focused beams and, again, the approximation gives an estimate that is too high. Better solutions are given in Sections 9.2.5 and 9.2.6. First, however, the question of phase matching must be discussed.

9.2.3. Phase Matching

The $(\sin x)/x$ term in Eq. (9.2.16) shows the importance of phase matching: If $\Delta k \neq 0$ the signal has a maximum whenever $\Delta kL/2$ is equal to an odd multiple of $\pi/2$, and is zero when $\Delta kL/2$ is equal to a multiple of π. The coherence length is the length necessary to obtain the first maximum

$$L_{\text{coh}} = \pi/\Delta k. \qquad (9.2.20)$$

Note that, if $\Delta k \neq 0$, the output never becomes larger than the output from one coherence length, no matter how large the actual interaction length is. Also note that the signal from one coherence length is smaller by $4\pi^{-2}$ than the signal from the same length of material would be if Δk were zero.

Several methods exist to obtain phase matching. The most common one[9] uses the fact that in a birefringent crystal the refractive index for a given wavelength depends on the direction of propagation and on the

[9] J. A. Giordmaine, *Phys. Rev. Lett.* **8**, 19 (1962); P. D. Maker, R. W. Terhune, M. Nisenoff, and C. M. Savage, *ibid.* p. 21.

direction of polarization of the wave.* For sum frequency generation in a negative uniaxial crystal the method is illustrated in Fig. 1. Here k_1 and k_2 are assumed to have ordinary polarization (polarization perpendicular to the optic axis) and k_3 is assumed to be extraordinary. Thus the lengths of the vectors k_1 and k_2 are independent of the direction of propagation in the crystal, but the length of k_3 changes as the angle θ between the wave normal and the optic axis is changed. The locus of the end of $k_1 + k_2$ is a circle, but the locus of the end of k_3 is an ellipse. At the intersection of the two $k_1 + k_2 = k_3$, and the interaction is phase-matched.

Obviously the method can also be used in positive uniaxial crystals—where k_1 and k_2 become extraordinary and k_3 ordinary—and for other interactions. If, as illustrated in Fig. 1, the two input waves in an interaction have the same polarization, the phase matching is called type I.[10] Clearly one can use a similar technique if the two input waves have oppo-

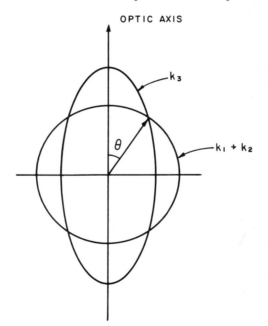

FIG. 1. Section of the indicatrix showing the elliptical section for the extraordinary output and the circular section for the sum of the two ordinary inputs. The intersection between the two determines the phase-matching angle.

[10] J. E. Midwinter and J. Warner, *Br. J. Appl. Phys.* **16**, 1135 (1965).

* The word polarization here is used in its normal "optical" meaning, such as in "polarized light." It should not be confused with the polarization mentioned earlier.

site polarizations. This is called type II phase matching. Many authors adhere to the convention that type I and type II phase matching are defined for SFG. Thus if in a DFG experiment the two lowest frequency beams have polarizations orthogonal to the highest frequency beam, these authors call the phase matching type I, even though the input beams have orthogonal polarizations.

The refractive index for an extraordinary beam whose wave normal makes an angle θ with the optic axis is given by

$$n(\theta) = \frac{n_e n_o}{(n_o^2 \sin^2 \theta + n_e^2 \cos^2 \theta)^{1/2}}, \qquad (9.2.21)$$

where n_e is the extraordinary index and n_o is the ordinary index. Also,

$$\sin \theta = \frac{n_e}{n(\theta)} \left(\frac{n_o^2 - n(\theta)^2}{n_o^2 - n_e^2} \right)^{1/2}. \qquad (9.2.22)$$

It is well known that the wave normal and the ray direction for the extraordinary beam are not parallel.[11] If ρ is the angle between the ray direction and the wave normal, one has

$$\tan(\theta + \rho) = (n_o/n_e)^2 \tan \theta. \qquad (9.2.23)$$

Consequently, two beams with parallel wave normals but orthogonal polarizations do not maintain a complete overlap if $\theta \neq 90°$. The walk-off angle ρ can often be on the order of a few degrees, and is largest for $\theta = 45°$.

If phase matching is obtained at an angle θ for a given combination of frequencies, and one of the input frequencies is changed, to tune the output frequency the angle θ has to be altered also. On the other hand, one can keep θ constant and change both input frequencies, *but always two parameters have to be changed to preserve phase matching.* It follows that, in many cases, one can find combinations of frequencies for which $\theta = 90°$. The walk-off then becomes zero, and the efficiency of the interaction can be maintained over a much larger length. Significantly the first optical parametric oscillator was demonstrated shortly after 90° phase matching was first used in $LiNbO_3$.[12,13] In addition, the change in Δk with angle, $d \Delta k / d\theta$ is smallest for $\theta = 90°$. For this reason, 90° phase matching is often referred to as noncritical phase matching.

In some crystals the birefringence $n_e - n_o$ is a strong function of temperature, and one can choose the temperature so that, for a given pair of input frequencies, phase matching is obtained for $\theta = 90°$. Then tuning of the output can also be obtained by changing one input frequency and

[11] M. Born and E. Wolf, "Principles of Optics." Pergamon, Oxford, 1959.
[12] J. A. Giordmaine and R. C. Miller, *Phys. Rev. Lett.* **14**, 973 (1965).
[13] R. C. Miller, G. D. Boyd, and A. Savage, *Appl. Phys. Lett.* **6**, 77 (1965).

the temperature of the crystal. A disadvantage is that the temperature of the crystal often has to be held to strict tolerances to maintain the phase-matched condition.

The section of the index ellipsoid shown in Fig. 1 is independent of the angle ϕ between the plane of the figure and the x axis of the crystal. However, the effective coefficient d_{eff} is almost always dependent on both θ and ϕ, as shown in Tables I and II, and so for some crystals $d_{\text{eff}} = 0$, even though the birefringence is such that 90° phase matching would be possible. A case in point is GaSe,[14] which has a birefringence large enough to allow 90° phase matching for infrared difference frequency generation between two visible beams traveling in opposite directions, but which belongs to space group 32 for which $d_{\text{eff}} = 0$ at $\theta = 90°$. There are indeed some cases where d_{eff} is only dependent on θ. In LiNbO$_3$, for example, frequency mixing is possible between two o rays and one e ray at $\theta = 90°$, but independent of the angle ϕ.

In biaxial crystals the refractive indices for both polarizations are dependent on the direction of propagation. However, for a given angle ϕ, the refractive index for the beam whose electric vector lies in the x,y plane does not change as a function of θ. Thus, for each angle ϕ one again has the situation shown in Fig. 1, but now the exact form of the figure is dependent on ϕ. In addition, for waves propagating in the crystal in a direction not parallel to the major axes (x, y, or z), the allowed directions of polarization are not parallel to the major planes and this has an additional effect on d_{eff}. Phase matching in birefringent crystals has been treated by Hobden[15] and by Ito et al.[16]

To determine the phase-matching angle for a given crystal, the refractive indices at the pertinent frequencies have to be known. Often the refractive indices are known only at a few wavelengths other than those desired. Methods for interpolation have been discussed by Bhar.[17]

The amount of birefringence needed to obtain collinear phase matching is always largest for second-harmonic generation. Also, for type II phase matching a larger birefringence is needed than for type I. Moreover, if the input beams have orthogonal polarizations and collinear wavenormals, they will walk off. This restricts the effective interaction length more than in type I phase matching. However, in certain cases type II may be preferable for other reasons. For SHG of 1.06 μm in ADP, for example, type

[14] G. B. Abdullaev, L. A. Kulevskii, A. M. Prokhorov, A. D. Savel'ev, E. Yu. Salaev, and V. V. Smirnov, Sov. Phys.—JETP Lett. (Engl. Transl.) 16, 90 (1972); Ph. Kupecek, E. Batifol, and A. Kuhn, Opt. Commun. 11, 291 (1974).

[15] M. V. Hobden, J. Appl. Phys. 38, 4365 (1967).

[16] H. Ito, H. Naito, and H. Inaba, Oyo Butsuri 41, 1201 (1972).

[17] G. C. Bhar, Appl. Opt. 15, 305 (1976).

TABLE II. Some Recent Parametric Oscillator Experiments

Tuning range (μm)	Pump laser	Crystal	Power in	Power out	Bandwidth (cm⁻¹)	Ref.
9.8–10.4	Q-sw. Nd:YAG, 1.833 μm	CdSe	5 kW	Up to 40%	2	[a]
4.3–4.5	HF laser,			Up to 800 W peak		[b]
8.1–8.3	2.7–3.0 μm	CdSe				
14.1–16.4	HF laser + amplifier, 2.87 μm	CdSe		100 μJ	1.8–9	[c]
15.5–16.5		CdSe		2 mJ	0.3	[d]
1.22–8.5	Q-sw. Nd:CaWO₄, 1.06 μm	Proustite		100 W	1	[e]

[a] R. L. Herbst and R. L. Byer, *Appl. Phys. Lett.* **21**, 189 (1972).
[b] J. A. Weiss and L. S. Goldberg, *Appl. Phys. Lett.* **24**, 389 (1974).
[c] R. G. Wenzel and G. P. Arnold, *Appl. Opt.* **15**, 1322 (1976).
[d] G. P. Arnold and R. G. Wenzel, *Appl. Opt.* **16**, 809 (1977).
[e] D. C. Hanna, B. Luther-Davies, and R. C. Smith, *Appl. Phys. Lett.* **22**, 440 (1973).

II shows less dispersion of Δk. This is of importance when the input beams have a wide bandwidth, such as in experiments with short pulses.

9.2.4. Other Phase-Matching Methods

The methods described above can only be used with birefringent crystals. This leaves out the cubic materials, such as GaAs and InSb, many of which have very large nonlinear coefficients. By using a stack of plates, each one coherence length thick, with odd-numbered plates having the opposite crystallographic direction from the even-numbered ones, a phase-matched interaction can still be obtained, even though $\Delta k \neq 0$. The efficiency of this interaction is $4/\pi^2$ of what it would be if Δk were zero. For example, in crystals of class $\overline{4}3m$ (such as GaAs), the nonlinear coefficients tensor for a coordinate system with the x axis along the body diagonal $\{111\}$ and the y axis along $\{\overline{1}, 1, 0\}$ is

$$\begin{vmatrix} \tfrac{1}{3}d\sqrt{3} & -\tfrac{1}{6}d\sqrt{3} & -\tfrac{1}{6}d\sqrt{3} & 0 & 0 & 0 \\ 0 & 0 & \tfrac{1}{3}d\sqrt{2} & -\tfrac{1}{6}d\sqrt{6} & 0 & -\tfrac{1}{6}d\sqrt{3} \\ 0 & -\tfrac{1}{6}d\sqrt{6} & \tfrac{1}{6}d\sqrt{6} & 0 & -\tfrac{1}{6}d\sqrt{3} & 0 \end{vmatrix}$$

where d is the nonlinear coefficient normally listed ($d = d_{14} = d_{25} = d_{36}$). A crystal cut with faces in the [1, 1, 1] crystallographic plane has yz faces in this coordinate system. The tensor shows that a beam incident perpendicular to the surface, and polarized along the y axis, generates a driven wave polarized parallel to the z axis. If the crystal is $2N + 1$ coherence lengths long, the free wave at the exit face will be 180° out of phase with the driven wave. Thus, at a given time when the polarization vector at the exit face is positive in the $+z$ direction, the free-wave electric field is positive in the $-z$ direction. In a second crystal, which is in contact with the first, but which has been rotated 180° around the y axis, the driven wave at the entrance face will point in the $+z$ direction of this crystal, but this is the same as the $-z$ direction in the first crystal, and so the driven wave and the free wave are in phase again. The efficiency of the interaction is of course largest if the plates are one coherence length thick.

This method, originally proposed by Armstrong et al.,[2] is not very useful in the visible, where the plates would have to be very thin, because the coherence length typically is on the order of 10 μm. In the infrared, however, where the dispersion is generally smaller, and where consequently the coherence length is longer, it has been used successfully.[18-21] Szilagyi

[18] A. Szilagyi, A. Hordvik, and H. Schlossberg, J. Appl. Phys. **47**, 2025 (1976).

[19] J. D. McMullen, J. Appl. Phys. **46**, 3076 (1975).

[20] M. S. Piltch, C. D. Cantrell, and R. C. Sze, J. Appl. Phys. **47**, 3514 (1976).

[21] D. E. Thompson, J. D. McMullen, and D. B. Anderson, Appl. Phys. Lett. **29**, 113 (1976).

et al.[18] have examined the method in detail, including the effects of transmission losses and errors in the thickness of the plates.

In some twinned crystals an enhancement of mixing efficiency can be observed due to the modulation of the nonlinear coefficient by the twins.[22-24] Hocker and Dewey have used this effect for SHG[23] and DFG[24] in twinned ZnSe crystals. A similar effect, where the SH was not collinear with the pump beam, was observed by Muzart *et al.*[25]

Another method of introducing a periodic phase change, also originally proposed by Armstrong *et al.*,[2] uses the phase change on total internal reflection. The waves are reflected between the top and bottom surfaces of a slab of crystal, and the phase mismatch accumulated in every pass is canceled by the difference in the phase change upon reflection for the different frequencies. This method has been used experimentally by Ashkin *et al.*[26] and by Boyd and Patel.[27] It is also fundamental to much recent work on SHG in optical waveguides.

In those cases where

$$|k_1| + |k_2| > |k_3| \qquad (9.2.24)$$

phase matching can be obtained by having the interaction take place between noncollinear beams.[28-30] This condition obtains in far-infrared difference frequency generation, when the input frequencies ω_3 and ω_2 are separated from the output frequency ω_1 by an absorption band, such that $n(\omega_1) > n(\omega_2), n(\omega_3)$. An additional advantage of noncollinear matching is that the input beams are not collinear outside the crystal either, so that no separate optical component is needed to combine the beams.

In birefringent crystals the condition given by Eq. (9.2.24) may be realized even if anomalous dispersion is not present. This is done in some cases where a noncollinear arrangement has advantages. Gerlach,[31] for example, used noncollinear matching between an e ray and an o ray in LiIO$_3$, making the angle between the two input wavevectors inside the crystal just equal but opposite in magnitude to the walk-off of the e ray, thus obtaining complete overlap of the input beams through the entire crystal.

[22] R. C. Miller, *Phys. Rev.* **134,** A1313 (1964).
[23] C. F. Dewey, Jr. and L. O. Hocker, *Appl. Phys. Lett.* **26,** 442 (1975).
[24] L. O. Hocker and C. F. Dewey, Jr., *Appl. Phys. Lett.* **28,** 267 (1976).
[25] J. Muzart, F. Bellon, C. A. Arguello, and R. C. C. Leite, *Opt. Commun.* **6,** 329 (1972).
[26] A. Ashkin, G. D. Boyd, and D. A. Kleinman, *Appl. Phys. Lett.* **6,** 179 (1965).
[27] G. D. Boyd and C. K. N. Patel, *Appl. Phys. Lett.* **8,** 313 (1966).
[28] V. T. Nguyen, J. Spalter, J. Manus, J. Ernest, and D. Kehl, *Phys. Lett.* **19,** 285 (1965).
[29] B. Lax, R. L. Aggarwal, and G. Favrot, *Appl. Phys. Lett.* **23,** 679 (1973).
[30] F. Zernike, *Bull. Am. Phys. Soc.* [2] **14,** 741 (1969); see also Zernike and Midwinter.[2b]
[31] H. Gerlach, *Opt. Commun.* **12,** 405 (1974).

Lee *et al.*[32] have used a folded geometry that uses both the phase change on reflection and noncollinear interaction to obtain phase matching. Their geometry is shown in Fig. 2.

Other methods of phase matching have been used. Patel and Nguyen[33] used magnetically induced optical rotatory dispersion to obtain phase matching. The theory of interactions using rotatory dispersion has been given by Rabin and Bey.[34] Phase matching using periodic laminar structures, with layers only a fraction of a wave thick, was proposed by Bloembergen and Sievers,[35] and treated theoretically by Tang and Bey.[36] Van der Ziel and Ilegems[37] verified this method experimentally using a periodic structure of alternating layers of GaAs and $Ga_{0.7}Al_{0.3}As$, grown by molecular beam epitaxy. Other phase-matching methods using periodic thin layers have also been used by van der Ziel and Ilegems.[37,38]

In optical waveguides[39] the propagation velocity for a given frequency depends, among other factors, on the dimensions of the guide and the particular mode that is used. Thus, in principle, one can obtain phase matching by choosing appropriate dimensions for the guide and different modes for the various frequencies. In addition, an optical waveguide has the advantage that a large energy density can be maintained over a length much longer than one can attain in a comparable bulk interaction. The theory of interactions in optical waveguides has been given by Conwell.[40] Though many experiments using this technique have been reported, low

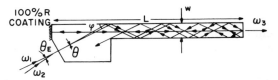

FIG. 2. Noncollinear folded geometry using the phase change on total internal reflection [after N. Lee *et al.*, *Opt. Commun.* **11**, 339 (1974)].

[32] N. Lee, R. L. Aggarwal, and B. Lax, *Opt. Commun.* **11**, 339 (1974).

[33] C. K. N. Patel and V. T. Nguyen, *Appl. Phys. Lett.* **15**, 189 (1969).

[34] H. Rabin and P. P. Bey, *Phys. Rev.* **156**, 1010 (1967); P. P. Bey and H. Rabin, *ibid.* **162**, 794 (1967).

[35] N. Bloembergen and A. J. Sievers, *Appl. Phys. Lett.* **17**, 483 (1970).

[36] C. L. Tang and P. P. Bey, *IEEE J. Quantum Electron.* **qe-9**, 9 (1973).

[37] J. P. van der Ziel and M. Ilegems, *Appl. Phys. Lett.* **28**, 437 (1976).

[38] J. P. van der Ziel and M. Ilegems, *Appl. Phys. Lett.* **29**, 200 (1976), and references therein.

[39] P. K. Tien, *Appl. Opt.* **10**, 2395 (1971).

[40] E. M. Conwell, *IEEE J. Quantum Electron.* **qe-9**, 867 (1973).

coupling between the different frequencies and very stringent dimensional tolerances have thus far limited the attainable efficiency.[41]

9.2.5. Large-Signal Conversion

When the photon conversion efficiency becomes larger than about 10%, the depletion of one or both of the input beams can no longer be neglected. If the inputs have frequencies ω_p and ω_i, if the output has a frequency ω_s, and if the interaction is perfectly phase-matched, one has the following equations[2]:

For SHG ($2\omega_p = \omega_s$),

$$I_s(z) = I_p(0) \tanh^2(\Gamma z), \tag{9.2.25}$$

$$I_p(z) = I_p(0) \operatorname{sech}^2(\Gamma z), \tag{9.2.26}$$

with

$$\Gamma = \tfrac{1}{2} K_p |E_p(0)|, \tag{9.2.27}$$

where K_p is given in Eq. (9.2.7) and $|E_p(0)|^2 = 8\pi I_p(0)/cn_p$.

For SFG ($\omega_p + \omega_i = \omega_s$), if $I_p(0) \approx I_i(0)$ and $I_s(0) = 0$,

$$I_p(z) = (\omega_p/\omega_i)I_i(z) = I_p(0) \operatorname{sech}^2(\Gamma z), \tag{9.2.28}$$

$$I_s(z) = (\omega_s/\omega_p)I_p(0) \tanh^2(\Gamma z). \tag{9.2.29}$$

If $I_p \gg I_i$, as in parametric upconversion of a weak signal at ω_i, the depletion of I_p can be neglected, and Eqs. (9.2.28) and (9.2.29) become

$$I_i(z) = I_i(0) \cos^2(\Gamma z), \tag{9.2.30}$$

$$I_s(z) = \frac{\omega_s}{\omega_i} I_i(0) \sin^2(\Gamma z). \tag{9.2.31}$$

In Eqs. (9.2.28)–(9.2.31),

$$\Gamma = (K_i K_s |E_p|^2/2)^{1/2}. \tag{9.2.32}$$

For DFG ($\omega_p - \omega_i = \omega_s$), if $I_p > I_i$, and the pump I_p is depleted,

$$I_p(z) = I_p(0) \operatorname{sn}^2(\Gamma z, \gamma), \tag{9.2.33}$$

$$I_i(z) = I_i(0) + \frac{\omega_i}{\omega_p} I_p(0)[1 - \operatorname{sn}^2(\Gamma z, \gamma)], \tag{9.2.34}$$

[41] P. K. Tien, R. Ulrich, and R. J. Martin, *Appl. Phys. Lett.* **17**, 337 (1970); Y. Suematsu, Y. Sasaki, and K. Shibata, *ibid.* **23**, 137 (1973); W. K. Burns and R. A. Andrews, *Appl. Opt.* **12**, 2249 (1973); H. Ito and H. Inaba, *Opt. Commun.* **15**, 104 (1975); B. U. Chen, C. L. Tang, and J. M. Telle, *Appl. Phys. Lett.* **25**, 495 (1974); J. P. van der Ziel, R. M. Mikulyak, and A. Y. Cho, *ibid.* **27**, 71 (1975); F. Zernike, *Dig. Top. Meet. Integr. Opt., 1976* p. WA3 (1976).

$$I_s(z) = I_s(0) + \frac{\omega_s}{\omega_p} I_p(0)[1 - \text{sn}^2(\Gamma z, \gamma)], \qquad (9.2.35)$$

where sn is the Jacobian elliptic function,[42] and

$$\gamma = 1 - \frac{1}{2} \frac{\omega_p}{\omega_i} \frac{I_i(0)}{I_p(0)}. \qquad (9.2.36)$$

If the depletion of I_p can be neglected, the equations for DFG become

$$I_i(z) = I_i(0) \cosh^2(\Gamma z), \qquad (9.2.37)$$

$$I_s(z) = \frac{\omega_i}{\omega_s} I_i(0) \sinh^2(\Gamma z), \qquad (9.2.38)$$

where Γ is given in Eq. (9.2.32).

Note that Γ has the dimension L^{-1}. For SHG, for example, Γ^{-1} is the length in which approximately 58% of the input power is converted to the second harmonic.

Depletion of the pump in a DFG experiment was shown clearly by Seymour and Choy.[43] They mixed two dye lasers (450.7 and 578.4 nm), pumped by the same N_2 laser, in $LiNbO_3$ using 90° phase matching. Figure 3 shows the pump pulse after passing through the crystal. In the upper curve the low-frequency laser, the idler, was detuned from phase matching. When this laser is tuned to the frequency that is matched, the pump is depleted, as shown in the lower curve.

Although the equations for DFG also describe parametric amplification, they do so only for perfect phase matching, and no losses. In parametric amplification it is also of interest to know the gain for frequencies that are not phase-matched. Assuming no losses ($\alpha = 0$) and no pump depletion, one finds for the single-pass power gain of a single frequency ω_i, incident on a parametric amplifier,

$$G_i(L) = \frac{I_i(L)}{I_i(0)} - 1 = \Gamma^2 L^2 \frac{\sinh^2(gL)}{(gL)^2}, \qquad (9.2.39)$$

where

$$g^2 = \Gamma^2 - (\Delta k/2)^2. \qquad (9.2.40)$$

At the same time, a frequency ω_s is of course also generated; however, the phase of this signal wave is automatically such that the gain is optimum. If, however, two frequencies ω_s and ω_i are incident on the amplifier, then

[42] L. M. Milne-Thomson, *in* "Handbook of Mathematical Functions" (M. Abramowitz and I. A. Stegan, eds.), p. 567. Natl. Bur. Stand., Washington, D. C., 1966.
[43] R. J. Seymour and M. M. Choy, *Opt. Commun.* **20**, 101 (1977).

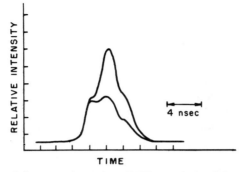

FIG. 3. Depletion of the pump beam in a DFG experiment [after R. J. Seymour and M. M. Choy, *Opt. Commun.* **20**, 101 (1977)].

their phases are predetermined, and they influence the gain. This is the case in the parametric oscillator, where either both the signal and the idler waves, or the idler wave only, are resonated in the crystal.* The two cases are known as doubly resonant oscillator (DRO) and singly resonant oscillator (SRO) (see also Chapter 9.3).

9.2.6. Focused Beams

Thus far, the interaction was assumed to take place between waves with plane wavefronts of infinite extent. In reality the beams are usually focused into the nonlinear crystal, to obtain a higher energy density. The amount of focusing of a Gaussian beam can be specified by its confocal parameter b, which is related to w_0, the radius to the e^{-1} point of the Gaussian spot at the focus, by

$$w_0 = (b/k)^{1/2}.$$

The phase mismatch, $\frac{1}{2}\,\Delta kL$, can also be expressed as a function of b by specifying a phase match parameter $\sigma = b\,\Delta k/2$. If the walk-off is zero, and the beam (power W_1) is focused to a spot with an area $A = \pi w_0^2$ the energy density at the focus is W_1/A. This energy density is more or less constant over a length $L = \xi b$, with the focus in the center, where $\xi \approx 1$. Substituting these values in Eq. (9.2.17) one finds

$$W_2 = \frac{128\pi^5 L d^2 W_1^2}{n_1 n_2 \lambda_2^3 c} h(\xi, \sigma), \qquad (9.2.41)$$

---

* The signal and the idler waves are fully interchangeable in the equations, so it is immaterial which wave is called the signal and which the idler.

where

$$h(\xi, \sigma) = \xi \left(\frac{\sin \sigma\xi}{\sigma\xi}\right)^2. \qquad (9.2.42)$$

Thus, if the length of the crystal is approximately equal to the confocal parameter b, the output power is proportional to the length, rather than to the square of the length.

A rigorous and detailed treatment of the parametric interaction of Gaussian beams, taking absorption, walk-off, and the location of the focus (but not the variation of Δk with angle) into account, has been given by Boyd and Kleinman.[44] They arrive at the same equation, Eq. (9.2.42), but with the function $h(\xi, \sigma)$ replaced by a function

$$h(\sigma, B, \kappa, \mu, \xi), \qquad (9.2.43)$$

where $B = \rho(Lk_1)^{1/2}/2$ is the walk-off parameter and where the parameters κ and μ give the absorption and the location of the focus, respectively. If there is no absorption, $\kappa = 0$ and the maximum efficiency occurs when the focus lies at $L/2$. In that case $\mu = 0$. An important result of the Boyd and Kleinman analysis is that for focused beams the efficiency is *not* optimum for $\Delta k = 0$, but rather for $\Delta k > 0$. This is due to the fact that the phase of a focused beam changes by 180° in the vicinity of the focus.[11] Thus Δk has to be adjusted to keep the phase difference between the polarization wave and the radiated wave close to $-90°$, the value for maximum energy transfer. The fact that Δk has to be positive can be understood by regarding the input beams as diverging pencils of plane waves. When $\Delta k > 0$ for the axial component of the pencil, there still exist mixing processes for which $\Delta \mathbf{k} = \mathbf{k}_1 + \mathbf{k}_2 - \mathbf{k}_3 = 0$. For $\Delta k < 0$ this is not the case.

For $\kappa = \mu = 0$, the function $h(\xi, \sigma, \kappa, \mu, B)$ can be optimized for the best phase-matching condition at every ξ. One then arrives at a function $h_m(B, \xi)$. For SHG in a negative uniaxial crystal using type I phase matching this function $h_m(B, \xi)$ is shown in Fig. 4 as a function of ξ for several values of B. For $B = 0$, the maximum value, $h = 1.068$, occurs for $\xi = 2.84$, and $\xi\sigma = \Delta kL/2 \approx 1.6$. Experimentally the exact value of $\xi\sigma$ for maximum conversion is not important, as it can normally be adjusted by a small change in the angle or the temperature of the crystal.

For $B = 0$ and $0.5 < \xi < 2$, the function $h(\xi, \sigma)$ of Eq. (9.2.42) gives fairly accurate values if the optimum values of σ are used.

The theory of Boyd and Kleinman does not take pump depletion into account. For 90° phase-matched SHG this was done by White *et al.*,[45]

[44] G. D. Boyd and D. A. Kleinman, *J. Appl. Phys.* **39**, 3597 (1968).
[45] D. R. White, E. L. Dawes, and J. H. Marburger, *IEEE J. Quantum Electron.* **qe-6**, 793 (1970).

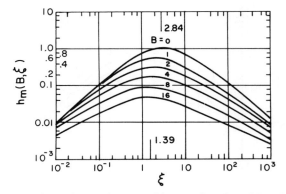

FIG. 4. SHG power for optimum phase matching as a function of focusing parameter for several values of double refraction parameter [after G. D. Boyd and D. A. Kleinman, *J. Appl. Phys.* **39**, 3597 (1968)].

who conclude that the optimum values of σ and ξ found by Boyd and Kleinman are very nearly independent of pump power to above 95% conversion. For the efficiency of the interaction at optimum σ they give an empirical relation

$$W_2 = W_1 \tanh^2 \left[\frac{128\pi^5 d^2}{n_1 n_2 \lambda_2^3 c} LW_1 h(\xi, \sigma) \right]^{1/2}, \qquad (9.2.44)$$

where $h(\xi, \sigma)$ is given by Eq. (9.2.42). . For small values of W_1, Eq. (9.2.44) agrees with Eq. (9.2.41).

It should be emphasized that the parameter b used above describes the beam in the crystal. That is to say, if the crystal—which has its entrance face at $z = 0$—were removed, the beam would have a confocal parameter b/n_1 and would be focused at a point $L/2n_1$ from $z = 0$.

Boyd and Kleinman also give a function $h_m(B, \xi)$ for three frequency interactions, i.e., interactions where $\omega_1 + \omega_2 = \omega_3$. Now

$$B = \rho(Lk_0)^{1/2}(n_3/n_0)^{1/2}/2,$$

where $\omega_0 \approx \omega_1 \approx \omega_2$. This function is shown in Fig. 5. The curve for $B = 0$ is the same as the one in Fig. 4.

For most practical applications, it is sufficient to use Eq. (9.2.18) with A for the average area of the spot sizes of the two input beams,

$$A = \tfrac{1}{2}\pi(w_p^2 + w_i^2).$$

The following example illustrates the effect of focusing. Suppose that the SH of 1.06 μm is generated in a 3-cm-long piece of ADP using type I angle phase-matching at room temperature. Then $\theta = 41°58'$ and

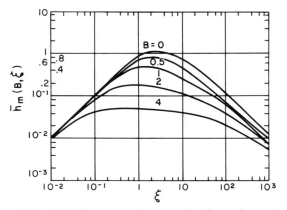

FIG. 5. Reciprocal threshold for parametric generation for optimum phase-matching as a function of focusing parameter for several values of B.

$\rho = 0.03$ rad, giving $B = 8$. By going from a beam with a radius of 1.3 mm ($\xi = 10^{-3}$) to an optimally focused beam ($\xi = 1.4$), one obtains a focused spot with radius 0.045 mm. Figure 4 shows that the increase in output power would be about 80 times, provided the small signal approximation is still valid. However, if LiNbO$_3$ is used in the same experiment, phase matching at 90° can be obtained by heating the crystal. Then $B = 0$, and so the optimum value of ξ is 2.84, giving a spot size of 0.025 mm. From Fig. 4 one finds an increase in output power of about 800 times in this case. A very thorough treatment of the interaction of focused beams in SHG for different phase-matching conditions has been given by Akhmanov et al.[46]

To focus a beam with radius w and wavelength λ to a spot with radius w_0, one needs a lens with focal length f, such that

$$f = \pi ww/\lambda. \qquad (9.2.45)$$

Equation (9.2.45) assumes that w is in the far field, i.e., that $w \gg w_0$.

9.2.7. Interactions with Short Pulses

When the inputs are generated by cw or Q-switched lasers, the bandwidths are, or can be made, small enough so that phase matching is maintained for all frequencies involved. When input pulses with durations of the order of picoseconds are used, however, this is no longer the case. The bandwidth of a pulse of duration t is $\Delta\nu = (tc)^{-1}$, where $\Delta\nu$ is in cm^{-1}.

[46] S. A. Akhmanov, A. I. Kovrygin, and A. P. Sukhorukov, in "Quantum Electronics: A Treatise" (H. Rabin and C. L. Tang, eds.), Vol. 1, p. 475. Academic Press, New York, 1975.

Thus a 1 psec pulse has a bandwidth of 33 cm^{-1}. For SHG of 1.06 μm in LiNbO$_3$, for example, the mismatch at the edges of such a band, assuming that the center is phase-matched, is $|\Delta k| \approx 5.0$ cm^{-1}. The important quantity here is the group velocity $\partial\omega/\partial k$, rather than the phase velocity. The fundamental pulse propagates with a group velocity $\alpha^{-1} = (\partial\omega/\partial k)_{\omega_1}$, and the SH has a group velocity $\beta^{-1} = (\partial\omega/\partial k)_{\omega_2}$. If these two velocities are matched, $\alpha \approx \beta$, continuous energy transfer between the pulses takes place. The envelope of the output pulse then is the square of the envelope of the input pulse, and as a result the output will in general have a shorter duration than the input.

If $\alpha \neq \beta$, the two pulses do not travel with the same velocity and if $(\alpha - \beta)L = t$, where L is the crystal length, the frequency components at the edge of the band are mismatched by 2π. The output pulse then is essentially square in time and does not increase in amplitude when L is increased, but instead the duration of the pulse increases.[47] Thus, the group velocity mismatch has an effect on the interaction of short pulses that is analogous to the effect of the walk-off in steady-state interactions. Akhmanov has shown that indeed the harmonic generation of wave packets is the temporal analog of harmonic generation of two-dimensional beams.[46]

The group velocity mismatch for the case given above (SHG of 1.06 μm in LiNbO$_3$) is large, $\alpha - \beta = 3.5 \times 10^{-12}$. For ADP, using type II phase matching in the same interaction, however, $\alpha - \beta = 9 \times 10^{-14}$.

In parametric amplification, a shortening of the pulse occurs that, in addition to being dependent on group velocity matching, is also dependent on the gain.[48,49]

9.2.8. Output Angle

In most cases the output is collinear with the inputs and has a divergence approximately equal to the convergence of the input. When a strong walk-off is present, the output beam becomes somewhat rectangular in cross section, but this effect is usually negligible. If noncollinear phase matching is used, the output exits at an angle to the inputs and this angle changes as the output frequency is tuned. Wellegehausen et al.[50] minimized this change by providing their mixing crystal with an exit face that is perpendicular to the mean direction of the output.

[47] J. Comly and E. Garmire, Appl. Phys. Lett. 12, 7 (1968); W. H. Glenn, IEEE J. Quantum Electron. qe-5, 284 (1965).

[48] W. H. Glenn, Appl. Phys. Lett. 11, 333 (1967).

[49] S. A. Akhmanov, A. P. Sukhorukov, and A. S. Chirkin, Sov. Phys.—JETP (Engl. Transl.) 28, 748 (1968).

[50] B. Wellegehausen, D. Friede, H. Vogt, and S. Shahdin, Appl. Phys. 11, 363 (1976).

In far-infrared generation the wavelength of the output can be of the same order of magnitude as the spot size of the focused input beams. From diffraction considerations it is obvious that the divergence of the output then has to be much larger than the convergence of the inputs. An approximate equation for the angular distribution of the output replaces the $(\sin x/x)^2$ function of Eq. (9.2.16) by

$$\left[\frac{2J_1(k_1w_0 \sin \theta)}{k_1w_0 \sin \theta}\right]^2 \left[\frac{\sin k_1L(1 - \cos \theta)/2}{k_1L(1 - \cos \theta)/2}\right]^2, \qquad (9.2.46)$$

where the interaction takes place in a cylinder of radius w_0 and length L. J_1 is a first-order Bessel function, k_1 is the magnitude of the wavevector of the output, and ϕ is the angle between the direction of observation and the axis of the input beams.[7] A rigorous treatment of the angular distribution of the output in far-infrared DFG has been given by Morris and Shen.[51]

9.2.9. Intracavity Generation

The output is proportional to the energy densities of the inputs. Thus one should always try to obtain the maximum input energy density, limited only by the damage threshold of the nonlinear material. If the material has no absorption at the input frequencies, the maximum available energy density can be utilized by placing the crystal inside the laser resonator, and equipping this resonator with mirrors that have maximum reflectivity for the laser wavelength. Zernike and Berman[7] used this method for far-infrared DFG and spatially separated the output from the input by using the fact that the output angle was much larger than the input angle [see Eq. (9.2.46)]. Geusic et al.[52] used a crystal of $Ba_2NaNb_5O_{15}$ for SHG inside the cavity of a Nd:YAG laser equipped with mirrors that had a reflectivity $>99.8\%$ at 1.064 μm. The SH was coupled out through one of the mirrors, which had a transmission of 98% at 532 nm. They pointed out that, if the loss in the nonlinear crystal is much smaller than the other losses in the laser, the maximum SH output power is equal to the maximum available fundamental power from the basic laser.

A detailed discussion of intracavity SHG using a dye laser has been given by Frölich et al.,[53] who carefully analyzed the losses in their system and found that the largest losses were due to walk-off losses (see Section 9.3.6) and thermally induced losses in the etalons used for frequency narrowing

[51] J. R. Morris and Y. R. Shen, *Phys. Rev. A* **15**, 1143 (1977).

[52] J. E. Geusic, H. J. Levinstein, S. Singh, R. G. Smith, and L. G. Van Uitert, *Appl. Phys. Lett.* **12**, 306 (1968).

[53] D. Frölich, L. Stein, H. W. Schröder, and H. Welling, *Appl. Phys.* **11**, 97 (1976).

of the dye laser. They note that these effects are smaller in ring lasers, and report an increase of a factor of 10 in power by going from a conventional laser to a ring laser.

Recently an enhancement of the efficiency of DFG and SFG experiments has been obtained by many authors by placing the crystal inside the cavity of one or the other of the input lasers.[54-56]

9.2.10. Effects That Reduce the Efficiency

A number of effects can reduce the efficiency of a given interaction. Most prominent among these are damage, absorption, and heating of the crystal.

Index damage[57] has been observed in a number of crystals, but most prominently in lithium niobate. It occurs at relatively low power densities. Heating of the crystal minimizes the damage.

Bulk damage and surface damage can occur in almost any material at high power densities. The surface damage threshold is usually lower than the bulk damage threshold, and depends very strongly on the perfection of the surface finish. The threshold for short pulses is normally higher than for cw operation.

Absorption can be accounted for by replacing the $(\sin x/x)^2$ function in Eq. (9.2.16) by the function

$$\frac{(1 + 2e^{-\alpha L} \cos \Delta kL + e^{-2\alpha L})}{[\alpha^2 + (\Delta k)^2]L^2} e^{-2\alpha_3 L},$$

where $\alpha = \alpha_1 + \alpha_2 - \alpha_3$, and α_i is the amplitude attenuation for the wave at frequency ω_i.

Heating of the crystal by the beams themselves can cause a deterioration of the phase-matching condition[58] (see also Table V). The efficiency can also be reduced by two-photon absorption at any of the three frequencies or by the production, by the generated second harmonic, of photo-induced carriers that absorb the pump radiation.[59]

In some cases more than one mixing interaction can be phase-matched simultaneously, reducing the efficiency of the desired interaction.[60,61]

[54] D. W. Meltzer and L. S. Goldberg, *Opt. Commun.* **5**, 209 (1972).

[55] L. S. Goldberg, *Appl. Opt.* **14**, 653 (1975).

[56] W. Lahmann, R. Tibulski, and H. Welling, *Opt. Commun.* **17**, 18 (1976).

[57] A. Ashkin, G. D. Boyd, J. M. Dziedzic, R. G. Smith, A. A. Ballman, J. J. Levinstein, and K. Nassau, *Appl. Phys. Lett.* **9**, 72 (1966).

[58] M. Okada and S. Ieiri, *IEEE J. Quantum Electron.* **qe-7**, 560 (1971).

[59] W. B. Gandrud and R. L. Abrams, *Appl. Phys. Lett.* **17**, 302 (1970).

[60] J. M. Yarborough and E. O. Amman, *Appl. Phys. Lett.* **19**, 145 (1971).

[61] R. A. Andrews, H. Rabin, and C. L. Tang, *Phys. Rev. Lett.* **25**, 605 (1970).

9.3. Parametric Oscillators

9.3.1. Introduction

The advantage of parametric oscillators is that the only input needed is the pump—the idler is generated from noise. Thus tunable signal frequencies can be generated from a single input. In the early years, parametric oscillators were the only practical tunable sources, and much work was done on the various possible configurations. This work has been described in detail in review articles by Smith[62] and by Byer.[63]

In recent years, frequency mixing using the experimentally much simpler tunable dye lasers has replaced the OPO for the generation of frequencies in the visible and the near infrared, except for cases where high average power is necessary. For the generation of frequencies in the mid-infrared, however, the requirement that the crystal be transparent at all three frequencies makes the OPO desirable, either as a direct source of the frequency to be generated, or as a source of two near-IR frequencies (signal and idler) from which the mid-IR frequency can be generated by DFG.

9.3.2. Threshold

In contrast to mixing interactions where two input frequencies are used, the OPO has a threshold. The threshold pump power density is that power density for which the gain is equal to the losses.

For small losses one has, for the DRO,

$$(\Gamma L \sin x/x)^2 \approx a_i a_s, \tag{9.3.1}$$

and for the SRO, assuming that the idler wave is resonated,

$$(\Gamma L \sin x/x)^2 \approx 2a_i, \tag{9.3.2}$$

where $x = \Delta k L/2$ and a_i, a_s are the single-pass power losses for the signal and the idler, respectively.

For the DRO the threshold power density in w/cm² is

$$I_{th}^{DRO} = 0.02 \frac{n_i n_s n_p \lambda_i \lambda_s a_i a_s}{d^2 L^2} \left(\frac{\sin x}{x}\right)^{-2}, \tag{9.3.3}$$

where n is the refractive index, a the loss at the frequency indicated by the subscript, and $x = \Delta k L/2$. The wavelength λ and the length of the crystal L are in cm, and d is in esu.

[62] R. G. Smith, *in* "Laser Handbook" (F. T. Arecchi and E. O. Schultz-Dubois, eds.), p. 837. North-Holland Publ., Amsterdam, 1972.

[63] R. L. Byer, *in* "Quantum Electronics: A Treatise" (H. Rabin and C. L. Tang, eds.), Vol. 1, Par B, p. 587. Academic Press, New York, 1975.

For the SRO one has

$$I_{th}^{SRO} = 0.04 \frac{n_i n_s n_p \lambda_i \lambda_s a_i}{d^2 L^2} \left(\frac{\sin x}{x}\right)^{-2}. \qquad (9.3.4)$$

Clearly the threshold power density for the SRO is a factor $2/a_s$ larger than for the DRO. However, in the DRO the cavity has to be resonant for both the signal and the idler, and the two frequencies that are phase-matched are not necessarily both resonant. As a result the oscillator then operates with signal and idler frequencies that are somewhere between perfect resonance and perfect phase matching, since the gain is influenced both by the Q of the cavity, and by the phase-matching condition. Thus smooth tuning cannot be achieved, because the signal and idler frequencies tend to jump from one cavity mode to the next. Even for a single frequency output this "mode hopping" can be eliminated only by careful control of the resonator, and by the use of a single-frequency pump source. For this reason the SRO is preferable for many applications in spite of its higher threshold.

9.3.3. Risetime

To obtain the necessary threshold power density, the oscillator is often pumped by a Q-switched laser with pulse lengths of 10 nsec to 1 μsec duration. It is of course necessary to reach threshold before the pump pulse is turned off; the shorter the pulse, the higher the peak power should be to obtain threshold quickly.

The risetimes are given by

$$\tau_R = \frac{L_c/c}{2a(\sqrt{N} - 1)} \ln \left(\frac{I_i(t)}{I_i(0)}\right), \qquad (9.3.5)$$

for the DRO, assuming $a_i = a_s = a$ and $\Delta k = 0$, and

$$\tau_R = \frac{L_c/c}{2a(N - 1)} \ln \left(\frac{I_i(t)}{I_i(0)}\right) \qquad (9.3.6)$$

for the SRO. Here N is the ratio between the pump power and the threshold power, and L_c the optical length of the cavity. Clearly the cavity length should be made as short as possible. The reason for this is of course that the shorter the cavity the larger the number of passes through the crystal in a given time.

In Eqs. (9.3.5) and (9.3.6) $I_i(t)$ is the "steady-state" power density in the cavity and $I(0)$ is the initial power density. The initial power density can be taken to be noise, i.e., one phonon per mode, $I_i = \hbar \omega_i c$. However, the risetime is very insensitive to the exact values of these two quantities as their ratio comes in as a logarithm. The value of this logarithm is usually about 30.

Equations (9.3.5) and (9.3.6) should be used for order of magnitude calculations only, because they assume a pump pulse with an infinitely short risetime and they neglect depletion of the pump. The risetime of an OPO has been treated in detail by Pearson *et al.*[64] and by Byer.[63]

9.3.4. Pump Power and Damage Threshold

At high-power operation the OPO is susceptible to damage, both of the nonlinear crystal itself and of the coatings used on the crystal and other cavity elements. Bey and Tang[65] have analyzed in detail the large-signal behavior of a SRO, pumped by a plane wave. An important finding of their work is that the photon flux of the resonated wave can be much higher than the pump photon flux, so that the power level that the oscillator has to withstand in that case is much higher than that of the pump (see Fig. 6). Arnold and Wenzel[65a] have observed this experimentally in a setup in which a grating is used both to couple in the pump radiation, and as a reflecting element of the resonator. They found that a replica grating, which was damage resistant to the pump beam alone, was destroyed after only a few pulses with the OPO operating.

In addition, the intensity distribution across the pump beam is usually not constant. Aperturing of the beam can lead to Fresnel diffraction with severe intensity variations.[66] This can be alleviated by the use of apodized apertures.[67]

9.3.5. Conversion Efficiency

Assuming a nearly uniform spatial distribution of the resonated waves, the conversion efficiency of the DRO is

$$\frac{I_s}{I_p} = \frac{\omega_s}{\omega_p} \frac{2}{N} (N^{1/2} - 1), \tag{9.3.7}$$

$$\frac{I_i}{I_p} = \frac{\omega_i}{\omega_p} \frac{2}{N} (N^{1/2} - 1). \tag{9.3.8}$$

The total efficiency, the sum of Eqs. (9.3.7) and (9.3.8), has a maximum of 50% for $N = 4$. The cause of this maximum is the wave generated in the backward direction as the sum frequency between the idler and the

[64] J. E. Pearson, U. Ganiel, and A. Yariv, *IEEE J. Quantum Electron.* **qe-8**, 433 (1972).

[65] P. P. Bey and C. L. Tang, *IEEE J. Quantum Electron.* **qe-8**, 361 (1972).

[65a] G. P. Arnold and R. G. Wenzel, *Appl. Opt.* **16**, 809 (1977).

[66] A. J. Campillo, J. E. Pearson, S. L. Shapiro, and N. J. Terrell, Jr., *Appl. Phys. Lett.* **23**, 85 (1973).

[67] A. J. Campillo, D. H. Gill, B. E. Newman, S. L. Shapiro, and J. Terrell, *IEEE J. Quantum Electron.* **qe-10**, 767 (1974).

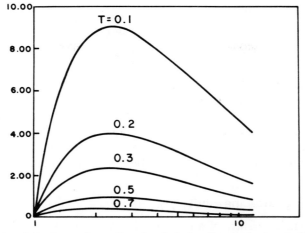

FIG. 6. Ratio of the forward-traveling signal photon flux at the input of a parametric oscillator to the pump photon flux as a function of the ratio of the product of the incident pump intensity and the square of the length of the nonlinear medium to the value of this product at threshold. $T = 1 - R_1 R_2$, where R_1 and R_2 are the power reflectivities of the input and the output mirrors, respectively [after P. P. Bey and C. L. Tang, *IEEE J. Quantum Electron.* **qe-8**, 361 (1972)].

signal. This reflected power, which has a frequency equal to that of the pump, effectively takes away a fraction of the incident pump power. Since it is reflected, it may also cause instabilities in the pump laser. Byer *et al.*[68] have shown that this reflected wave can be avoided by using a ring resonator, which has an efficiency of 100% at $N = 4$.

If the intensity distribution across the beam is Gaussian, the efficiency for a DRO becomes[69]

$$\frac{I_i + I_s}{I_p} = \frac{4}{N}(N^{1/2} - 1 - \ln N^{1/2}) \qquad (9.3.9)$$

and reaches a maximum of 41% at $N = 12.5$.

For the SRO the efficiency is

$$\frac{I_s + I_i}{I_p} = \frac{\kappa^2}{\beta^2}\sin^2(\beta L), \qquad (9.3.10)$$

where β is found from

$$\left(\frac{\sin \beta L}{\beta L}\right)^2 = \frac{1}{N}\left(\frac{\sin x}{x}\right)^2, \qquad (9.3.11)$$

[68] R. L. Byer, A. Kovrigin, and J. F. Young, *Appl. Phys. Lett.* **15**, 136 (1969).
[69] J. E. Bjorkholm, *IEEE J. Quantum Electron.* **qe-5**, 293 (1969).

where $x = \Delta k L/2$ and

$$\kappa^2 = \beta^2 - (\tfrac{1}{2} \Delta k)^2. \qquad (9.3.12)$$

For $\Delta k = 0$, i.e., at line center, the SRO reaches 100% efficiency for $N = (\pi/2)^2$.

Bjorkholm[69] has given an integral solution for the efficiency of a SRO with a Gaussian intensity distribution. This efficiency reaches a maximum of 71% at $N = 6.5$.

The conversion efficiencies given here are internal efficiencies. To convert to external efficiencies, the transmission of the resonator mirrors at the appropriate frequencies has to be taken into account.

9.3.6. Tuning and Linewidth

The output frequency of an OPO is determined by its phase-matching condition, and can thus be tuned by changing the angle, or the temperature of the crystal, or by changing the pump frequency. In general, the range over which one can tune in a particular crystal depends on the pump frequency also. Tuning curves for a number of crystals have been given by Byer.[63] Since the frequency of the output is determined by the phase-matching condition, it follows that the bandwidth is a function of the dispersion of this condition. The gain is appreciable only over a bandwidth for which $|\Delta k L| < \pi$, so that, to first order

$$\Delta \omega = 2\pi/bL, \qquad (9.3.13)$$

where

$$b = (k_s - k_i)/\omega.$$

The gain is also influenced by the resonator. In general, a number of modes of the resonator lie within the bandwidth $\Delta \omega$. In the transient state, when the oscillation is building up from noise, a number of modes are present. However, if all modes within $\Delta \omega$ are equally lossy, then only the mode with the highest unsaturated gain will oscillate in the steady-state condition. In general, the oscillator behaves like a homogeneously broadened medium, that is to say, if the resonator is made very selective, so that only one mode can oscillate in it, then this mode will have the same power as the entire bandwidth would have with a non-selective cavity, providing of course that the losses of the two cavities are the same.

The cavity can be made selective by incorporating frequency-selective elements such as gratings and etalons. To attain the necessary finesse, the beam diameter at these elements should be as large as possible.

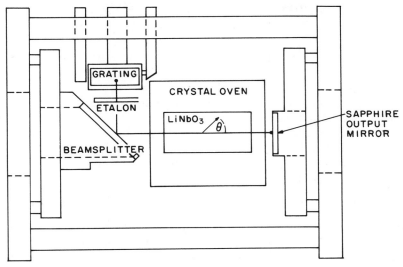

FIG. 7. Folded cavity of a singly resonant oscillator [after R. L. Byer, *in* "Nonlinear Optics" (P. G. Harper and P. S. Wherrett, eds.),p. 47. Academic Press, New York, 1977.]

However, the selective element should operate only on one of the frequencies inside the resonator. Arnold and Wenzel[65a] achieved this by using a grating as one mirror of the resonator of a SRO driven by one line of a HF laser amplifier combination. This grating performs three functions: (1) it couples the pump wave in, (2) it reflects the resonant wave back into the cavity by being placed at the Littrow angle for this resonant frequency, and (3) it acts as a specular reflector for the much longer wavelength signal wave, which therefore is reflected out of the cavity. As mentioned, a replica grating could not be used with the high energies of this OPO. Even a gold-coated stainless steel master grating showed damage effects after 10^3 to 10^4 pulses.

Byer *et al.*[70] use a dichroic beam splitter to fold the cavity of a SRO (see Fig. 7). The beam splitter transmits the pump beam, but reflects the resonated beam, which is then incident on a grating that acts as the other reflector. The output of the oscillator can be further narrowed by placing an etalon between the grating and the beam splitter. In this particular configuration, the output at 1.95 μm was narrowed to 1.1 cm^{-1} by the grating and to 0.08 cm^{-1} by the combination of grating and etalon.

If an etalon is placed perpendicular to the axis of the cavity, its reflec-

[70] R. L. Byer, *in* "Nonlinear Optics" (P. G. Harper and P. S. Wherrett, eds.) p. 47. Academic Press, New York, 1977.

tions combine with the cavity's mirrors to form separate cavities. To avoid this, the etalon is tilted over a small angle (typically 1°) from the perpendicular. This leads to a nonoverlapping (a "walk-off") of consecutive reflections in the etalon, which introduces a loss and causes a slight deterioration of the finesse of the etalon (see also Section 9.2.9).

Some representative recent OPO experiments are listed in Table II. Note the change in tuning range when the pump wavelength of a CdSe oscillator is changed from 1.833 to 2.87 μm.

Recently, Weisman and Rice[71] have demonstrated a synchronously pumped mode-locked parametric oscillator that generates high-power tunable picosecond pulses in the near infrared. The pump originates as a train of 6 psec pulses in a passively mode-locked Nd–glass laser. This pulse train is amplified and frequency doubled to a total energy of 15 mJ and is then incident on a SRO whose cavity round-trip time is exactly equal to the separation of successive pumping pulses. The nonlinear material is $LiIO_3$ and a slightly noncollinear mixing of the pump and the resonated wave is used. This causes the much longer wavelength output to be coupled out directly. The tuning ranges are 11,600 to 16,200 cm^{-1} (resonated wave), and 2600 to 7200 cm^{-1}. The output energy is 150 μJ (1% conversion) at 5400 cm^{-1}. The key to the proper operation of this OPO is the length of the resonator: If this length is detuned by 10 μm, a decrease in output of an order of magnitude results.

9.4. Generation by Mixing of Two Inputs

9.4.1. UV and Near IR

As mentioned in Chapter 9.1, optical frequency mixing has been greatly simplified by the development of tunable dye lasers. These lasers provide a simple method to obtain two input frequencies, one or both of which can be tuned. Frequencies in the near UV can be generated by SFG and SHG of these inputs, and by DFG one can generate frequencies in the IR.

Experimentally, these interactions are very simple. Since the mixing process does not start from noise (as it does in a parametric oscillator), there is no threshold, and an initial output signal can usually be obtained even with an experimental setup that is far from optimum. After that it is a fairly simple task to increase the output by small adjustments of the crystal angle and or its temperature, and by optimization of the temporal and spatial coincidence and the focusing of the inputs. It is often advan-

---

[71] R. B. Weisman and S. A. Rice, *Opt. Commun.* **19**, 28 (1976).

tageous to focus the two input frequencies separately. In general it is best to start with a weakly focused beam. This minimizes the possibility of damaging the crystal. If damage occurs, it is usually possible to use a different area of the crystal at a lower input power. Usually the same apparatus used to generate the inputs for SFG can also be used for DFG.

In addition to this experimental simplicity, the bandwidth of the output is determined by the bandwidths of the inputs and can thus be made very narrow. Pine,[72] for example, has generated extremely narrow band outputs in the infrared by mixing the frequency-narrowed output of an argon-laser-pumped dye laser with the output of a second argon laser, operating in a single longitudinal mode. He used 90° phase matching in $LiNbO_3$, heated to above 100°C to avoid optical damage, and achieved a tuning range of 2.2 to 4.2 μm with a resolution of 5×10^{-4} cm and a scan accuracy of 2×10^{-3} cm^{-1}. Output powers of approximately 1 μW were obtained. It is of interest to note that both inputs in Pine's experiment were generated by commercially available lasers. In addition, Pine gives a useful comparison of his method to other methods of IR generation.

As mentioned, either one or both input frequencies can be tunable. One can use a laser-pumped dye laser and mix its output with the pump radiation[73]; two flashlamp or laser-pumped dye lasers[74]; or two frequencies obtained from the same dye laser.[75-77]

Royt *et al.*[78] found that if a dye laser is pumped by a mode-locked laser, the dye laser pulses can be made to have picosecond duration and to be in temporal coincidence with the pump pulses if the cavities of the pump and the dye laser are matched in length to within a fraction of a millimeter. Using this synchronous pumping method with a frequency-doubled repetitively pulsed Nd:YAG laser as the pump,[79] Moore and Goldberg[80] were able to generate picosecond pulses in the UV and the near-IR by mixing the dye laser output with the pump. Their experimental setup is shown in Fig. 8. Angle tuning of the phase-matching condition was used, with the crystals at room temperature. The spectral coverage of the experi-

[72] A. S. Pine, *J. Opt. Soc. Am.* **64**, 1683 (1974).

[73] C. F. Dewey and L. O. Hocker, *Appl. Phys. Lett.* **18**, 58 (1971).

[74] R. J. Seymour and F. Zernike, *Appl. Phys. Lett.* **29**, 705 (1976).

[75] C. Kittrell and R. A. Bernheim, *Opt. Commun.* **19**, 5 (1976).

[76] B. R. Marx, G. Holloway, and L. Allen, *Opt. Commun.* **18**, 437 (1976).

[77] K. H. Yang, J. R. Morris, P. L. Richards, and Y. R. Shen, *Appl. Phys. Lett.* **23**, 669 (1973).

[78] T. R. Royt, W. L. Faust, L. S. Goldberg, and C. H. Lee, *Appl. Phys. Lett.* **25**, 514 (1974).

[79] L. S. Goldberg and C. A. Moore, *Appl. Phys. Lett.* **27**, 217 (1975).

[80] C. A. Moore and L. S. Goldberg, *Opt. Commun.* **16**, 21 (1976).

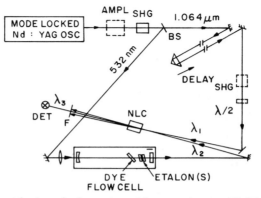

FIG. 8. Experimental setup of a frequency-mixing experiment. NLC is the mixing crystal [after C. A. Moore and L. S. Goldberg, *Opt. Commun.* **16**, 21 (1976)].

ment and the dyes used in the different ranges are shown in Figs. 9 and 10.

Though LiNbO$_3$ has been used successfully for many nonlinear interactions, the strong temperature dependence of its birefringence requires precise temperature control. Pine, for example, controlled the temperature of his crystal to better than 0.05°C. In cw interactions this limits the amount of input power because nonuniform heating of the crystal by too high an input power would destroy the phase matching. In addition, LiNbO$_3$ is susceptible to index damage. For these reasons many workers have used LiIO$_3$, a material that has a high damage threshold and a birefringence that is only very weakly dependent on temperature. Its nonlinear coefficients are slightly higher than those of LiNbO$_3$. Some rep-

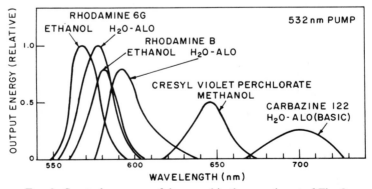

FIG. 9. Spectral coverage of dyes used in the experiment of Fig. 8.

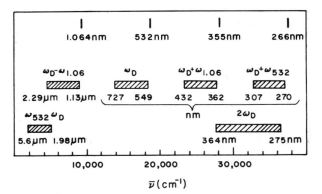

FIG. 10. Spectral coverage of the experiment of Fig. 8.

resentative experiments using LiIO$_3$ in near infrared DFG are listed in Table III.

Though many other crystals have been developed, ADP and its isomorphs have been found to be the most useful crystals for SFG and SHG in the visible and the near-UV, despite their small nonlinear coefficients. This is due to their high damage resistance, the availability of large single crystals of good optical quality, and the fact that, by picking different isomorphs, one can phase-match at 90° over a large part of the range,[81] thus obtaining large efficiencies. This is illustrated, for example, in an experiment by Massey[82] to generate the fifth harmonic of a Nd:YAG laser in a cascade process. He generated the SH of 1.064 µm by 90° matching in CD*A and doubled this frequency by 90° matching in ADP. This fourth harmonic was then mixed with the fundamental in a second crystal of ADP, cooled to −55°C to obtain 90° matching. Table IV shows some experimental results obtained with these crystals.

Though LiIO$_3$ and ADP and its isomorphs are water soluble and hygroscopic, deterioration of the polished surfaces can be prevented, or at least retarded, by a number of methods, such as flowing dry nitrogen over the surfaces, equipping the crystal or the heating stage with permanently bonded windows, or always keeping the crystal at a slightly elevated temperature. Frölich et al.,[53] for example, used an index-matching liquid to bond antireflection-coated fused-silica flats on their crystal (ADA).

In Tables III and IV the tuning range reported for each individual experiment is often rather small. This does not mean that a larger range is

[81] I. M. Beterov, A. A. Chernenko, V. I. Samarin, V. I. Stroganov, and V. I. Trunov, Opt. Commun. 19, 329 (1976).
[82] G. A. Massey, Appl. Phys. Lett. 24, 371 (1974).

TABLE III. Representative Mixing Experiments Using LiIO$_3$[a]

Tuning range (μm)	Input lasers	Input power	Output power	Bandwidth	Comments
4.1–5.2	Amplified ruby DTTC	1.6 MW 70 kW intracavity	100 W peak (800 W theory)	6 Å in dye laser	No damage at 50 MW/cm² [b]
2.8–3.4	2 × Q-sw.Nd:YAG Rhodamine B + cresyl violet	3.8 kW 3.6 KW	80 mW at 3 μm (1.9 W theory)	3 Å in dye laser	20 p.p.s [c]
3.4–5.65	2 × Q-sw.Nd:YAG Rhodamine B	3.5 MW/cm² 2.3 kW intracavity	0.5 W at 5.02 μm	0.1 cm⁻¹	[d]
1.25–1.6	Q-sw. Nd:YAG Rhodamine B	3.1 MW/cm² 4.7 kW intracavity	70 W at 1.35 μm	0.08 cm⁻¹	[d]

1.5–4.8	Coumarin Rhodamine 6G	14 kW 30 kW	0.4 W		Damage at 10MW/cm^2, flashlamp pump[e]
1.28–1.62	Nd:YAG cw Rhodamine 6G (Ar laser pumped)	30 W intra-cavity 200 mW	35 μW (100 μW theory)	200 GHz	Noncollinear mixing[f]
2.3–4.6	Argon (cw) Rhodamine 6G	(intracavity)	4 μW multimode 0.5 μW single mode	5 cm^{-1} <50 MHz	Noncollinear mixing[g]
2.2–4.2	Argon Rhodamine 6G (AR laser pumped)	25–50 mW	0.5 μW at 3 μm (4.6 μm theory)	5 × 10^{-4} cm	[h]

[a] Also shown for comparison are Pine's data for LiNbO$_3$.
[b] D. W. Meltzer and L. S. Goldberg, *Opt. Commun.* **5**, 209 (1972).
[c] H. Tashiro and T. Yajima, *Opt. Commun.* **12**, 129 (1974).
[d] L. S. Goldberg, *Appl. Opt.* **14**, 653 (1975).
[e] H. Gerlach, *Opt. Commun.* **12**, 405 (1974).
[f] W. Lahmann, K. Tibulski, and H. Welling, *Opt. Commun.* **17**, 18 (1976).
[g] B. Wellegehausen, D. Friede, H. Vogt, and S. Shahdin, *Appl. Phys.* **11**, 363 (1976).
[h] A. S. Pine, *J. Opt. Soc. Am.* **64**, 1683 (1974).

179

TABLE IV.   Representative Experiments in ADP and Isomorphs

Crystal	Input laser	Input power	Output power	Range (nm)	Comments
Ammonium dihydrogen phosphate (ADP)	cw Rhod.6G laser intra-cavity	40 W	85 mW peak[a]	285–315	[b]
Ammonium dihydrogen arsenate (ADA)	cw dye laser intracavity		50 mW multi-mode, 4 mW single mode	285–315	[c]
Cesium dihydrogen arsenate (CDA)	Nd:YAG	50 MW	28 MW	532.1	[d]
Cesium dideuterium arsenate CD*A	Nd:YAG	50 MW	23 MW	532.1	[e]
Rubidium dihydrogen phosphate		7 MW	3.6 MW	305–335	[f]
Rubidium dihydrogen arsenate	Nd:YAG	50 MW	27 MW 6 MW	532.1 354.7	[g]

[a] Peak power when the fundamental is chopped at 1:50 duty cycle. At cw temperature, detuning occurs and the output power reduces to 20 mW.
[b] A. I. Ferguson, M. H. Dunn, and A. Maitland, *Opt. Commun.* **19**, 10 (1976).
[c] D. Frölich, L. Stein, H. W. Schröder, and H. Welling, *Appl. Phys.* **11**, 97 (1976).
[d] K. Kato, *IEEE J. Quantum Electron.* **qe-10,** 616 (1974).
[e] K. Kato, *J. Appl. Phys.* **46,** 271 (1975).
[f] K. Kato, *Opt. Commun.* **13,** 93 (1975).

not possible, but rather that this is the range that was covered by the use of one dye.   In almost all cases, the range could easily be extended by changing the dyes in the input laser(s).   This is amply illustrated in Fig. 9.

Table III also shows the difference between the experimentally observed and the theoretically expected output powers.   In some cases the disparity is large.   Note, however, that for those experiments where the input beams are well defined, with a close to Gaussian distribution, such as in Pine's experiment, the disparity is small.

In Massey's experiment[82] the ADP crystal had to be cooled, which makes the experiment more complicated.   For room temperature operation the shortest wavelength that can be phase-matched in ADP and its isomorphs is approximately 262 nm.   Room temperature operation down to 230 nm has been achieved in lithium formate-monohydride.[83] Potassium pentaborate[84-87] has been used down to 208 nm. An interesting method to tune the dye laser and the input angle to the SHG crystal simultaneously has been described by Saikan.[88]

9.4.2. Mid- and Far-Infrared

While for generation of frequencies in the UV the highest frequency obtainable is usually determined by the phase matching in the crystal, for generation of difference frequencies in the IR the constraints of the experiment are more often determined by the transmission range of the material. A few birefringent crystals have a transmission range that is large enough to allow generation of wavelengths up to about 20 μm from inputs in the visible. Two of these crystals are proustite and $AgGaS_2$. However, in both these crystals a two-phonon absorption severely limits the efficiency of the interaction at around 14 μm. $GaSe$[14] in principle can be phase-matched for generation out to past 20 μm from input sources in the red or the near-IR, but this has not been demonstrated.

If input sources in the infrared are used, the efficiency of the interaction can be higher, and more materials are available. Alignment becomes more of a problem, but often a visible laser, such as a He–Ne laser, can be used as an alignment tool before the crystal is put in place.

Table V gives some results of experiments in mid-IR generation. In the far-IR the situation is, in some ways, easier again in that many crystals have one or more Reststrahlen bands in the mid-IR range and are transparent again in the far-IR. In addition, the anomalous dispersion across the Reststrahlen can allow phase matching even in isotropic materials.

Yang et al.[77] have demonstrated far-IR generation in the 20–190 cm^{-1} range, using a dual-frequency dye laser system to generate the inputs. They employed collinear forward and backward phase matching, and also noncollinear phase matching in $LiNbO_3$. In addition they used forward collinear matching in ZnO, ZnS, CdS, and CdSe, obtaining as much as 14 mW at 190 cm^{-1} using ZnO.

[83] F. B. Dunning, F. K. Tittel, and R. F. Stebbings, *Opt. Commun.* **7**, 181 (1973).

[84] C. F. Dewey, Jr., W. R. Cook, Jr., R. T. Hodgson, and J. J. Wynne, *Appl. Phys. Lett.* **26**, 714 (1975).

[85] F. B. Dunning and R. F. Stickel, Jr., *Appl. Opt.* **15**, 3131 (1976).

[86] H. Zacharias, A. Anders, J. B. Halpern, and K. H. Welge, *Opt. Commun.* **19**, 116 (1976).

[87] K. Kato, *Appl. Phys. Lett.* **29**, 562 (1976).

[88] S. Saikan, *Opt. Commun.* **18**, 439 (1976).

TABLE V. Some Representative Experiments in Mid-IR Generation

Range (μm)	Input sources	Input power	Output power	Crystal	Comments
4.6	Ruby + ruby-pumped dye	140 kW 13 kW	300 mW at 11.1 μm (3 W theory)	$AgGaS_2$	Damage threshold 140 kW in 10 nsec pulses[a]
5.5–18.3	N_2-pumped dye lasers 8–10 nsec	4.7 kW 5.7 kW	4 W, 4 nsec at 11.8 μm (20 W theory)	$AgGaS_2$	Two-phonon absorption at 14 μm[b]
10.1–12.7	Ruby + ruby-pumped dye	45 kW 500 W	100 mW at 10.6 μm (750 mW theory)	Proustite	[c]
18.6	Ruby + raman-shifted ruby			Proustite	Backward-wave generation[d]
11–23	Two ruby-pumped dye lasers		3 W at 19 μm	Proustite	Two-phonon absorption at 14.5 μm[e]
9.4–24.3	Proustite OPO		10 W at 10 μm, 100 mW at 22 μm	CdSe	[f]
4–21	Ruby + ruby-pumped dye		5 W at 4 μm	ZnSe	Twinned crystal[g]
15.5–16.5	CO CO_2	1 kW 200 kW	2.5 kW at 16 μm	$CdGeAs_2$	10 mW/cm² damage threshold[h]

[a] D. C. Hanna, V. V. Rampal, and R. C. Smith, *Opt. Commun.* **8**, 151 (1973).

[b] R. J. Seymour and F. Zernike, *Appl. Phys. Lett.* **29**, 76 (1976).

[c] D. C. Hanna, R. C. Smith, and C. R. Stanley, *Opt. Commun.* **4**, 300 (1971).

[d] D. Cotter, D. C. Hanna, B. Luther-Davies, R. C. Smith, and A. J. Turner, *Opt. Commun.* **11**, 54 (1974).

[e] L. O. Hocker and C. F. Dewey, Jr., *Appl. Phys.* **11**, 137 (1976).

[f] D. C. Hanna, B. Luther-Davies, R. C. Smith, and R. Wyatt, *Appl. Phys. Lett.* **25**, 142 (1974).

[g] L. O. Hocker and C. F. Dewey, Jr., *Appl. Phys. Lett.* **28**, 267 (1976).

[h] M. S. Piltch, J. Rink, and C. Tallman, *Opt. Commun.* **15**, 112 (1975).

Various workers[29,33,89,90] have used a pair of CO_2 lasers to generate the input beams for far-IR generation. Since a CO_2 laser can be tuned over a large number of discrete frequencies, this provides stepwise tuning of the output through a large range. Lax *et al.*,[29] for example, tuned from 70 μm to 2 mm in steps 0.01 cm^{-1} apart. Boyd *et al.*[90] used CO_2 lasers to tune from 70 to 110 cm^{-1} using birefringent phase matching in the ternary semiconductor $ZnGeP_2$. They obtained 1.7 μW at 83.37 cm^{-1} from input powers of 360 and 50 W.

Alignment of the CO_2 lasers, both temporally and spatially, can be facilitated by first observing SHG and SFG in a different material. Polycrystalline CdSe, for example,* can be used. The advantage here is that the second harmonic of both inputs is always observed but only if the alignment is correct is the sum frequency signal equal in magnitude to the second harmonic.

9.5. Upconversion

As mentioned in the introduction, weak signals in the infrared can be upconverted to the visible by SFG with a strong visible signal. By employing noncollinear phase-matching techniques, some image information can also be upconverted. The extensive work done in this field has been reviewed by Warner.[91] Though the technique is interesting, it is in general much more complicated than the use of normal infrared detectors, and except in some very special cases does not offer an advantage. A more recent application of upconversion is in the measurement of very short pulses in the IR. To do this, the pulse to be measured is mixed with a shorter, visible pulse, that is incident on the crystal via a variable time delay, and the sum frequency signal is measured as a function of this delay. Using 20 psec pulses from a mode-locked dye laser as the probe, Mahr and Hirsch[92] have mapped the pulse shape of pulses with a width on the order of 150 psec.

[89] F. Zernike, *Phys. Rev. Lett.* **22**, 931 (1969).

[90] G. D. Boyd, T. J. Bridges, C. K. N. Patel, and E. Buehler, *Appl. Phys. Lett.* **21**, 553 (1972).

[91] J. Warner, *in* "Quantum Electronics: A Treatise" (H. Rabin and C. L. Tang, eds.), Vol. 1, Part B, p. 703. Academic Press, New York, 1975.

[92] H. Mahr and M. D. Hirsch, *Opt. Commun.* **13**, 96 (1975).

---

* Manufactured by the Eastman Kodak Company under the trade name IRTRAN 6.

10. EXAMPLES OF LASER TECHNIQUES AND APPLICATIONS

10.1. Picosecond Spectroscopy*

10.1.1. Introduction

Not long after the first demonstration of the laser it was discovered that lasers could be used to generate ultrashort optical pulses. It was also apparent that these short optical pulses could provide a means of investigating the properties of matter on a time scale three orders of magnitude faster than had been previously possible. In the last decade a great deal of activity has taken place in this field, developing a large body of knowledge in pulse generation and measurement techniques. Optical pulses as short as a few tenths of a picosecond have been generated and high-power systems are being developed that will produce peak optical intensities in the terawatt range.

The applications of picosecond pulse techniques run a wide range of diversity. Chemists are studying photochemical reactions and energy relaxation processes in molecules. Biologists are investigating fundamental biological processes on a picosecond time scale, which provides new understanding in such areas as vision and photosynthesis. Physicists have opened an entirely new area of nonlinear optics: the study of matter with intense optical fields. Short optical pulses provide a unique means of supplying high-peak-intensity optical fields with the least energy to minimize catastrophic material failure. With terawatt optical pulses attempts are being made to compress matter to the point of thermonuclear fusion.

In this chapter we cover the present state of the art in generation of short optical pulses and some techniques for their application to measurements of ultrafast phenomena.

10.1.2. Picosecond Pulse Sources

For purposes of application to picosecond spectroscopy, pulse sources may be divided into two categories: flashlamp-pumped systems and continuously operated systems.

---

* Chapter 10.1 is by E. P. Ippen and C. V. Shank.

METHODS OF EXPERIMENTAL PHYSICS, VOL. 15B

Historically, mode-locked flashlamp-pumped lasers provided the framework for the early development of picosecond techniques. The two most important flashlamp-pumped systems for picosecond work have been the Nd:glass laser and flashlamp-pumped dye laser. Until very recently, these systems were the only ones to provide pulses with durations less than 10 psec. They produce sufficient optical intensity for single-pulse measurements and for studies of highly nonlinear processes. Unfortunately, they also suffer the inherent disadvantages of low repetition rate and a certain lack of reproducibility. For purposes of comparison we shall include brief reviews of the important characteristics of these two types of lasers.

Continuously operated, mode-locked lasers make possible the generation of ultrashort pulses in a controlled, highly reproducible manner. Pulses as short as 0.3 psec in duration have been generated. Perhaps even more important is the very high pulse repetition rate of these systems. Extensive signal-averaging techniques can be employed and experiments can be performed at low excitation levels. Recent experiments in our laboratory also indicate that a reliable cw system can provide the best foundation for a high-power pulse system. Pulses selected from a cw subpicosecond oscillator are amplified to high power by a series of dye amplifiers. Subpicosecond pulsewidths can be maintained and powers sufficient for the study of highly nonlinear processes can be achieved. It will become increasingly difficult for conventional flashlamp lasers to compete with such a system.

In our discussion we have restricted ourselves to those few lasers that have produced pulses with durations of a few picoseconds. Mode-locked ruby lasers, Nd:YAG lasers, and several gas lasers can be used to generate pulses in the range 20–100 psec, but these provide insufficient temporal resolution for the type of studies we are interested in here. A review of many aspects of these other systems has been given elsewhere.[1]

10.1.2.1. Nd:Glass Lasers. Picosecond optical pulses first became a reality with the passive mode-locking of the flashlamp-pumped Nd:glass laser.[2] In the intervening years the pulse output characteristics of this laser have also been the most extensively investigated.[3,4] These investigations have to a certain extent been necessitated by the variety and variability of pulse outputs obtained. Pulse characteristics vary greatly

[1] P. W. Smith, M. A. Duguay, and E. P. Ippen, *Prog. in Quantum Electron.* **3**, Part 2 (1974).

[2] A. J. DeMaria, D. A. Stetster, and H. Heyman, *Appl. Phys. Lett.* **8**, 174 (1966).

[3] For a review, see D. J. Bradley, *in* "Ultrashort Light Pulses" (S. L. Shapiro, ed.), Chapter 2. Springer-Verlag, Berlin and New York, 1977.

[4] E. P. Ippen and C. V. Shank, *in* "Ultrashort Light Pulses" (S. L. Shapiro, ed.), Chapter 3. Springer-Verlag, Berlin and New York, 1977.

with differences in laser design, from shot to shot in a particular system, and with position in the train of pulses produced by a single flash. Much of this behavior can be explained in terms of transient laser build-up from an initially statistical distribution of fluctuations.[5] Selection of a single, short, isolated pulse from this condition requires careful control of a variety of parameters. It is especially important that the laser be operated in a low-order transverse mode, near threshold, and that spurious reflections and beam irregularities be avoided. In order to avoid satellite pulses the saturable absorber is generally placed in direct contact with one of the end mirrors of the laser.[6] Furthermore, a fast (picosecond recovery time) saturable absorber is required.

The build-up of Nd:glass laser pulse trains and the variation of pulse characteristics during this build-up have been studied by many workers.[3] During the first part of the flashlamp pumping, spontaneous-emission fluctuations are amplified linearly. When they reach sufficient intensity to begin bleaching the saturable absorber, the largest fluctuation peak grows most rapidly and takes over. Although gain saturation plays a role in emphasizing the largest peak,[7] it cannot have a dramatic effect on pulse shape. (This is in direct contrast to dye lasers, where strong gain depletion and gain recovery can occur within a cavity round-trip time.) As the main intensity peak (now a "mode-locked pulse") continues to grow in amplitude, nonlinearities of the glass laser medium (self-phase modulation and self-focusing) start to introduce new spectral broadening and temporal substructure.[8,9] If the gain is too high, the pulse can literally be torn apart.[10]

Because of this great variation between pulses within a train, accurate picosecond measurements must be made with single, properly selected pulses.[11,12] Pulse selection is accomplished by passing the pulse train through an optical Kerr shutter that is activated by a high-voltage spark gap.[13] When the laser output reaches a predetermined level, the spark gap fires and the next pulse is transmitted. Usually the pulse is chosen to be near the beginning of the train (as viewed on an oscilloscope). At this

[5] P. G. Kryukov and V. S. Letokhov, *IEEE J. Quantum Electron.* **qe-8,** 766 (1972).

[6] D. J. Bradley, G. H. C. New, and S. J. Caughey, *Opt. Commun.* **2,** 41 (1970).

[7] W. H. Glenn, *IEEE J. Quantum Electron.* **qe-11,** 8 (1975).

[8] M. A. Duguay, J. W. Hansen, and S. L. Shapiro, *IEEE J. Quantum Electron.* **qe-16,** 725 (1970).

[9] V. A. Korobkin, A. A. Malyutin, and A. M. Prokhorov, *JETP Lett.* (*Engl. Transl.*) **12,** 150 (1970).

[10] R. C. Eckhardt, C. H. Lee, and J. N. Bradford, *Opto. Electron.* **6,** 67 (1974).

[11] D. von der Linde, *IEEE J. Quantum Electron.* **qe-8,** 328 (1972).

[12] W. Zinth, A. Laubereau, and W. Kaiser, *Opt. Commun.* **22,** 161 (1977).

[13] D. von der Linde, O. Bernecker, and A. Laubereau, *Opt. Commun.* **2,** 215 (1970).

point the laser is already in the regime where the primary fluctuation peak has been isolated and subsequent broadening has not yet come strongly into play. In a carefully designed system it is possible in this way to obtain pulses with an average pulsewidth of about 5 psec and an rms width fluctuation of about 0.5 psec.[11,12] These pulses can be in a low-order transverse mode and have energies of several millijoules. Thus peak powers are around 10^9 W, already sufficient for efficiently driving a variety of nonlinear processes. In nonlinear experiments, however, the fluctuations of pulse characteristics become increasingly problematic. Care must be taken to characterize each pulse used.

Some applications require even higher powers than can be obtained with a single Nd:glass oscillator. One or more amplifiers can then be used. Picosecond pulses saturate amplifier gain with pulse energy. In the case of Nd:glass the saturation energy density is about 5 J/cm². This means that Nd:glass amplifiers can only be operated in a small-signal-gain, low-efficiency configuration. Even with 5 psec pulses, operation near saturation would result in peak powers of 10^{12} W/cm². At this point destructive nonlinear effects in the glass become very severe. Although pulse energies of several joules can be achieved,[14] such very high power systems are complicated and expensive, and are needed only for a few laboratory applications such as very short wavelength generation.[15]

10.1.2.2. Flashlamp Dye Lasers.
Passive mode-locking of a dye laser was first demonstrated by Schmidt and Schäfer[16] with a system employing rhodamine 6G as the gain medium and DODCI as the saturable absorber. Since then a variety of different dye combinations have been used to obtain wavelength-tunable picosecond pulses from 575 to 700 nm.[17,18] It may be possible to extend these results into the blue[3] and into the near-infrared.[19]

Pulse generation by flashlamp-pumped dye lasers has been studied extensively[3,20] with high-resolution streak-cameras as well as with more conventional nonlinear techniques. It is apparent that the build-up of short pulses in these systems is fundamentally different from that in Nd:glass lasers. The reason is that the dye gain medium (under conditions of continued pumping) can recover in a time comparable to a cavity round-trip time. Dye recovery times are on the order of several nanosec-

[14] H. Kuroda, H. Masuko, and S. Mackawa, *Opt. Commun.* **18**, 169 (1976).

[15] J. Reintjes, C. Y. She, R. C. Eckhardt, N. E. Karangelen, R. Andrews, and R. C. Elton, *Appl. Phys. Lett.* **30**, 480 (1977).

[16] W. Schmidt, and F. P. Schäfer, *Phys. Lett. A* **26**, 558 (1968).

[17] E. G. Arthurs, D. J. Bradley, and A. G. Roddie, *Appl. Phys. Lett.* **19**, 420 (1971).

[18] E. G. Arthurs, D. J. Bradley, and A. G. Roddie, *Appl. Phys. Lett.* **20**, 125 (1972).

[19] A. Hirth, K. Volrath, and D. J. Lougnot, *Opt. Commun.* **8**, 318 (1973).

[20] D. J. Bradley and G. H. C. New, *Proc. IEEE* **62**, 313 (1974).

onds, while that for Nd:glass is several hundred microseconds. This means that gain saturation can play a large role not only in pulse selection but in pulse shaping. In turn, this allows the use of saturable absorbers, like DODCI, which recover slowly compared to the pulsewidth. New[21] first described how the saturation of a "slow" absorber in combination with gain saturation can produce rapid and continuing pulse shortening. More recently, Haus[22] has obtained analytical solutions for pulseshape and stability in this type of system. In essence, the absorber acts to steepen the leading edge of the pulse while depletion of the gain carves away the trailing edge. Thus, the ability of the system to form short pulses rapidly (or at all) depends on the relative absorption and emission cross sections of the two dyes as well as geometrical focusing factors in the laser. This leads to different operating characteristics as a function of wavelength and particular dye combination.

An optimized flashlamp-pumped dye laser is operated near threshold and has the saturable absorber in a thin cell in direct contact with one of the end mirrors.[6] It is tuned with (and has its bandwidth restricted by) a thin Fabry–Perot etalon. Pulse durations between 1.5 and 4 psec can be obtained between 580 and 700 nm.[17,18] Pulse energies can be on the order of 50 μJ. Some pulse variation still occurs within a single-pulse train as well as from flash to flash, and single pulses should be selected for quantitative measurements.

Flashlamp dye amplifiers have been used to amplify 2 psec pulses to pulse energies of about 7 mJ.[23] Because of the much lower saturation energy (~ 3 mJ/cm^2) dye amplifiers can be operated in a regime of saturation. Small-signal gain, however, is correspondingly larger and considerable care must be taken to suppress amplified spontaneous emission. Because of the relatively short (nsec) energy storage time with dyes, only a small fraction of flash energy can go into amplification of a single pulse. In this regard, laser-pumped amplifiers have a considerable advantage (see Section 10.1.4). Schmidt[24] has used a nitrogen-laser-pumped amplifier to amplify single pulses from a train of flashlamp-pumped dye laser pulses. Another possibility for efficient flashlamp amplification of a pulse train is the technique demonstrated by Moses et al.[25] They synchronously mode-locked a high-power oscillator to a low-power pulse train and achieved a gain of 3×10^4.

[21] G. H. C. New, *IEEE J. Quantum Electron.* **qe-10,** 115 (1974).
[22] H. A. Haus, *IEEE J. Quantum Electron.* **qe-11,** 736 (1975).
[23] R. S. Adrain, E. G. Arthurs, D. J. Bradley, A. G. Roddie, and J. R. Taylor, *Opt. Commun.* **12,** 136 (1974).
[24] A. J. Schmidt, *Opt. Commun.* **14,** 287 (1975).
[25] E. I. Moses, J. J. Turner, and C. L. Tang, *Appl. Phys. Lett.* **28,** 258 (1976).

10.1.2.3. Passively Mode-Locked cw Dye Lasers. Significant progress in picosecond spectroscopy has been made possible by the invention of the passively mode-locked cw dye laser.[26-28] Pulses as short as a few tenths of a picosecond can be generated.[29,30] Perhaps even more important is the high pulse repetition rate that allows powerful signal averaging techniques to be applied to the study of weak processes. With continuous operation, laser mode-locking is in the steady-state regime and pulse reproducibility improves greatly. Pulse formation in a passively mode-locked cw dye laser is understood in approximately the same terms used above to describe pulse build-up in a flashlamp dye laser except that the variabilities and uncertainties of flashlamp and high-power pulses have been removed. The conditions for pulse shortening and stable operation given by New[21] and Haus[22] appear to hold. Exponential behavior in the wings of steady-state pulses is observed in accordance with theory.[30,31]

A continuously operated system that has been used successfully in a variety of picosecond studies is shown in Fig. 1. It is pumped by the cw (4–5 W) output of a commercial argon ion laser. The gain medium, rhodamine 6G in a free-flowing stream of ethylene glycol, is located at a focal point approximately in the center of the resonator. Near one end of the resonator is a second, free-flowing stream of ethylene glycol containing the saturable absorber dyes for mode-locking. The beam waist in this stream is about a factor of two smaller than that in the gain medium. Subpicosecond pulses can be generated with only the dye DODCI in the mode-locking stream, but addition of a second dye, malachite green, improves reproducibility and stability.[29] The recovery time of DODCI in this system has been measured to be 1.2 nsec,[32] confirming its behavior as a "slow" absorber. Malachite green provides some absorption recovery on a picosecond time scale[33] and facilitates subpicosecond pulse operation in a stable regime well above threshold.

At the other end of the resonator is a commercially available cavity dumping arrangement. An acousto-optic deflector with a risetime somewhat less than the round-trip time of the cavity can dump single pulses from the laser at a rate of more than 10^5 pps. This arrangement is preferable to taking the pulse output through a partially transmitting mirror.

[26] E. P. Ippen, C. V. Shank, and A. Dienes, *Appl. Phys. Lett.* **21,** 348 (1972).
[27] C. V. Shank and E. P. Ippen, *Appl. Phys. Lett.* **24,** 373 (1974).
[28] F. O'Neill, *Opt. Commun.* **6,** 360 (1972).
[29] E. P. Ippen and C. V. Shank, *Appl. Phys. Lett.* **27,** 488 (1975).
[30] I. S. Ruddock and D. J. Bradley, *Appl. Phys. Lett.* **29,** 296 (1976).
[31] H. A. Haus, C. V. Shank, and E. P. Ippen, *Opt. Commun.* **15,** 29 (1975).
[32] C. V. Shank and E. P. Ippen, *Appl. Phys. Lett.* **26,** 62 (1975).
[33] E. P. Ippen, C. V. Shank, and A. Bergman, *Chem. Phys. Lett.* **38,** 611 (1976).

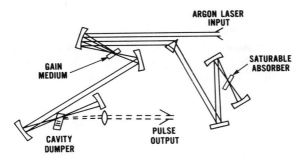

FIG. 1. Continuously operated, subpicosecond pulse generator.

The peak power is at least an order of magnitude greater, and the lower repetition rate can be adjusted to allow for complete sample recovery (or replacement) between pulses.

The output pulses of the system shown in Fig. 1 have been well characterized. Subpicosecond pulses are generated near 615 nm. Energy in each pulse is about 5×10^{-9} and the peak power exceeds 5 kW. In a focused beam, energy densities of several millijoules per square centimeter are easily achieved. When the highest temporal resolution is required, pulses can be compressed and filtered[29] to produce bandwidth-limited pulses with a duration of 0.3 psec.

10.1.2.4. Other cw Dye Laser Systems. Several other techniques have also been used to produce ultrashort pulses with continuously operated systems. Most prominent of these has been synchronous pumping.[34-39] In this technique an actively mode-locked argon laser is used to pump a dye laser of matching cavity length. The major advantage is that synchronously pumped lasers can be operated at cw dye laser wavelengths, where a proper combination of gain and saturable absorber dyes may not be available. Pulse generation in the near-infrared is an example.[40] Conversion efficiency and average output power can be relatively high because of the low loss and higher peak pumping intensities. Most systems have produced pulses in the 5–10 psec range, considerably longer than those obtained from the passively mode-locked system; but very recent results[41] indicate that subpicosecond pulses may be produced

[34] C. V. Shank and E. P. Ippen, *in* "Dye Lasers" (F. P. Schafer, ed.), p. 121. Springer-Verlag, Berlin and New York, 1972.

[35] H. Mahr and M. D. Hirsch, *Opt. Commun.* **13**, 96 (1975).

[36] J. M. Harris, R. W. Chrisman, and F. E. Lytle, *Appl. Phys. Lett.* **26**, 16 (1975).

[37] C. K. Chan, S. O. Sari, and R. E. Foster, *J. Appl. Phys.* **47**, 1139 (1976).

[38] J. de Vries, D. Bebelaar, and J. Langelaar, *Opt. Commun.* **18**, 24 (1976).

[39] A. Scavennec, *Opt. Commun.* **17**, 14 (1976).

[40] J. Kuhl, R. Lambrich, and D. von der Linde, *Appl. Phys. Lett.* **31**, 657 (1977).

[41] R. K. Jain and C. P. Ausschnitt, *Opt. Lett.* **2**, 117 (1978).

directly in a synchronously mode-locked system if laser bandwidth and cavity length are very carefully controlled.

Synchronous pumping also opens the possibility of operating two mode-locked dye lasers simultaneously for experiments requiring pulses at two wavelengths. The two lasers can be operated in parallel,[42] pumped by the same argon laser, or in tandem,[43] where the first dye laser pumps the second. The latter method has been shown capable of producing 0.8 psec pulses in a rhodamine B laser pumped by the pulses of a rhodamine 6G laser.

Another interesting scheme still in the early stages of development is the doubly mode-locked system due originally to Runge[44] in which the saturable absorber is allowed to lase in synchronism with the laser it is mode-locking. Such a passively mode-locked rhodamine 6G–cresyl-violet laser has been operated simultaneously at two wavelengths.[45] Pulses as short as 0.6 psec have been produced in this manner.[46] This technique may be limited to particular dye combinations but could be most useful if reliability is demonstrated. Synchronous locking has also been observed in a double-wavelength system, where two lasers compete for the same gain medium.[47]

All of these systems offer the principal advantages of continuous operation: steady-state control over pulse shape and high repetition rate.

10.1.2.5. High-Power, Subpicosecond Amplification. Although high repetition rate and pulse reproducibility make it possible to perform a great variety of picosecond experiments with continuously operated lasers, there are several important reasons for pushing toward higher power. Some time-resolved experiments (especially in low-density systems) simply require more pulse energy. Higher power facilitates the conversion of pulse energy to other wavelengths for more experimental versatility. Finally, well-controlled subpicosecond pulses make possible some very high power experiments not amenable to longer pulse durations.

An amplifier system used in conjunction with subpicosecond pulses from a passively mode-locked cw dye laser is shown in Fig. 2.[48] A frequency-doubled, Q-switched Nd:YAG pump laser is fired in synchronism with the selection of a single pulse from the cw dye laser. Successively larger fractions of the resulting high-power pulse at 5300 Å

[42] J. P. Heritage and R. Jain, *Appl. Phys. Lett.* **32**, 101 (1978).

[43] R. Jain and J. P. Heritage, *Appl. Phys. Lett.* **32**, 41 (1975).

[44] P. K. Runge, *Opt. Commun.* **5**, 311 (1972).

[45] Z. A. Yasa and O. Teschke, *Appl. Phys. Lett.* **27**, 446 (1975).

[46] Z. A. Yasa, A. Dienes, and J. R. Whinnery, *Appl. Phys. Lett.* **30**, 24 (1977).

[47] B. Couillaud and A. Ducasse, *Appl. Phys. Lett.* **29**, 665 (1976).

[48] E. P. Ippen and C. V. Shank, *in* "Picosecond Phenomena" (Shank *et al.*, eds.), p. 103. Springer-Verlag, Berlin and New York, 1978.

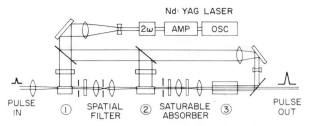

FIG. 2. Three-stage dye amplifier system for subpicosecond pulses. The Q-switched Nd:YAG pump laser is fired at 10 Hz in synchronism with single-pulse selection from the subpicosecond pulse generation.

are split off to pump three dye amplifiers. The pumping duration is about 10 nsec, synchronization with the subpicosecond pulse is within 2 nsec, and the whole system can be operated at a rate of 10 pps.

Considerable care must be taken to prevent excessive losses due to amplified spontaneous emission. Geometrical design can be optimized to avoid self-dumping of a single stage Spectral filters, spatial (pinhole) filters, and a saturable absorber stream are used to isolate stages from each other. With the system illustrated in Fig. 2, 615 nm pulses, 0.5 psec in duration, have been amplified to peak powers approaching 1 GW. Autocorrelation measurements of pulses before and after amplification are shown in Fig. 3. It is clear that the pulse duration has been maintained. Amplification to even higher powers is expected in the near future. This combination of reliable cw oscillation and high-power amplifier is perhaps the most versatile tool available to date for picosecond spectroscopy.

10.1.2.6. Frequency Conversion. For wide-ranging spectroscopic applicability it is obviously desirable to have a source of wavelength-tunable subpicosecond pulses. Even the dye laser sources described

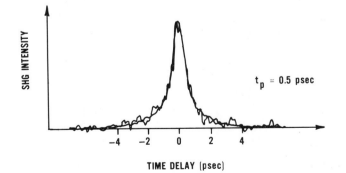

FIG. 3. Two autocorrelation traces show pulse duration (~0.5 psec) (a) before and (b) after high-power amplification.

above are limited in this regard.   Fortunately, with sufficient peak power
it is possible to convert energy from a fixed-frequency picosecond source
to wavelengths ranging from the infrared to the vacuum ultraviolet.   We
simply review here briefly the different nonlinear optical techniques that
can be used for such frequency conversion.   A more extensive discussion
has been given recently by Auston.[49]   Most of these techniques were originally demonstrated with pulses from Nd lasers.   New, amplified subpicosecond pulse systems such as described above will also be able to take advantage of these techniques, offering at the same time shorter pulse durations and higher repetition rates.   Some early results with such a system
are included below.

10.1.2.6.1. SHG. Perhaps the simplest and most efficient means of extending wavelength capability is second harmonic generation.   Energy
conversion efficiencies of 80–85% can be achieved.[50,51]   Ultraviolet
pulses at 266 nm have been generated in two steps from 1.06 μm pulses
with an overall conversion efficiency of about 60%.[51]   Achieving such
high conversion efficiency while maintaining short pulse duration requires
some attention to detail.   Optimization of beam focusing for different
crystal parameters has been described in detail by Boyd and Kleinman.[52]
The importance of group velocity and dispersion when dealing with ultrashort pulses has been pointed out by several authors.[53-56]   Recently,
Shank et al.[57] have demonstrated that subpicosecond pulse resolution can
be maintained with the use of tight focusing and a thin crystal.   Subpicosecond ultraviolet pulses at 307 nm can be generated with an efficiency of
15% using pulses directly from a mode-locked cw dye laser.   Even more
conversion will be obtained with amplified pulses.

10.1.2.6.2. SHORTER WAVELENGTHS. Third-harmonic generation has
been demonstrated and studied under a variety of conditions.[49]   The most
important are those which allow the generation of wavelengths shorter
than one can reach with SHG.   Since generation of odd harmonic does
not require material acentricity, it can be accomplished in the gaseous
media that are needed for transmission of short wavelengths.   Clever use

[49] D. H. Auston, in "Ultrashort Light Pulses" (S. L. Shapiro, ed.), Chapter 4.
Springer-Verlag, Berlin and New York, 1977.

[50] A. H. Kung, J. F. Young, G. C. Bjorklund, and S. E. Harris, Phys. Rev. Lett. 29, 985
(1972).

[51] J. Reintjes and R. C. Eckardt, Appl. Phys. Lett. 30, 91 (1977).

[52] G. D. Boyd and D. A. Kleinman, J. Appl. Phys. 39, 3597 (1968).

[53] R. U. Orlov, T. Usmanov, and A. S. Chirkin, Sov. Phys.—JETP (Engl. Transl.) 30, 584
(1970).

[54] R. C. Miller, Phys. Lett. A 26, 177 (1968).

[55] J. Comly and E. Garmire, Appl. Phys. Lett. 12, 7 (1968).

[56] W. H. Glenn, IEEE J. Quantum Electron. qe-5, 284 (1969).

[57] C. V. Shank, E. P. Ippen, and O. Teschke, Chem. Phys. Lett. 45, 291 (1977).

of alkali metal vapor absorption resonances and buffer gases for phase matching[58] led first to the generation of pulses at 177, 152, and 118 nm by three-photon processes in mixtures of cadmium and argon.[50] In argon and xenon, conversion from 355 to 118 nm has been achieved with a conversion efficiency of 2.8%.[59] Pulses from a Xe_2 laser at 171 nm have been tripled in argon to 57 nm.[60] Several authors have analyzed the optimization[61,62] of these third-order processes and have discussed the various factors limiting conversion efficiency.[63,64]

The shortest wavelengths achieved so far are 53.2 and 38 nm, obtained by fifth[65] and seventh[15] harmonic conversion of 266 nm pulses. These VUV results were produced without phase matching in helium and required focused intensities of 10^{15} W/cm^2. The seventh harmonic, 38 nm, is already well into the continuum of He.

A means of generating VUV picosecond pulses by spontaneous anti-Stokes scattering has recently been suggested by Harris.[66] This scheme would make use of energy storage in high-lying atomic metastable levels and offers the potential of bright tunable emission near 60 nm (He) and 20 nm (Li).

10.1.2.6.3. PARAMETRIC EMISSION. Three photon ($\omega_p \rightarrow \omega_0 + \omega_i$) parametric fluorescence and amplification provides the capability for tunable picosecond pulse generation.[67-69] It also allows conversion to wavelengths longer than that of the original source. With intense picosecond pulses, angle or temperature control of a crystal phase-matching angle can provide selective wavelength emission on a single pass. Selectivity depends upon divergence of the pump beam and the length of coherent interaction. Several authors have reported efficient tunable emission.[70-72]

[58] S. E. Harris and R. B. Miles, *Appl. Phys. Lett.* **19**, 385 (1971).

[59] A. H. Kung, J. F. Young, and S. E. Harris, *Appl. Phys. Lett.* **22**, 301 (1973).

[60] M. H. R. Hutchinson, C. C. Ling, and D. J. Bradley, *Opt. Commun.* **18**, 203 (1976).

[61] J. F. Ward and G. H. C. New, *Phys. Rev.* **185**, 57 (1969).

[62] G. C. Bjorklund, *IEEE J. Quantum Electron.* **qe-11**, 287 (1975).

[63] D. M. Bloom, G. W. Bekkers, J. F. Young, and S. E. Harris, *Appl. Phys. Lett.* **26**, 687 (1975).

[64] S. E. Harris, *Phys. Rev. Lett.* **31**, (1973).

[65] J. Reintjes, R. C. Eckardt, C. Y. She, N. E. Karangelen, R. C. Elton, and R. A. Andrews, *Phys. Rev. Lett.* **37**, 1540 (1976).

[66] S. E. Harris, *Appl. Phys. Lett.* **31**, 498 (1977).

[67] W. H. Glenn, *Appl. Phys. Lett.* **11**, 333 (1967).

[68] S. A. Akhmanov, A. S. Chirkin, K. N. Drabovich, A. I. Koorigin, A. S. Piskarskas, and A. P. Sukhorukov, *IEEE J. Quantum Electron.* **qe-4**, 598 (1968).

[69] A. G. Akhmanov, S. A. Akhmanov, R. V. Khokhlov, A. I. Koorigin, A. S. Piskaraskas, and A. P. Sukhorukov, *IEEE J. Quantum Electron.* **qe-4**, 828 (1968).

[70] A. H. Kung, *Appl. Phys. Lett.* **25**, 653 (1974).

[71] P. G. Kryukov, Yu. A. Matveets, D. H. Nikigosyan, A. V. Sharkov, E. M. Gordeev, and S. D. Fanchenko, *Sov. J. Quantum Electron. (Engl. Transl.)* **7**, 127 (1977).

[72] A. Laubereau, L. Greiter, and W. Kaiser, *Appl. Phys. Lett.* **25**, 87 (1974).

Kung[70] obtained a 10% conversion efficiency with 1 mJ, 266 nm pump pulses and a tuning range of 420–720 nm. Two temperature-tuned ADP crystals were used: one primarily as a fluorescence source and the other as a selective amplifier. Kryukov et al.[71] achieved 5% conversion with single, 530 nm, 4 mJ pulses. Two angle-tuned $LiIO_3$ crystals were used in this case. Smooth tuning occurred over the range 660 nm to 2.75 μm, limited by absorption of the idler at the long-wavelength end. Laubereau et al.,[72] using 1.06 μm pulses, found a tuning range with $LiNbO_3$ from about 1.45 to 4 μm. They pointed out the advantage of using single pulses to avoid crystal damage and were able to obtain conversion efficiencies of several percent with their peak input powers of about 10^9 W. In an extension of this work,[73] two crystals have been employed to reduce the emission bandwidth close to the pulse transform limit. This extension of picosecond sources into the infrared provides an important source for time-resolved molecular vibrational spectroscopy.

10.1.2.6.4. CONTINUUM GENERATION. When picosecond pulses with enough intensity are focused into almost any dielectric medium, a dramatic super-broadening of the spectrum occurs. Such "continuum" generation was initially demonstrated[74,75] in liquid argon and krypton, sodium chloride, calcite, quartz, and various glasses. It has since been generated efficiently and studied in a variety of other materials including water (H_2O and D_2O),[76–82] isopropyl alcohol,[76] CCl_4,[76,83,84] and polyphosphoric acid.[85] The use of liquids over solids has the practical advantage that permanent damage to the medium is avoided.

A variety of different and related phenomena may contribute to continuum generation. Experimental results have been analyzed in terms of

[73] A. Seilmeier, K. Spanner, A. Laubereau, and W. Kaiser, Opt. Commun. **24**, 237 (1978).

[74] R. R. Alfano and S. L. Shapiro, Phys. Rev. Lett. **24**, 584 (1970).

[75] R. R. Alfano and S. L. Shapiro, Chem. Phys. Lett. **8**, 631 (1971).

[76] N. N. Il'ichev, V. V. Korobkin, V. A. Korshunov, A. A. Malyutin, T. G. Okroashvili, and P. P. Pashinin, JETP Lett. (Engl. Transl.) **15**, 133 (1972).

[77] G. E. Busch, R. P. Jones, and P. M. Rentzepis, Chem. Phys. Lett. **18**, 178 (1973).

[78] W. Werncke, A. Lau, M. Pfeifer, K. Leng, J. H. Weigman, and C. D. Thag, Opt. Commun. **4**, 413 (1972).

[79] H. Tashiro and T. Yajima, Chem. Phys. Lett. **25**, 582 (1974).

[80] A. Penzkofer, A. Laubereau, and W. Kaiser, Phys. Rev. Lett. **14**, 863 (1973).

[81] D. K. Sharma, R. W. Yip, D. F. Williams, S. E. Sugamori, and L. L. T. Bradley, Chem. Phys. Lett. **41**, 460 (1976).

[82] W. L. Smith, P. Liu, and N. Bloembergen, Phys. Rev. A **15**, 2396 (1977).

[83] R. R. Alfano, L. L. Hope, and S. L. Shapiro, Phys. Rev. A **6**, 433 (1972).

[84] D. Magde, B. A. Bushaw, and M. W. Windsor, Chem. Phys. Lett. **28**, 263 (1974).

[85] N. Nakashima and N. Mataga, Chem. Phys. Lett. **35**, 350 (1975).

self-focusing and self-phase modulation,[83,86] phase-matched four-photon parametric amplification,[74,87] and rapid plasma formation in avalanche breakdown.[88] Each of these effects can become relatively more or less important depending upon experimental parameters such as beam quality, pulse duration, material dispersion, and power level. With a series of experiments designed to guard against self-focusing and breakdown, Penzkofer and co-workers[87,89,90] arrived at the conclusion that super-broadening may be explained entirely by parametric conversion. On the other hand, with slightly longer pulses, Smith *et al.*[82] performed experiments that argue that superbroadening in water only occurs at power levels that are also sufficient for catastrophic self-focusing and avalanche breakdown.

An example of the continuum generated by focusing amplified subpicosecond pulses[48] in a cell of water is shown in Fig. 4. Measurements of selected parts of this broadened spectrum show that it has approximately the same duration as the input pulse. This 0.5 psec continuum will allow study of superbroadening on a new, shorter timescale and already provides the shortest-duration source for tunable or broadband picosecond spectroscopy.

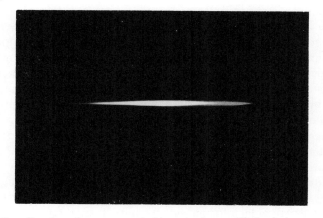

FIG. 4. Broadband continuum generated in water by amplified subpicosecond pulses extends from the near-UV (left) to the near-infrared (right).

[86] Y. R. Shen, *Prog. Quantum Electron.* **4**, 1 (1975).
[87] A. Penzkofer, A. Laubereau, and W. Kaiser, *Phys. Rev. Lett.* **14**, 863 (1973).
[88] N. Bloembergen, *Opt. Commun.* **8**, 285 (1973).
[89] A. Penzkofer, *Opt. Commun.* **11**, 265 (1974).
[90] A. Penzkofer, A. Seilmeier, and W. Kaiser, *Opt. Commun.* **14**, 363 (1975).

10.1.3. Pulse Measurement Techniques

In a division similar to that outlined above for pulse sources, we shall discuss picosecond techniques in terms of their usefulness to either single-pulse measurements or repetitive-pulse measurements.

Single-pulse measurements are generally made with pulses selected from flashlamp-pumped laser outputs and can only be repeated on a time scale of minutes. This requires a means of detecting and recording the entire event with each pulse. Any further averaging or plotting of data must be done on a stepwise basis and can be very tedious. Averaging over the multiple pulses in a flashlamp pulse train is not recommended procedure under any circumstances. Pulse characteristics vary during the measurement and cumulative pumping effects can completely distort the result.

By repetitive-pulse measurements we mean reproducible, single-pulse events repeated at high rate. The high rate is made possible by continuous operation of the picosecond pulse source. With low-energy subpicosecond pulses this is done at rates up to 10^5 pps, and with high-power amplification of these pulses at rates of 10 pps. At these repetition rates lower excitation intensities can be used in conjunction with sensitive detection methods, and extensive signal averaging is possible.

With the invention of the passively mode-locked Nd:glass laser[2] it became immediately apparent that conventional electronics would not provide sufficient time resolution either for measuring pulses or for studying the dynamics of their interactions with matter. The rapid development of pulse measurement techniques[4] that followed relied mostly on the use of nonlinear optical interactions. In a nonlinear material one picosecond pulse can be used to probe either itself or another pulse, and the resulting measurement is a function of autocorrelation or cross correlation. The most practical and widely used of these schemes are TPF, used primarily in conjunction with single-pulse systems, and SHG, used with cw or high-repetition-rate systems. Picosecond streak camera techniques have been less widely used because of the investment involved; but they have also provided valuable information on pulse formation in mode-locked lasers, pulse isolation, and pulse-to-background ratio. An excellent review of these applications and current streak camera capabilities has been given by Bradley.[3]

10.1.3.1. TPF. Pulse measurement by two-photon fluorescence[91] is the most convenient means for monitoring the output of single-pulse lasers.

---

[91] J. A. Giordmaine, P. M. Rentzepis, S. L. Shapiro, and K. W. Wecht, *Appl. Phys. Lett.* **10**, 16 (1967).

It is commonly accomplished in a triangular arrangement[92] where equal intensities, derived from a single pulse, propagate in opposite directions past each other in a (linearly transparent) two-photon-absorbing dye solution. There is nonlinearly enhanced two-photon absorption (and subsequent fluorescence) in the region where the two pulses overlap. The entire fluorescence track is photographed from the side. Its spatial distribution, averaged over several wavelengths, is given theoretically by[93]

$$f(\tau) = 1 + 2G^2(\tau),$$

where $\tau \cong 2nz/c$ is the relative delay between the arrival times at the point z. $G^2(\tau)$ is the second-order autocorrelation function of the field with intensity $I(t)$:

$$G^2(\tau) = \frac{\langle I(t)I(t + \tau)\rangle}{\langle I^2(t)\rangle}.$$

Several comments can be made on the interpretation of experimental results. $G^2(\tau)$ is by definition equal to unity at $\tau = 0$ so that a well-isolated pulse gives a contrast ratio $f(0):f(\infty)$ of $3:1$. Accurate measurement of this contrast ratio is important since even a purely random signal has an autocorrelation peak at $\tau = 0$ and a variety of statistical behavior is possible.[94] $G^2(\tau)$ is also by definition symmetric in τ. This is a fundamental limitation on the information one can derive about the pulse shape $I(t)$. Higher order correlation measurements and other probing methods are necessary to determine pulse asymmetry.[95,96] Nevertheless, if a proper measure of $G^2(\tau)$ is obtained, one can estimate the pulse duration to better than a factor of two.[4] A recent, careful comparison of autocorrelation by TPF and by SHG have shown agreement between the two techniques to better than 10^{-13} sec.[97] The primary difficulties with TPF are experimental. Photographic exposures are nonlinear and must be accurately calibrated. Proper contrast is very difficult to get and the inherent background level prevents accurate measurement.

10.1.3.2. SHG. Autocorrelation measurement by second-harmonic generation preceded the use of TPF historically,[98-100] but was not well

[92] G. Kachen, L. Steinmetz, and J. Kysilka, *Appl. Phys. Lett.* **13**, 229 (1968).
[93] J. R. Klauder, M. A. Duguay, J. A. Giordmaine, and S. L. Shapiro, *Appl. Phys. Lett.* **13**, 174 (1968).
[94] H. E. Rowe and T. Li, *IEEE J. Quantum Electron.* **qe-6**, 49 (1970).
[95] D. H. Auston, *Appl. Phys. Lett.* **18**, 249 (1971).
[96] D. Von der Linde, A. Laubereau, and W. Kaiser, *Phys. Rev. Lett.* **26**, 954 (1971).
[97] E. P. Ippen, C. V. Shank, and R. L. Woerner, *Chem. Phys. Lett.* **46**, 20 (1977).
[98] H. P. Weber, *J. Appl. Phys.* **38**, 2231 (1967).
[99] M. Maier, W. Kaiser, and J. A. Giordmaine, *Phys. Rev. Lett.* **17**, 1275 (1966).
[100] J. A. Armstrong, *Appl. Phys. Lett.* **10**, 16 (1967).

suited for use with low-repetition-rate systems. It has become much more important with the development of mode-locked cw lasers. SHG is much more efficient than TPF and is easily used with low-intensity pulses. Measurements are made photoelectrically and relative delays are accurately calibrated in terms of variable mechanical position. Finally, with SHG it is possible to measure the pulse autocorrelation function $G^2(\tau)$ without background by proper choice of polarization[98,99] or geometry.[31,100] A review of the variety of SHG techniques and measurements reported in the literature has been given recently elsewhere.[4]

A particular experimental arrangement that has been used successfully with high-repetition-rate systems is shown in Fig. 5. The picosecond pulse train is divided into two beams that traverse different path lengths in a modified interferometer. A stepper-motor-driven translation state provides a variable path length in one of the arms and allows digitally controllable scan of the desired delay. The two beams, with relative time delay, leave the interferometer parallel but noncollinear. A simple lens focuses both beams to the same spot in an angle-phase-matched nonlinear crystal. SHG, proportional only to the product of the two beams, is detected at an angle bisecting that of the individual beams. The signal as a function of relative delay is stored in a multichannel analyzer and averaged by repeated scanning over the desired range. Good autocorrelation traces are obtained in a matter of minutes; the same system provides averaging over hours for applications where the signal is small.

A pulse measurement made by this background-free autocorrelation system using SHG is shown in Fig. 6. The pulses, from a passively mode-locked cw dye laser, have a duration (FWHM) of about 0.3 psec. Other careful studies of the pulses produced by this laser have completely characterized their dynamic amplitude and spectral behavior.[29,31]

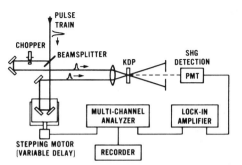

FIG. 5. Experimental set-up for noncollinear, background-free correlation using SHG.

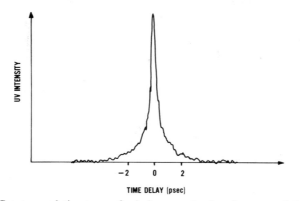

FIG. 6. SHG autocorrelation trace of subpicosecond pulses from a mode-locked cw dye laser.

10.1.4. Applications

The availability of reliable, short optical pulses has resulted in a number of innovative techniques for studying dynamical processes in matter. Diverse applications have taken place in important areas of physics, chemistry, and biology. Nevertheless, there is a common thread that extends through all the measurement techniques. A short optical pulse induces some perturbation to the system under measurement. At a later time an optical probe is used to measure this disturbance induced by the first pulse. The time resolution depends upon the reliability of the pulse source, accurate characterization of the pulses, and careful measurement of the delay between the pumping pulse and the probing pulse.

Techniques for data accumulation and averaging depend upon the source. With flashlamp lasers, a single pulse is selected from the pulse train and as much data as possible are collected from a single event. Averaging of data from successive events must be done with care to account for fluctuations in pulse width and amplitude. With highly repetitive systems, data from many events can be more easily collected and averaged. Perturbations as small as 10^{-4} can be measured with high accuracy, and distorting nonlinear, high-power effects can be avoided.

In the first reported measurements with picosecond optical pulses Shelton and Armstrong[101] used a beam splitter to pick off a small portion of the pumping beam and a delay path was provided by a mirror corner cube arrangement on a translation stage. The probe beam transmission through the sample was measured at different relative path delays to mon-

[101] J. W. Shelton and J. A. Armstrong, *IEEE J. Quantum Electron.* **qe-3**, 302 (1967).

itor the absorbent recovery.   A novel technique introduced by Topp *et al.*[102] using an echelon delay provides a means of measuring a complete absorption recovery with a single pulse.   As shown in Fig. 7, a strong pump beam that bleaches the sample travels almost parallel to the interrogating beam, which is divided up into a number of delayed pulses by means of echelon.   The echelon is formed by a stack of glass plates arranged in a step sequence.   It is assumed in this technique that the beams are very nearly spatially uniform, a condition that can be strictly met only if the laser operates in its lowest order mode.   The echelon technique is most useful for making single-pulse recordings with high-power lasers. With highly repetitive sources, a system with continuously variable delay such as that in Fig. 5 is easier to implement and provides greater accuracy.

Often it is desired to measure an entire absorption spectrum in a single shot using a picosecond interrogating pulse with a broad optical band-

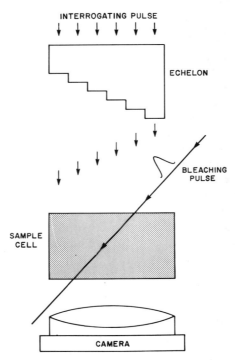

FIG. 7. An echelon provides an array of probe pulses with different delay for single-shot measurement.

[102] M. R. Topp, P. M. Rentzepis, and R. P. Jones, *Chem. Phys. Lett.* **9,** 1 (1971).

width. Alfano and Shapiro[74] first suggested that a picosecond continuum, generated when an intense picosecond laser pulse is focused into a nonlinear medium, could be used for monitoring absorption spectra of transient species on a picosecond time scale. Malley[103] has described the use of a continuum in a configuration that allows one to take both time- and wavelength-resolved data with a single-pulse system. An experimental arrangement that utilizes the continuum to measure time and wavelength resolved absorption spectra of excited molecular statements is shown in Fig. 8. In this technique a laser pulse of a given frequency is used to pump the molecules of interest into an excited state and a series of picosecond continuum pulses is used to measure the absorption. Care should be taken to measure the duration of the continuum pulses because the pulses can be appreciably broadened by group delay in passage through the generating medium. A spectrograph followed by a photodiode array is used to image a three-dimensional recording of intensity time and wavelength. Light intensity versus time is recorded parallel to the entrance slit. This information is easily recorded and read out electronically from the diode array. Amplified pulses from the mode-locked cw dye laser, and the subpicosecond continuum they can be

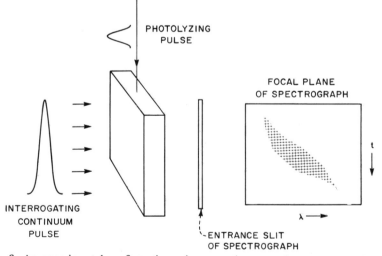

FIG. 8. An experimental configuration using a continuum probe to measure time- and wavelength-resolved spectra. Intensity vs. wavelength is plotted perpendicular to the slit; intensity vs. time is plotted parallel to the slit.

[103] M. Malley, in "Creation and Detection of the Excited State" (W. R. Ware, ed.), Vol. 2. Dekker, New York, 1974.

used to generate, now allow time- and wavelength-resolved spectra to be obtained with extensive averaging and high resolution.

10.1.4.1. Molecular Dynamics. The study of nonradiative relaxation in organic molecules is a natural application of picosecond optical pulses. The experimental arrangement shown in Fig. 5 for measuring the optical pulse autocorrelation measurement is readily adapted for nonradiative relaxation study. The harmonic generation crystal can be replaced by a cell containing a solution of a molecule that is to be measured. The molecule azulene is of particular interest in the field of chemistry because of its unusual fluorescence properties.[104] The molecular structure and a schematic energy diagram of azulene are given in Fig. 9. Of particular interest is the lifetime of the lowest excited singlet S_1. A very low quantum efficiency of fluorescence indicates that this level is rapidly depopulated by nonradiative decay. Excitation to the lowest excited singlet state produces negligible fluorescence, but excitation to higher states causes an easily detectable fluorescence. The fact that fluorescence can be ob-

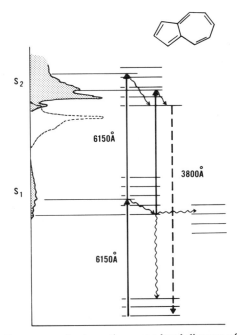

FIG. 9. Structure and schematic energy level diagram of azulene.

[104] J. B. Birks, *Chem. Phys. Lett.* **17**, 320 (1972).

served following excitation to higher states suggests a method for monitoring the population in S_1.[105-107]

A single pulse is used to excite molecules to S_1, where they remain for some short time τ. A second delayed pulse arrives within the singlet lifetime and excited molecules are further elevated to a state from which fluorescence can occur. Each individual pulse can also induce a two-step excitation, but this process only gives a background independent of the relative pulse delay. By measuring the fluorescence as a function of relative pulse delay one can deduce the singlet S_1 lifetime. It is possible to accurately measure the response of the system by substituting for the azulene sample a two-photon fluorescing dye (α-NPO) that has no real intermediate state. The experiment then is simply a measurement of the autocorrelation function of the pulses by two-photon fluorescence.

The experimental result of the measurement of S_1 relaxation of azulene in solution is shown in Fig. 10.[107] The two reference curves are also shown plotted on top of one another, one obtained by second-harmonic generation autocorrelation and the other by two-photon fluorescence. These two curves are indistinguishable on a 50 fsec timescale. The third curve shows clearly the response of the azulene and the broadening yields a measurement of the S_1 lifetime. Analysis and deconvolution yield a simple exponential behavior with a recovery time constant of 1.9 ± 0.2 psec.

A second type of measurement can be performed by placing a detector

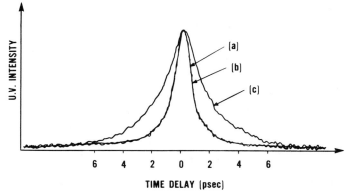

FIG. 10. Experimental traces of (a) pulse correlation by SHG, (b) pulse correlation by TPF, and (c) time-resolved excited-state absorption in azulene.

[105] P. M. Rentzepis, *Chem. Phys. Lett.* **2**, 117 (1968).
[106] J. P. Heritage and A. Penzkofer, *Chem. Phys. Lett.* **44**, 76 (1976).
[107] E. P. Ippen, C. V. Shank, and R. L. Woerner, *Chem. Phys. Lett.* **46**, 20 (1977).

at the weak probing beam and detecting a small modulation on the probe beam induced by the strong, chopped, pumping beam in the experimental setup in Fig. 5. In this case, the detector monitors directly the transmission of the probe beam. Any modulation of the probe transmission due to the chopped pump beam can be detected, averaged, and plotted as a function of time delay. The result of such a measurement for the dye malachite green in a methanol solution is shown in Fig. 11.[33] The pulse induces an increase in transmission due to absorption saturation, which recovers with an exponential time constant 2.1 psec. The same experimental arrangement can be extended to studies of induced absorption, induced dichroism,[108] and even modifications of surface reflectivity.[109]

10.1.4.2. Biological Processes. Picosecond pulses have also found their way into the study of biological processes.[110] Although conventional measurement techniques can determine that a biological reaction has taken place, the complexity of biological molecules often masks actual atomic or molecular rearrangements. Recent experiments indicate that many biological processes occur on a picosecond timescale. Dynamic measurements can be used to distinguish different reaction pathways as for example in the measurement we describe here for the ultrafast dynamics of bacteriorhodopsin photochemistry.

Bacteriorhodopsin has been the object of much research because of its similarity to visual pigment rhodopsins and because of its biological role as an energy converter.[111] The absorption of light is known to induce

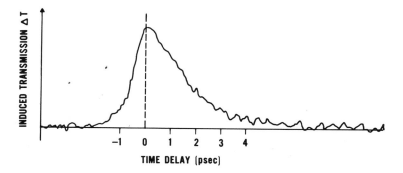

FIG. 11. Picosecond absorption saturation and recovery in the dye malachite green.

[108] C. V. Shank and E. P. Ippen, *Appl. Phys. Lett.* **26,** 62 (1975).

[109] D. H. Auston, S. McAfee, C. V. Shank, E. P. Ippen, and O. Teschke, *Solid State Commun.* **21,** 147 (1978).

[110] A. J. Campillo and S. L. Shapiro, *in* "Ultrashort Light Pulses" (S. L. Shapiro, ed.), Chapter 7. Springer-Verlag, Berlin and New York, 1977.

[111] D. Oesterhelt and W. Stoeckenius, *Nature (London), New Biol.* **223,** 149 (1971); R. A. Blaurock and W. Stoeckenius, *ibid.* p. 152.

very rapid formation of a more energetic species, with a red shift in absorption maximum, called bathobacteriorhodopsin.[112] The pump-probe technique previously described can be applied to resolve the dynamics of the primary step, bacteriorhodopsin → bathobacteriorhodopsin. The bacteriorhodopsin is excited at 615 nm and an increase in absorption due to the formation of red shifted bathobacteriorhodopsin is probed at the same frequency. The results of such a measurement are shown in Fig. 12.[113] The solid curve is the experimentally measured response of the system to an instantaneous formation time. The measured curve can be compared with a dotted line that shows an ideal 1.0 psec response. The actual curve deviates somewhat from this simple curve, but nevertheless approaches its final value within a time constant of 1.0 ± 0.5 psec. This result is the first resolved measurement of the primary photochemical step in bacteriorhodopsin. The rapidity with which this formation takes place argues against any major change in molecular configuration.

The field of biology will certainly provide a wide range of opportunities for investigation by researchers with short optical pulses. The technique described above has been applied to the study of photodissociation of hemoglobin compounds.[114] A great deal of other work on photosynthesis, pigments in the eye, and DNA has also been reported.[110]

10.1.4.3. Semiconductor Physics. The application of short optical pulses to the study to the properties of the semiconductors is an important topic of current study. Short optical pulses have revealed how high-density electron–hole plasmas diffuse[115] and recombine[116] at extremely

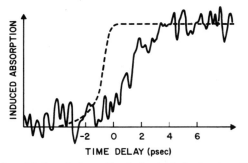

FIG. 12. Time-delayed induced absorption in bacteriorhodopsin. The solid curve indicates an instantaneous response and the dashed curve is computed for a 1.0 psec exponential rise.

[112] K. J. Kaufman, P. M. Rentzepis, W. Stoeckenius, and A. Lewis, *Biochem. Biophys. Res. Commun.* **68,** 1109 (1976).
[113] E. P. Ippen, C. V. Shank, A. Lewis, and M. Marcus, *Science* **200,** 1279 (1978).
[114] C. V. Shank, E. P. Ippen, and R. Bersohn, *Science* **193,** 50 (1976).
[115] D. H. Auston and C. V. Shank, *Phys. Rev. Lett.* **32,** 1120 (1974).
[116] D. H. Auston, C. V. Shank, and P. LeFur, *Phys. Rev. Lett.* **35,** 1022 (1975).

high densities.   Also at high densities it is possible to study nonlinear recombination processes, for example, Auger recombination in Ge.[116] An electron–hole plasma was excited in germanium by an energetic short 1.06 μm picosecond pulse.   By measuring the free-carrier absorption with a picosecond probe pulse at 1.55 μm, it was possible to monitor the recombination of carriers in the high-density plasma.   If equal concentrations of electrons and holes are assumed, the rate of change of electron concentration is given by

$$dn/dt = -\gamma_3 n^3,$$

where γ_3 is the Auger recombination constant.   By monitoring the carrier recombination as a function of time, it is possible to determine the Auger recombination constant.   In Fig. 13 the induced free-carrier absorption is plotted as a function of probe delay for the experiment of Auston et al.[116] From the observed initial free-carrier decay rate, the Auger recombination rate was calculated to be 1.1×10^{-31}.   At the high densities in this experiment, the Auger process dominates free-carrier recombination, allowing an unambiguous interpretation.

Short pulses also provide a unique way to directly observe the dynamics of hot-carrier distributions in optically excited semiconductors.   In recent experiments, a short optical pulse was used to generate a distribution of hot carriers on the surface of a semiconductor by band-to-band absorption of an optical pulse of frequency greater than the bandgap.   A second less-intense pulse then probed the incremental reflectivity change at a de-

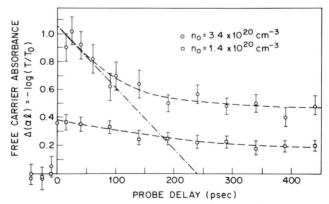

FIG. 13. Induced free-carrier absorption in germanium as a function of delay.   At the high experimental carrier densities, the rapid recovery is determined by Auger recombination.

layed time.   As the carrier distribution relaxes by optical phonon emission, the reflectivity is altered due to the continuously changing occupa-
tion probabilities of the different electronic states until a carrier distribution is in thermal equilibrium with the lattice.   Details of the relationship
between the optical reflectivity and the carrier distribution depend on the
frequency of the probing pulse and the band structure of the particular
semiconductor.

In Fig. 14 we plot the results for the case of gallium arsonide excited
with a short pulse at 4 eV and probed with a 2 eV pulse.[117]   In this example
the sign of the reflectivity change goes from negative to positive as the
carriers relax through the probing pulse energy.   The energy loss rate for
hot carriers can be estimated from the zero crossing to be 0.4 eV/psec.

Numerous applications are on the horizon for the use of short optical
pulses to provide new understanding of ultrafast phenomena in semiconductors.   The ability to resolve short-term events may ultimately provide
the physics necessary for understanding the whole new generation of ultrafast semiconductor devices.

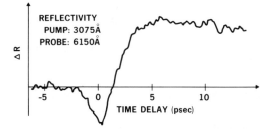

FIG. 14. Time-resolved reflectivity dynamics of an optically excited GaAs surface.   As
the initially hot carriers relax, the induced reflectivity change goes from negative to positive.

[117] C. V. Shank, D. H. Auston, E. P. Ippen, and O. Teschke, *Solid State Commun.*
26, 567 (1978).

10.2. VUV Spectroscopy†*

10.2.1. Introduction

The age of the laser has made many new types of spectroscopic studies possible. Before the advent of widely tunable lasers, material systems with tunable resonances were the subjects of laser "spectroscopic" investigations. In addition, effects such as stimulated Raman scattering (SRS) were used to produce a number of laser lines that were relatively closely spaced and could be used for spectroscopy. Outstanding early examples are the works of Soref and Moos,[1] who worked with ZnS–CdS and CdS–CdSe alloys as a way to tune the band gap through the photon energy of the frequency-doubled Nd:glass laser; Maker and Terhune,[2] who used SRS to produce photon energies with which they performed CARS experiments on benzene and benzene derivatives; and Chang, Ducuing, and Bloembergen,[3] who used SRS to produce photon energies with which they measured the dispersion of second harmonic generation from III–V semiconductors.

Currently, tunable lasers, most notably dye lasers, have permitted a dramatic growth in the number of laser spectroscopic studies. We are now seeing the emergence of the age of laser spectroscopy.

Laser spectroscopy is presently limited by the tuning ranges of lasers. But the history of coherent sources of electromagnetic radiation and their application is characterized by a steady trend toward shorter wavelengths. Laser spectroscopy is no exception. Dye lasers pumped by ultraviolet (UV) light from a KrF laser at 249 nm have operated at wavelengths as short as 335 nm.[4] The Xe_2 excimer laser has been tuned from 169.5 to 174.5 nm[5] and represents a truly tunable, vacuum ultraviolet

[1] R. A. Soref and H. W. Moos, *J. Appl. Phys.* **35**, 2152 (1964).

[2] P. D. Maker and R. W. Terhune, *Phys. Rev.* **137**, A801 (1965).

[3] R. K. Chang, J. Ducuing, and N. Bloembergen, *Phys. Rev. Lett.* **15**, 415 (1965).

[4] D. G. Sutton and G. A. Capelle, *Appl. Phys. Lett.* **29**, 563 (1976).

[5] S. C. Wallace and R. W. Dreyfus, *Appl. Phys. Lett.* **25**, 498 (1974).

† Supported by the U. S. Army Research Office.

* Chapter 10.2 is by J. J. Wynne.

METHODS OF EXPERIMENTAL PHYSICS, VOL. 15B

(VUV) source for spectroscopy. But the present state of laser-assisted VUV spectroscopy is that it relies on multiphoton processes and uses near-infrared (IR), visible, or near-UV lasers.

This chapter will deal with several methods of laser spectroscopy that allow an investigator to study material excitations with energies $\geqslant 50,000$ cm^{-1}, the low-energy cutoff for the VUV region of the spectrum. The VUV region covers the wavelength ranges from 200 to 0.2 nm and corresponds to the optical radiation emitted by outer electrons of atoms and ions. It overlaps the soft X-ray spectral region, which corresponds to radiation emitted by inner electrons when they change energy state. The name VUV is given because of the opacity of air between 200 and 0.2 nm.[6] The main emphasis will be on techniques for laser spectroscopy in the Schumann UV (between 200 and 100 nm), although some mention will be made of extending the techniques to the XUV (< 100 nm) region.

Readily available tunable lasers have output powers well below that of nontunable laser sources such as ruby, Nd:glass, and Nd:YAG. The nontunable sources are powerful enough to produce strong VUV at fixed wavelengths via nonlinear processes. But the limitations on the output power of the tunable lasers necessitate resonant enhancement of the nonlinear processes used to reach the VUV. Thus, this chapter first deals with the question of multiphoton processes and resonant enhancement. Next, methods of generating VUV by optical sum mixing will be discussed, followed by a description of the use of this generated VUV light for spectroscopic studies. Then, the subject of multiphoton absorption and ionization spectroscopy, which access states $\geqslant 50,000$ cm^{-1} above the ground state, will be discussed. The final section presents concluding remarks and offers some predictions as to future directions.

10.2.2. Multiphoton Processes and Resonant Enhancement

In the absence of readily available tunable lasers in the VUV, use may be made of nonlinear optical processes such as third-harmonic generation (THG) and optical sum mixing to produce coherent, tunable VUV. Alternatively, one may use a multiphoton process to reach energy states that could be reached by the one photon only if it were a VUV photon.

The field of nonlinear optics emerged rapidly after the invention of the laser. Early experiments involved second-harmonic generation (SHG) in acentric media, THG in crystalline media, and two-photon absorption

[6] J. A. R. Samson, "Techniques of Vacuum Ultraviolet Spectroscopy," pp. 1–2. Wiley, New York, 1967.

studies in crystalline media.[7]   However, while SHG is an extremely simple and relatively efficient method of extending coherent output wavelengths into the UV, it is restricted by the available acentric media.   In the absence of strong DC electric fields, the only practical acentric media are acentric crystals.   Despite a large effort by workers in the field, the short-wavelength limit for SHG is 217 nm in the crystal $KB_5O_8 \cdot 4H_2O$.[8] No other acentric crystals have been found with sufficient transparency and birefringence to allow efficient SHG into the VUV.   Light has been generated in many crystals by THG using a ruby laser at 694.3 nm to generate 231.4 nm.   Among those crystals tried have been LiF[2] the crystal with the shortest wavelength transmission limit at 104 nm.   This cubic crystal has a small nonlinear susceptibility and is not phase matchable, so that THG is relatively weak.   No one has reported using LiF or any other transparent crystalline material to generate VUV by THG.   Early work in gases involved measurements of THG in rare gas atomic vapors and in some common diatomic and triatomic molecular gases.[9]   These experiments were also done with a ruby laser.   Again, the nonlinearities are weak.

Harris and co-workers[10-12] were the first to exploit the possibilities of resonant enhancement as a way to increase the efficiency of THG.   They used alkali metal atomic vapors with resonance lines near the frequency of the input laser to obtain nonlinear susceptibilities 10^6 greater than that of He.[9]   Kung et al.[13] utilized Cd vapor to extend this technique into the VUV, reaching as far as 118.2 nm.   However, this VUV radiation was not tunable since the input laser was derived from the 1060 nm fundamental of a Nd:YAG laser and its harmonics.   Kung, Young, and Harris[14] subsequently used Xe vapor to produce strong VUV at 118.2 nm by THG.

True tunability came about with the recognition that one could tune very close to a two-photon resonance and have strong resonant enhance-

[7] N. Bloembergen, "Nonlinear Optics." Benjamin, New York, 1965.   The early experimental work is covered in this book.   Particular attention should be given to Chapter 5 which discusses experimental results.

[8] C. F. Dewey, Jr., W. R. Cook, Jr., R. T. Hodgson, and J. J. Wynne, *Appl. Phys. Lett.* **26**, 714 (1975).

[9] J. F. Ward and G. H. C. New, *Phys. Rev.* **185**, 57 (1969).

[10] S. E. Harris and R. B. Miles, *Appl. Phys. Lett.* **19**, 385 (1971).

[11] J. F. Young, G. C. Bjorklund, A. H. Kung, R. B. Miles, and S. E. Harris, *Phys. Rev. Lett.* **27**, 1551 (1971).

[12] R. B. Miles and S. E. Harris, *IEEE J. Quantum Electron.* **qe-9**, 470 (1973).

[13] A. H. Kung, J. F. Young, G. C. Bjorklund, and S. E. Harris, *Phys. Rev. Lett.* **29**, 985 (1972).

[14] A. H. Kung, J. F. Young, and S. E. Harris, *Appl. Phys. Lett.* **22**, 301 (1973).

ment without the deleterious effects of strong linear absorption and dispersion. Hodgson, Sorokin, and Wynne (HSW)[15] utilized this resonant enhancement from a two-photon transition in Sr vapor to produce VUV tunable between 200 and 150 nm, starting with visible dye lasers. Wallace and co-workers have recently used Mg vapor[16] to extend the tuning range to less than 140 nm and have also reported the first use of molecular vapors[17] for tunable VUV generation to 130 nm.

The key to all of these recent methods is the use of resonant enhancement of the nonlinear susceptibilities, occurring when a laser photon energy or sum of photon energies is equal to the transition energy from the ground state of the atomic or molecular vapor to an excited state. The excited state must be connected to the ground state by allowed transitions utilizing the correct number of photons.

The effect of resonant enhancement may be quantitatively seen by considering the expressions for the polarizability of a nonlinear atomic vapor. Consider the process depicted in Fig. 1. This is one of many processes that contribute to the nonlinear polarization at frequency ν_4 when the vapor is irradiated with laser beams at frequencies ν_1, ν_2, and ν_3. The nonlinear polarization is given by (see Bloembergen[7])

$$P^{\mathrm{NL}}(\nu_4) = \chi^{(3)}(\nu_4)\mathscr{E}(\nu_1)\mathscr{E}(\nu_2)\mathscr{E}(\nu_3), \qquad (10.2.1)$$

where $\chi^{(3)}(\nu_4)$ may be calculated by fourth-order perturbation theory,[7] and $\mathscr{E}(\nu_i)$ is the electric field at frequency ν_i. The contribution of the process shown in Fig. 1 to $\chi^{(3)}(\nu_4)$ is given by

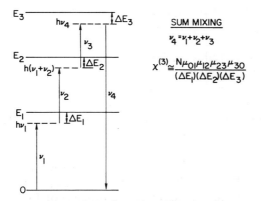

FIG. 1. Four-wave sum mixing process from simple four-level system. $\chi^{(3)}$ is the contribution of this process to the nonlinear susceptibility.

[15] R. T. Hodgson, P. P. Sorokin, and J. J. Wynne, *Phys. Rev. Lett.* **32,** 343 (1974).
[16] S. C. Wallace and G. Zdasiuk, *Appl. Phys. Lett.* **28,** 449 (1976).
[17] K. K. Innes, B. P. Stoicheff, and S. C. Wallace, *Appl. Phys. Lett.* **29,** 715 (1976).

$$\chi^{(3)}(\nu_4) \cong \frac{N\mu_{01}\mu_{12}\mu_{23}\mu_{30}}{(\Delta E_1)(\Delta E_2)(\Delta E_3)}, \tag{10.2.2}$$

where N is the volume density of atoms, μ_{ab} is the electric dipole matrix element between states a and b, and ΔE_i is the amount of energy detuning from resonance as shown in Fig. 1.

Other processes involving different orders of the four vertical arrows of Fig. 1 also contribute to $\chi^{(3)}(\nu_4)$. However, the particular process that has been singled out is one that shows the resonant enhancement resulting from reducing ΔE_1, ΔE_2, and ΔE_3, i.e., tuning near to resonance. It is often possible to pick out one important term in the expression for $\chi^{(3)}(\nu_4)$ and to ignore other contributions that are so far from resonance as to be negligible.

In the early work of Ward and New,[9] the energy levels were far in excess of the photon energies. Thus no terms were strongly resonantly enhanced, and the nonlinear susceptibilities were small. Harris and co-workers[10-14] utilized atomic vapors having energy levels that provided resonant enhancement by reducing ΔE_3 and ΔE_1. But the detuning from resonance was sufficient to avoid the associated problems of linear dispersion and linear absorption that occur when tuning close to resonance. Linear dispersion becomes a problem when there is enough dispersion to cause the various waves to get out of phase. It is necessary to match the phase velocities of the various waves (i.e., phase-match) for efficient VUV generation by sum mixing. The Stanford group[10-14] found that it was possible to correct for linear dispersion by adding a buffer gas with compensating linear dispersion characteristics. Linear absorption is a problem that cannot be corrected when tuning too close to single-photon resonances. Either the input lasers or the generated VUV output will be absorbed, seriously reducing VUV output. HSW[15] recognized that the way around these problems was to minimize ΔE_2. Then there is just as much resonant enhancement of $\chi^{(3)}(\nu_4)$ [Eq. (10.2.2)] without the problems of linear dispersion and linear absorption. Limitations do occur due to two-photon absorption and the resulting saturation of the nonlinearity,[16,18,19] but these occur at higher powers than the comparable one-photon effects.

In the process depicted in Fig. 1, the input frequencies ν_1, ν_2, and ν_3 need not all be different. When $\nu_1 = \nu_2 = \nu_3$, $\nu_4 = 3\nu_1$ and one has the case of THG. One may generalize the process of Fig. 1 to higher-order processes such as five-wave sum mixing of the form $\nu_1 + \nu_2 + \nu_3 + \nu_4 + \nu_5 = \nu_6$. There will be five energy denominators that may be resonantly

[18] S. E. Harris and D. M. Bloom, *Appl. Phys. Lett.* **24**, 229 (1974).
[19] C. C. Wang and L. I. Davis, *Phys. Rev. Lett.* **35**, 650 (1975).

enhanced. If there are sets of energy levels with large allowed matrix elements for electric dipole transitions, then such processes may generate useful amounts of VUV radiation. Harris[20] has theoretically considered higher-order processes but no experimental results have been published. Reintjes et al.[21] have reported fifth-harmonic generation in Ne and He producing radiation at 53.2 nm. The input wave was the fourth harmonic of a Nd:YAG laser at 266.1 nm. Four photons of this wavelength were nearly resonant with the $3p[1\frac{1}{2}]$ $J = 2$ level in Ne ($\Delta\tilde{\nu} = 12$ cm^{-1}). The four-photon resonance was much less exact in He ($\Delta\tilde{\nu} \approx 1600$ cm^{-1}). So far no tunability has been demonstrated in these higher-order processes. The effects have been seen only with very high power, nontunable lasers.

As an alternative to actually generating VUV light for spectroscopic purposes, experiments may be done using multiphoton transitions to populate energy states $>50,000$ cm^{-1} higher than the ground state. As an example, consider the three-photon process depicted in Fig. 2. The transition probability for such a process will be resonantly enhanced by tuning near resonance with the intermediate states. Either the one-photon resonance to level 1 or the two-photon resonance to level 2 or both may be utilized to make the three-photon transition probability large and thereby significantly populate level 3. Such techniques have been used extensively in the visible and near-UV[22,23] and are currently being extended

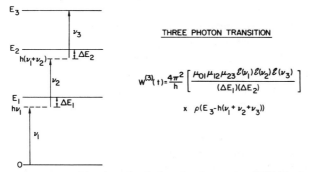

FIG. 2. Three-photon transition in a simple four-level system. $W^{(3)}(t)$ is the contribution of this process to the transition rate.

[20] S. E. Harris, *Phys. Rev. Lett.* **31**, 341 (1973).

[21] J. Reintjes, R. C. Eckardt, C. Y. She, N. E. Karangelen, R. C. Elton, and R. A. Andrews, *Phys. Rev. Lett.* **37**, 1540 (1976).

[22] J. E. Bjorkholm and P. F. Liao, *Phys. Rev. Lett.* **33**, 128 (1974).

[23] P. Esherick, J. A. Armstrong, R. W. Dreyfus, and J. J. Wynne, *Phys. Rev. Lett.* **36**, 1296 (1976).

into the VUV.[24] Laser-assisted, excited-state spectroscopy has already been used to study high-lying states $>50,000$ cm^{-1} above the ground state in Mg vapor.[25]

10.2.3. VUV Generation

This section will describe the experimental technique of optical sum mixing of three input waves to generate tunable VUV radiation. Brief mention will also be made of the methods of fixed-frequency VUV generation.

Generation of VUV by parametric sum mixing depends on many factors: the nonlinearity of the nonlinear medium, the strength of the input lasers, linear dispersion (which affects phase-matching), linear absorption, two-photon absorption, and photoionization. The interplay of all these factors has been discussed by Miles and Harris.[12] Our main concern will be the nonlinearity and its resonant enhancement.

Of the various approaches taken toward producing tunable VUV generation, that of HSW has proven most fruitful experimentally and will now be reviewed in some detail. For the input lasers they have used nitrogen-laser-pumped dye lasers. These lasers have the advantage of relatively high peak power, short pulse width to avoid saturation,[16,19] wide tunability, low shot-to-shot variation in output power, and good repetition rates. Using a Molectron UV-1000 nitrogen laser with 1 MW output at 337 nm in 10 nsec long pulses at repetition rates to 50 pps, two dye lasers may be simultaneously pumped, which will emit at least 10 kW and be tunable throughout the visible region of the spectrum. The dye lasers incorporate optics to narrow the spectral linewidth and are designed along lines described by Hänsch.[26] A schematic diagram of such a dye laser is given in Fig. 3. Linewidths of ~ 0.01 nm are routinely achieved with such a design. The laser is tuned by tilting the diffraction grating. The output beam from such a laser contains many transverse modes. For higher brightness an oscillator–amplifier configuration may be used.[27] Higher brightness will result in higher intensity when the beam is focused into a nonlinear medium, resulting in greater VUV output in general.

The beams from the two dye lasers are combined into a collinear beam, which is focused into the nonlinear gas. HSW used a Glan prism to combine beams with orthogonal linear polarization. This is shown in Fig. 4,

[24] J. J. Wynne, unpublished.

[25] D. J. Bradley, P. Ewart, J. V. Nicholas, J. R. D. Shaw, and D. G. Thompson, *Phys. Rev. Lett.* **31,** 263 (1973).

[26] T. W. Hänsch, *Appl. Opt.* **11,** 895 (1972).

[27] I. Itzkan and F. Cunningham, *IEEE J. Quantum Electron.* **qe-8,** 101 (1972).

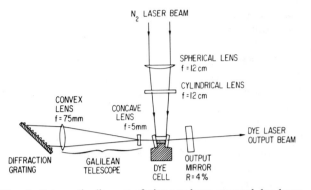

FIG. 3. Schematic diagram of nitrogen-laser-pumped dye laser.

which also shows the experimental configuration for detecting VUV produced in Sr vapor. The procedure for maximizing the tunable VUV out-
put is first to look for resonant enhancement of the VUV output as $\nu_{vuv} = \nu_4 = 3\nu_1$ when a single dye laser (ν_1) is focused into the nonlinear medium.
When the laser is tuned off resonance, weak THG may be observed. As
the laser is tuned so that the condition $2\nu_1 = \nu_0$ (where ν_0 is the frequency
of a two-photon transition) is approached, THG gets stronger. On resonance, THG increases by many orders of magnitude. In Sr vapor, HSW
observed a THG enhancement of six orders of magnitude when a rhodamine 6G dye laser was tuned to 575.7 nm so that two photons were resonant with the two-photon transition from the $5s^2\ {}^1S_0$ ground state to the
$5s5d\ {}^1D_2$ excited state at 34727 cm^{-1}. There are many other even-parity
states, connected to the ground state by two-photon transitions, which
may be used for resonant enhancement.

Once strong THG is observed, the next step is to make the VUV output
tunable. This is accomplished by combining the beam from the second
laser (ν_3) and then generating output at $\nu_{vuv} = \nu_4 = 2\nu_1 + \nu_3$ ($\nu_2 = \nu_1$ in
the terminology of Fig. 1). Since the only resonance needed is that at
$2\nu_1 = \nu_0$, ν_1 can be left fixed and ν_3 tuned in order to tune ν_{vuv}. Then the

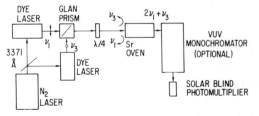

FIG. 4. Experimental configuration for generating $\nu_{vuv} = 2\nu_1 + \nu_3$ in Sr. The monochromator may be removed when all other VUV signals are suppressed (see text).

tuning rate and range of ν_{vuv} is the same as that of ν_3. One has, in effect, up-conversion from ν_3 to ν_{vuv}.

The VUV output emerging from the nonlinear medium will, in general, consist of both the THG light at $3\nu_1$ and the desired tunable component at $2\nu_1 + \nu_3$. If selection rules allow it, the THG may be eliminated while preserving the sum mixing by circularly polarizing the light at ν_1. It has been shown that frequency tripling in isotropic media is not allowed with circularly polarized light.[28] The fundamental reason is that angular momentum must be conserved among the photons. In THG, three input photons are destroyed and one frequency-tripled photon is created. With circular polarization, three units of angular momentum would be destroyed, and the created photon would at most restore one unit of angular momentum, making it impossible to conserve angular momentum. Using this idea, one starts with linearly polarized light at ν_1, tunes it onto resonance ($2\nu_1 = \nu_0$) to maximize THG, then inserts a $\lambda/4$ plate to circularly polarize the beam and null out the THG signal. Thus the only VUV output remaining is that at $2\nu_1 + \nu_3$. The selection rule just mentioned requires that the two-photon transition from the ground state to the resonant intermediate state be a $\Delta M = 2$ transition. When the ground state is $^1S_0$ as in the alkaline earths, then the resonant intermediate state must be $J = 2$ and of the same parity (even) as the ground state. For the case of Sr, HSW used a series of $^1D_2$ intermediate states, which when coupled with various dyes for ν_3, covered the tuning range 150–200 nm in the VUV. Table I shows the tuning ranges that can be covered using only four different, readily available dyes.

One may use a VUV monochromator to verify the VUV wavelength and also to discriminate against the visible dye laser light. But the VUV light is much easier to use for spectroscopy if the monochromator can be eliminated. For the case of Sr, where dye laser wavelengths are longer than 465 nm, one may use a solar-blind photomultiplier to detect the VUV light with no additional filtering. Such a photomultiplier has a photocathode that will not emit electrons from one-photon photoemission processes when the photon is in the visible or near-UV region of the spectrum. However, in intense laser light, two-photon photoemission processes become important. The EMR 541GX-08-18 photomultiplier used by HSW detected laser light below 450 nm due to two-photon processes, but had no response to laser light at wavelengths longer than 450 nm.

Figure 5 shows the VUV signal as a function of frequency when ν_3 comes from the Rhodamine 6G laser and ν_1 is set at the half-frequency of the 5s5d $^1D_2$ level (34727 cm^{-1}). The signal is not normalized in any way.

[28] P. P. Bey and H. Rabin, *Phys. Rev.* **162**, 794 (1967).

TABLE I. Tuning Ranges in the VUV Obtainable by Sum Frequency Mixing in Sr with the Use of Four Dye Solutions

Dye for laser at λ_1: Resonant state ($^1D_2$): λ_1 (nm):	7-Diethylamino-4-methylcoumarin 5s7d 477.9	7-Diethylamino-4-trifluoro-methyl-coumarin 5s6d 503.2	Sodium fluorescein $5p^2$ 540.9	Rhodamine 6G 5s5d 575.7
Dye for laser at λ_3 and its tuning range (nm)				
7-Diethylamino-4-methylcoumarin 464.8–492.5 nm	157.8–160.9	163.2–166.6	171.0–174.6	177.8–181.7
7-Diethylamino-4-trifluoro-methyl-coumarin 490.0–545.0 nm	160.7–166.2	166.2–172.2	174.3–180.8	181.4–188.4
Sodium Fluorescein 533.7–571.0 nm	165.7–168.5	171.0–174.7	179.5–183.6	187.0–191.4
Rhodamine 6G 568.0–611.1 nm	168.2–171.8	174.4–178.3	183.3–187.5	190.7–195.7

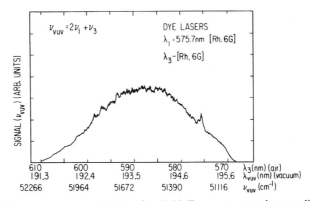

FIG. 5. VUV signal generated in Sr using 5s5d $^1D_2$ as resonant intermediate state and scanning the laser (Rhodamine 6G) at λ_3.

The fall-off at the ends of the scan corresponds to the fall-off of laser power (ν_3) as the laser is scanned over its fluorescent bandwidth. In this tuning range, with laser powers of ~20 kW focused to ~100 μm diameter with ~5 Torr Sr vapor, conversion efficiencies were ~10^{-6}. With pulse durations of ~10 nsec, this corresponds to 10^8 photons/pulse of VUV radiation. Since the laser linewidths are ~0.2 cm^{-1}, this represents a spectrally bright source of tunable VUV radiation.

This method of producing tunable VUV radiation has been extended to shorter wavelengths by Wallace and co-workers. Wallace and Zdasiuk[16], using Mg vapor, reported VUV generation from 140 to 160 nm. Wallace has recently extended the tuning in Mg to Lyman α at 121.6 nm.[29] Wallace and Zdasiuk have also compared the power conversion efficiencies of Sr and Mg. They find that Mg has a 10^3 higher conversion efficiency with incident powers of 10 kW. They attribute this increased efficiency to a larger $\chi^{(3)}(\nu_{VUV})$ in Mg, originating in the constructive contributions of the bound-continuum matrix elements as one integrates over the continuum. With input laser powers of 30 kW they saw a power conversion efficiency of 2×10^{-3} at 143.6 nm.

More recently, Innes et al.[17] observed sum mixing in molecular vapors NO, Br$_2$, and CH$_6$. The VUV was tuned from 130 to 150 nm in NO. At low pressures, various rotational lines resonantly enhanced the VUV output. However, by going to high pressures (10 atm) the rotational substructure was eliminated by pressure broadening. Thus it was possible to use only one laser and see resonantly enhanced THG smoothly tunable

[29] S. C. Wallace, T. J. McKee, and B. P. Stoicheff, *Int. Quantum Electron. Conf. 10th, 1978* Paper L.6 (1978).

over ~ 100 cm^{-1} range with conversion efficiencies of 10^{-7} (corresponding to outputs of $\sim 10^7$ photons/pulse). This represents a great simplification in the method of VUV generation. Only one laser is needed and the experimental problems associated with spatial and temporal overlap of two lasers are eliminated. The extension of this method to higher pressures (and densities) to further pressure broaden the resonant intermediate state will eventually be limited by dispersion and absorption problems.

Using picosecond pulses, Kung[30] reported tunable VUV radiation from Xe vapor. The input waves were produced from an ADP parametric generator. Wavelengths as short as 117.3 nm were observed. Interestingly, when this source was tuned to emit at or near Lyman α, no output was observable. The problem is probably related to phase matching difficulties. This VUV source is more difficult to realize experimentally, and much work remains to be done before it can be used for spectroscopic studies.

Mention must now be made of phase matching. In general, for highest efficiency generation by parametric mixing processes, one wants to keep the driving waves and the generated wave in phase over the region of strong nonlinear interaction. If they get out of phase, generation efficiency is impaired. But attempting to maintain exact phase matching as the output is tuned puts a severe constraint on the ease of tuning. In practice, one may give up some efficiency by not requiring exact phase matching. This enables one to tune smoothly the VUV output without doing anything else to try to maintain phase matching. Miles and Harris[12] have discussed in detail the method of phase matching in which another gas with desirable linear dispersion properties is added. In working with Sr, HSW found that by adding the correct amount of Xe the VUV output power could be increased by a factor 5 over that in the absence of Xe. However, as the output was tuned or as the Sr pressure varied due to temperature variations, the output power rapidly changed due to changing phase mismatch. With no Xe the changes of phase mismatch with temperature or wavelength were not significant, and smoothly tunable VUV output, such as shown in Fig. 5, could be produced. With proper engineering, it should be possible to have both high efficiencies due to phase matching and smooth tuning capabilities.

The concluding remarks of this section on VUV generation concern methods of generating XUV radiation. The work of Reintjes et al.[21] in frequency quintupling 266 nm radiation to 53.2 nm has already been mentioned. More recently, Reintjes and his colleagues obtained 38 nm out-

[30] A. H. Kung, *Appl. Phys. Lett.* **25,** 653 (1974).

put as the seventh harmonic of 266 nm generated in He.[31] These sources are not tunable since the fundamental derives from a fixed-frequency Nd:YAG laser. In contrast, Hutchinson, Ling, and Bradley[32] have frequency tripled the tunable output of a Xe_2 laser. The Xe_2 laser is tunable in a modest range about 171 nm,[5] and hence its third harmonie is tunable. Using the $3p^6$–$3p^55p$ two-photon transition in Ar for resonant enhancement, Hutchinson et al.[32] tuned the XUV output signal around 57 nm. These methods are just first steps toward the eventual development of a widely tunable, coherent XUV source.

10.2.4. Spectroscopic Studies Using VUV Generation

Once VUV light is generated it may be used for studies such as absorption or ionization spectroscopy. Thus the spectrally narrow VUV light may be used to directly irradiate a sample, the transmission may be monitored to study absorption, or ionization may be monitored. Another important application will be to monitor small densities of atomic or molecular species via the technique of resonance fluorescence.[33] An atom like H has a very high cross section for scattering light at Lyman α. If a tunable, spectrally bright source of Lyman α with enough power can be produced by the techniques discussed in the previous section, it will allow detection of very low ($< 100/cm^3$) H atomic densities, with the added possibility of velocity discrimination via the Doppler shift. With a VUV source of 0.1 cm^{-1} bandwidth tunable around Lyman α (82,259 cm^{-1}), the resolution of a Doppler shift as small as one part in 10^6 is achievable. This allows velocity resolution of $\Delta v/c = 10^{-6}$ or $\Delta v = 3 \times 10^4$ cm/sec. For H atoms, this corresponds to resolving the kinetic energy to ~5 cm^{-1} (10^{-15} erg). This high resolution could be very useful in monitoring chemical kinetics in crossed molecular beam experiments.[34]

An attractive alternative to using the generated VUV light on an external sample is to study the VUV spectroscopic properties of the nonlinear medium via nonlinear VUV generation. In this section most of the attention will be given to this alternative. But first some results on VUV absorption spectroscopy will be presented.

In the experimental geometry where the monochromator can be eliminated, the solar blind photomultiplier can look directly at the collimated

[31] J. Reintjes, C. Y. She, R. C. Eckardt, N. E. Karangelen, R. A. Andrews, and R. C. Elton, Appl. Phys. Lett. 30, 480 (1977).

[32] M. H. R. Hutchinson, C. C. Ling and D. H. Bradley, Opt. Commun. 18, 203 (1976).

[33] W. M. Fairbank, Jr., T. W. Hänsch, and A. L. Schawlow, J. Opt. Soc. Am. 65, 199 (1975).

[34] R. N. Zare, Lec. Notes Phys. 43, 112 (1975).

VUV light. Typically the nonlinear medium is contained in a cell with a VUV transmitting output window, such as LiF. A vacuum-tight sample chamber encloses the space between the output window and the photo-multiplier. If the sample is gaseous it can simply be admitted directly into the chamber. A liquid or solid sample may be placed in the chamber so that the VUV beam passes through it. A simple test that was done by HSW to verify the capability of this system to operate as a spectrometer was to observe absorption in CH_3I at 192.7 nm[35] as the VUV output wavelength was swept through this line. Figure 6 shows the results of ad-mitting ~5 Torr air into the ~10 cm long path between the output window and the photomultiplier. The observed absorption bands are due to the Schumann–Runge band system of atmospheric O_2.[36] This demon-strates the utility of the VUV generation method as a VUV spectrometer.

Although further published results are yet to come, it is clear how to use this VUV source for spectroscopic studies on external samples. The use of VUV generation itself for studying the nonlinear medium needs more explanation. Referring to Eq. (10.2.2) and Fig. 1, note that re-ducing ΔE_3 resonantly enhances the VUV signal. Thus as one tunes the

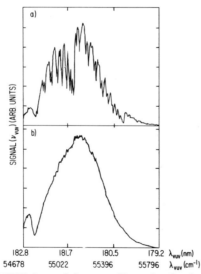

Fig. 6. Absorption of VUV due to Schumann–Runge bands of atmospheric O_2. (a) Scan taken with ~5 Torr of air in a ~10 cm path length; (b) reference scan with <0.1 Torr air (i.e., almost no O_2).

[35] G. Herzberg and G. Scheibe, Z. Phys. Chem., Abt. B 7, 390 (1930).

[36] R. W. B. Pearse and A. G. Gaydon, "The Identification of Molecular Spectra," p. 239. Wiley, New York, 1963.

laser at ν_3 to tune the VUV output (ν_4), the signal should show a dependence on the detuning from resonance with level 3. This is experimentally verified in Sr as shown in Fig. 7. As the VUV is tuned through odd-parity levels that are connected to the ground state by one-photon transitions, the VUV signal is strongly enhanced. The solid line in Fig. 7 shows the VUV output signal. The dashed line shows the one-photon absorption signal in the same spectral range.[37]

Although there is a one-to-one correspondence between absorption and VUV generation spectra, the lineshapes are definitely different. This difference is not surprising since different combinations of matrix elements contribute to the absorption and VUV generation spectra, respectively. For the situation appropriate to VUV generation in Sr, level 3 (Fig. 1) is an autoionizing level. The lineshape for absorption from a discrete bound state to an isolated autoionizing level has been discussed by Fano.[38] The absorption cross section is proportional to $|\mu_{30}|^2$ [notation defined just after Eq. (10.2.2)]. Fano gave the frequency dependence of $|\mu_{30}|^2$ and found that the lineshape could be described completely in terms of a single parameter q. It has become fashionable to describe observed autoionization line profiles by Fano q values.[37] On the other hand, the lineshape for VUV generation depends on $\chi^{(3)}(\nu_4)$ as given in Eq. (10.2.2). Thus, the lineshape involves the matrix elements μ_{23} and μ_{30}, which are both frequency dependent ($\nu_1 = \nu_2$ is assumed fixed so that μ_{01} and μ_{12} may be taken as constant as ν_3 and ν_4 are tuned). It has been shown that

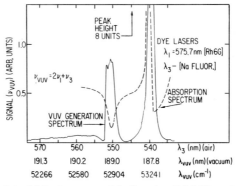

FIG. 7. VUV signal (solid line) generated in Sr using 5s5d $^1D_2$ as resonant intermediate state and scanning the laser (Na Fluorescein) at λ_3. The dashed line gives the one-photon absorption spectrum in the same VUV frequency range.[37]

[37] W. R. S. Garton, G. L. Grasdalen, W. H. Parkinson, and E. M. Reeves, *J. Phys. B* **1**, 114 (1968).

[38] U. Fano, *Phys. Rev.* **124**, 1866 (1961).

the lineshape for VUV generation depends on two q parameters, one involving the transition from the ground state (level 0) to the autoionizing level (level 3), and the other involving the transition from the resonant intermediate state (level 2) to the autoionizing level.[39,40] Thus, by using VUV generation as a method of studying the autoionizing level, one may learn more about that level than by only doing one-photon ionization or absorption studies.

A specific case has been studied by Armstrong and Wynne.[39] They looked at the autoionizing level of Sr labeled $4d(^2D_{5/2})4f[3/2]_1^0$ by Garton et al.[37] This level is seen centered at 186.8 nm in one-photon photoabsorption studies.[37] This level resonantly enhances the VUV signal generated when the first laser (ν_1) sits on a two-photon resonance. In particular, Armstrong and Wynne[39] studied the VUV line shape when the $5p^2$ $^1D_2$ or the $5s5d$ $^1D_2$ was used as the resonant intermediate level (level 2 in Fig. 1). Their results are shown in Fig. 8, where it is clear that the line shapes from the two different intermediate levels are different. Furthermore, these lineshapes differ from the one-photon absorption lineshapes.[37]

Before comparing these experimental results to theory, the analytical formulas that express the lineshape in terms of Fano q parameters are pre-

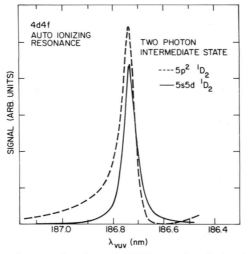

FIG. 8. VUV signal as a function of VUV wavelength in the vicinity of the $4d(^2D_{5/2})4f[3/2]_1^0$ state.[37] The two scans correspond to two different intermediate states, the $5s5d$ $^1D_2$ state at 34727.5 cm^{-1} and the $5p^2$ $^1D_2$ state at 36960.9 cm^{-1}.

[39] J. A. Armstrong and J. J. Wynne, *Phys. Rev. Lett.* **33**, 1183 (1974).
[40] L. Armstrong, Jr. and B. Beers, *Phys. Rev. Lett.* **34**, 1290 (1975).

sented. Interested readers should refer to the original papers[38-40] for details of the derivations. The lineshape for one-photon absorption from the ground state (level 0) to the autoionizing state (level 3) is given by

$$(q_{03} + x)^2/(1 + x^2), \tag{10.2.3}$$

where q_{03} is the Fano q parameter for the transition from 0 to 3, and $x = (\nu - \nu_{30})/(\Gamma/2)$ is the amount of detuning from resonance normalized to the half-width $\Gamma/2$ of the line. The frequency of the transition is given by ν_{30}. In contrast, the lineshape for the VUV-generated signal is given by

$$[(q_{03}q_{23})^2 + (x + q_{03} + q_{23})^2]/(1 + x^2), \tag{10.2.4}$$

where q_{23} is the Fano q parameter for the transition from 2 to 3 and $x = (\nu_{VUV} - \nu_{30})/(\Gamma/2)$.

From fitting the lineshape of the one-photon photoabsorption signal to Eq. (10.2.3), Garton et al.[37] found $q_{03} = -3.5$ with a linewidth $\Gamma = 20.2$ cm^{-1}. Using this value for q_{03}, Armstrong and Beers[40] fit the VUV lineshapes of Armstrong and Wynne[39] to Eq. (10.2.4) and found $q_{23} = 2.1$ when level 2 was the 5s5d $^1D_2$ and $q_{23} = -0.6$ when level 2 was the 5p^2 $^1D_2$. In each case the linewidth was taken as $\Gamma = 17$ cm^{-1}, in good agreement with Garton et al.[37] The results of these fits are not reproduced here. The fits are so good that an illustration does not show any difference between theory and experiment.

Thus, in the case of isolated, autoionizing level resonances, where the model of Fano[38] is expected to apply, VUV generation spectroscopy is another tool for determining Fano q parameters and characterizing the autoionizing levels. The use of many intermediate levels of correct parity and J can provide many separate values of q, thereby aiding in the identification of the autoionizing line. The use of this method by Armstrong and Wynne[39] is just the first example of what should become a widely used tool for VUV spectroscopy.

10.2.5. Multiphoton Absorption and Ionization Spectroscopy

When the sum of input photon energies equals the energy of an allowed multiphoton transition from a populated (ground) state to an excited state, this excited state will become populated. The presence of such a state can be detected by looking for decreased transmission of the light (absorption spectroscopy) or by detecting an ionization signal when the excitation of the state leads to ionization. If the energy of the excited state is $>50,000$ cm^{-1} above the ground state, one has a method of doing VUV spectroscopy. If the lasers are individually tuned onto resonance with one-photon transitions, the method of multiphoton excitation becomes a method of excited-state spectroscopy from well-defined excited states.

Laser-assisted, excited-state spectroscopy has become a widely used experimental method. The spectrum of the 6snd $^1D_2$ states of Ba has been studied by Bradley *et al.*[41] They excited the 6s6p $^1P_1{}^0$ level of Ba with a tunable dye laser and then looked at absorption originating from this state via transitions to the higher 6snd $^1D_2$. Rubbmark, Borgström, and Bockasten[42] have extended this method to observe the 6sns $^1S_0$ and 6snd $^3D_2$ levels. The extension of this technique to the VUV has been done for the case of Mg by Bradley *et al.*,[25] who observed absorbing transitions from the 3s3p $^1P_1{}^0$ state to the 3snd $^1D_2$ levels. They also saw an autoionizing state identified as 3p^2 $^1S_0$. This technique was capable of reaching states as high as 68,000 cm^{-1} above the ground state.

In the technique of absorption spectroscopy from excited states, it is necessary to prepare enough atoms in the excited state to give rise to a detectable amount of absorption. In contrast, detection by ionization requires far fewer excited atoms. In the ideal case, there is no background ionization in the absence of laser excitation. With good ion detectors and a proper geometry, ions may be counted one at a time. This has formed the basis for the technique of laser-assisted ionization spectroscopy. Using a space-charge ionization detector and a single dye laser, Popescu *et al.*[43] measured the two-photon excitation spectrum of Cs by three-photon ionization, resonantly enhanced by two-photon resonances. More recently, Esherick *et al.*[23] have used a simple ionization probe at low collecting voltage to measure the two-photon excitation spectrum of Ca. The ionization that follows excitation of bound states is thought to proceed via chemi-ionization, alternatively called associative ionization.[44] The same technique has been extended to Sr, where, in addition to bound states, an even-parity autoionizing state has been seen.[45] The extension of this technique to the VUV has been carried out by Wynne,[24] who has looked at even-parity autoionizing states of Ca between 50,000 and 63,000 cm^{-1} above the ground state.

The technique of multiphoton ionization spectroscopy is relatively simple to implement and capable of uncovering all sorts of new spectroscopic data. A brief description of the experimental method used to study even-parity autoionizing states in Ca will be presented along with some

[41] D. J. Bradley, P. Ewart, J. V. Nicholas, and J. R. D. Shaw, *J. Phys. B* **6**, 1594 (1973).
[42] J. R. Rubbmark, S. A. Borgström, and K. Bockasten, *J. Phys. B* **10**, 421 (1977).
[43] D. Popescu, C. B. Collins, B. W. Johnson, and I. Popescu, *Phys. Rev. A* **9**, 1182 (1974).
[44] A. Fontijn, *Prog. React. Kinet.* **6**, 75 (1971).
[45] P. Esherick, *Phys. Rev. A* **15**, 1920 (1977); P. Ewart and A. F. Purdie, *J. Phys. B* **9**, L437 (1976).

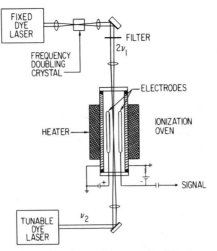

FIG. 9. Experimental configuration for studying autoionizing states. The filter blocks the laser (ν_1) and transmits the second harmonic ($2\nu_1$).

results.   The basic experimental setup is depicted in Fig. 9.   A dye laser (ν_1) is frequency doubled in a phase-matched ADP crystal.   The second harmonic frequency ($2\nu_1$) is tuned near a single-photon transition from the $4s^2\ ^1S_0$ ground state to either the $4s5p\ ^1P_1^0$ or $3d4p\ ^1P_1^0$ level.   This second-harmonic beam is focused into the pipe containing hot Ca vapor and electrodes for detecting ionization.[46]   A second dye laser (ν_2) is focused into the oven from the opposite end.   The beam is adjusted to give good spatial overlap with the first beam.   Then the wavelength of the second dye laser is scanned and the ionization spectrum is recorded.   With the first laser set so its second harmonic was near 227.6 nm ($43933.3\ cm^{-1}$), the second laser was scanned over a wide range from 14,000 to 20,000 cm^{-1}.   Strong peaks in the ionization signal were seen. A sample spectrum is shown in Fig. 10.   The peaks are due to enhanced ionization when the sum of the two input frequencies is resonant with even-parity, $J = 0, 1$, and 2 autoionizing states.   The full details of the assignments of these peaks have not been worked out although it is known that they correspond to 3dns, 3dnd, and 4pnp configurations interacting with $4s\epsilon s$ and $4s\epsilon d$ continua.

   Figure 10 dramatically shows how the technique of multiphoton ionization can be used to measure spectra in the VUV region.   The extension of this technique to three and more photon ionization spectroscopy is rela-

[46] For a more detailed description of the oven, see J. A. Armstrong, P. Esherick, and J. J. Wynne, *Phys. Rev. A* **15**, 180 (1977).

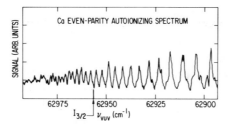

FIG. 10. The even-parity autoionizing spectrum of Ca. Data are the ionization signal resulting from photoionizing Ca atoms that have been photoexcited to the 4s5p $^1P_1^0$ state at 36731.6 cm^{-1}. The ionization limit labeled $I_{3/2}$ is the energy of the lowest excited state of the Ca II ion, the 3d $^2D_{3/2}$ state at 62956 cm^{-1}.

tively straightforward. The technique has the advantage of utilizing the narrow spectral linewidth of the lasers to achieve high resolution well into the VUV.

10.2.6. Conclusion

This chapter has tried to give the reader a perspective on the techniques of laser-assisted VUV spectroscopy. The methods discussed are new and only beginning to be fully exploited. The techniques of VUV generation and multiphoton excitation have given the experimental spectroscopist a new handle on the VUV region. The combination of spectral brightness and well-defined path of excitation, which are germane to laser-assisted spectroscopy, has made possible the detection and identification of many atomic energy levels that were not identifiable by methods of conventional spectroscopy.

While little has been said about the study of molecules and ions by these techniques, it is expected that the same positive results will follow when these tools are applied more extensively to molecular and ionic media. Recently, Provorov, Stoicheff, and Wallace[47] have carried out fluorescence studies of CO using VUV generated by optical sum mixing in Mg vapor. They tuned the VUV to selectively excite individual rotational levels within the (0, 0) vibrational manifold the lowest allowed electronic transition in CO. Much work has been done on two-photon excitation of states in molecules.[48] Faisal, Wallenstein, and Zacharias[49] have studied three-photon excitation of CO by observing the decay of the excited states by VUV fluorescence.

In the near future we expect to see the development of more tunable

[47] A. C. Provorov, B. P. Stoicheff, and S. Wallace, *J. Chem. Phys.* **67**, 5393 (1977).
[48] See, for example, P. M. Johnson, *J. Chem. Phys.* **62**, 4562 (1975).
[49] F. H. M. Faisal, R. Wallenstein, and H. Zacharias, *Phys. Rev. Lett.* **39**, 1138 (1977).

lasers in the near-UV and VUV. For example, the ArF laser output is centered at 193.3 nm and should be tunable over 1 nm.[50] The rare gas halide lasers should provide good pumps for fluorescent compounds such as organic dyes and scintillators, which will be tunable in the near-UV.

As a tool for VUV spectroscopic studies, tunable laser-generated VUV will have to compete with synchotron radiation, which provides widely tunable VUV. However, the laser techniques have the great advantage of being implemented in a modest size laboratory at a relatively modest cost. Synchotron radiation requires the large-size installation and cost of a synchotron, which limits its availability to large, well-funded, national-type laboratories.

10.2.7. Further Developments

Since this chapter was originally written, several developments worth mentioning have occurred.

In the area of sources, coherent VUV radiation, tunable to 185 nm, has been generated by sum mixing in the crystal $KB_5O_8 \cdot 4H_2O$.[51] Also various fifth-order mixing processes in He and Ne have produced step-tunable VUV near 60 nm.[52] Sum-mixing in Sr vapor has been considerably enhanced by careful choice of the intermediate state and by phase matching,[53] permitting observation of cw coherent VUV generation for the first time.[54] A new type of source has recently been realized. This source is based on laser-induced emission on a two-photon transition and has produced strong emission from He near 60 nm.[55] This source is spectrally narrow and should be widely tunable, making it useful for spectroscopic purposes.

In the area of spectroscopy, Freeman and Bjorklund[56] have studied the effects of electric fields on autoionizing states in Sr using three-photon ionization techniques. They saw significant changes in line shapes and energies, controlled by external electric fields. Using mode-locked dye

[50] For a recent review of rare-gas halide lasers, see J. J. Ewing, *Phys. Today,* 32 (1978).

[51] K. Kato, *Appl. Phys. Lett.* **30,** 583 (1977); R. E. Stickel, Jr. and F. B. Dunning, *Appl. Opt.* **17,** 981 (1978).

[52] C. Y. She and J. Reintjes, *Appl. Phys. Lett.* **31,** 95 (1977).

[53] G. C. Bjorklund, J. E. Bjorkholm, R. E. Freeman, and P. Liao, *Appl. Phys. Lett.* **31,** 330 (1977).

[54] R. R. Freeman and G. C. Bjorklund, *Int. Quantum Electron. Conf. 10th, 1978* Paper L.2 (1978).

[55] L. J. Zych, J. Lukasik, J. F. Young, and S. E. Harris, *Phys. Rev. Lett.* **40,** 1493 (1978).

[56] R. R. Freeman and G. C. Bjorklund, *Phys. Rev. Lett.* **40,** 118 (1978).

lasers, Royt and Lee[57] have carried out time resolved spectroscopic studies on autoionizing states in Sr. Finally, atomic H and O have been detected by three-photon ionization spectroscopy for concentrations as low as 4×10^9 atoms/cm³ in the presence of 10^{17} atoms/cm³ buffer gas.[58]

[57] T. R. Royt and C. H. Lee, *Appl. Phys. Lett.* **30,** 332 (1977).

[58] G. C. Bjorklund, C. P. Ausschnitt, R. R. Freeman, and R. H. Storz, *Appl. Phys. Lett.* **33,** 54 (1978).

10.3. Doppler-Free Laser Spectroscopy[*]

10.3.1. Introduction

Since the invention of the laser[1] in 1960, many new spectroscopic techniques that take advantage of the laser's unique capabilities have been developed. It is now possible to produce tunable, narrow-band laser radiation over nearly all of the optical spectrum, and applications of lasers to atomic and molecular spectroscopy are extremely diverse. We shall limit our brief discussion to three methods of laser spectroscopy that enable one to obtain optical spectra of atomic or molecular vapors without Doppler broadening. These methods are saturation spectroscopy, two-photon spectroscopy, and laser-induced line narrowing. We shall describe some of the experimental and theoretical aspects of these techniques with emphasis on the experimental methods that have been utilized. No attempt will be made to completely cover the vast amount of experimental and theoretical work that has been done. Each technique makes use of the laser's high intensity, low divergence, and spectral purity. The high intensity allows one to perform the nonlinear processes implicit in the words saturation, two-photon, and laser-induced. The low divergence enables one to precisely define the direction of propagation of the laser beams as is required for the elimination of the first-order Doppler effect. Spectral purity is required in these techniques, since the spectra are obtained by scanning the laser frequency over the atomic resonances. Hence, highly stable, narrow-band lasers are necessary to obtain high-resolution spectra. The atomic energy levels are probed directly with the laser. In other techniques, such as rf and level-crossing spectroscopy, the laser acts merely as a source of pump light to populate certain states. These methods can provide extremely accurate information about splittings and fine structure without a narrow-band requirement on the laser. Direct laser spectroscopy, on the other hand, offers the ability to make measurements associated with the absolute energies of atomic levels

[1] T. H. Maiman, *Nature* (*London*) **187**, 493 (1960).

---

* Chapter 10.3 is by P. F. Liao and J. E. Bjorkholm.

METHODS OF EXPERIMENTAL PHYSICS, VOL. 15B

rather than the differences in energy between sublevels of a given state. In addition, the wide tuning ranges of many lasers make it possible to much more rapidly survey the structure of a state than is possible with rf or level-crossing techniques.

10.3.2. Saturation Spectroscopy

In 1964 W. E. Lamb published a theory of a gas laser[2] that predicted a decrease in the power output of the laser at the frequency corresponding to the peak of the atom's transition. This power dip has a width that is the order of the atomic linewidth, inclusive of collision broadening, but not Doppler broadening; and hence observation of such "Lamb dips" provides a method of obtaining high-resolution spectra without Doppler broadening.

The basic principle leading to the formation of the Lamb dip is very simple. Consider the velocity distribution of the atoms in a gas laser as shown in Fig. 1. The standing wave in the laser cavity can be decomposed into two oppositely directed running waves. Atoms with longitudinal velocities v^+ such that $|\omega_0 - (1 - v^+/c)\omega| \leq \Delta\omega$ are stimulated to emit into the forward traveling wave, while atoms with velocities v^- such that $|\omega_0 - (1 + v^-/c)\omega| \leq \Delta\omega$ will emit into the backward wave; where ω is the laser frequency, ω_0 is the atomic resonance frequency and $\Delta\omega$ is the width of the atomic resonance line exclusive of Doppler broadening. The dashed lines labeled a and b in Fig. 1 represent these two velocity groups. As ω approaches ω_0, v^+ and v^- both approach zero and the two groups overlap into one velocity group (the group centered around zero longitudinal velocity); hence the number of atoms that can emit into the laser fields decreases by approximately a factor of two. This decrease produces the Lamb dip. The dip occurs at the atomic resonance frequency ω_0 and its width is independent of Doppler broadening. The original Lamb dip technique was restricted to studies of laser transitions; how-

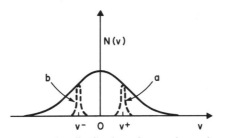

FIG. 1. Doppler distribution of atoms in gas laser.

[2] W. E. Lamb, *Phys. Rev.* **134**, A1429 (1964).

ever, in 1967 Lee and Skolnick[3] reported another version of the technique in which they placed an absorption cell into the laser cavity. The absorption of the gas in the cell has a dip similar to the Lamb dip at absorption line center ω_a. This dip in absorption leads to an increase in laser emission at this frequency and gives rise to the "inverted Lamb dip."

The Lamb dip technique has finally evolved into what is commonly known as saturation spectroscopy. In these experiments the gas to be studied is placed outside the laser cavity. A schematic diagram of a typical experimental setup is shown in Fig. 2. A strong saturating beam "burns a hole" in the Doppler-broadened line of the absorbing gas. The attenuation of a weaker probe beam that travels in the opposite direction is monitored as the frequency of the laser is varied. When tuned to line center, both laser beams interact with the same velocity group (zero longitudinal velocity) and the saturation produced by the strong beam produces a decrease in absorption of the probe beam.

The simple hole-burning model predicts that the change $\Delta\alpha_{ab}$ in the absorption coefficient α_{ab} for the probe beam is given by[4]

$$\Delta\alpha_{ab} = -\frac{1}{2}\,\alpha_{ab}\,\frac{I}{I_{sat}}\,\frac{\gamma_{ab}^2}{\gamma_{ab}^2 + (\Omega_{ab} - \omega)^2}, \qquad (10.3.1)$$

for $I \ll I_{sat}$; I is the strong saturating beam intensity, α_{ab} is the unsaturated absorption coefficient, γ_{ab} is the natural linewidth of the transition, $\hbar\Omega_{ab}$ is the energy separation of the two levels involved, and I_{sat} is the saturation intensity for the transition.

The saturating beam is usually chopped at audio frequencies and its effect on the probe beam is then monitored with phase-sensitive detection. Either the laser can be tuned or the sample resonance tuned[5] by application of magnetic or electric fields. Both pulsed and cw lasers have been used, with the greatest frequency resolution obtained with carefully stabi-

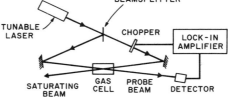

FIG. 2. Schematic diagram of an experimental setup for saturated absorption spectroscopy.

[3] P. H. Lee and M. L. Skolnick, *Appl. Phys. Lett.* **10**, 303 (1967).

[4] T. W. Hansch, M. D. Levenson, and A. L. Schawlow, *Phys. Rev. Lett.* **26**, 946 (1971).

[5] R. G. Brewer, M. J. Kelly, and A. Javan, *Phys. Rev. Lett.* **23**, 559 (1969).

lized cw lasers. In the case of a pulsed-laser experiment it is also possible to study linewidths as a function of time delay between the saturating and probe pulses and thereby obtain a lifetime associated with spectral diffusion.[6]

The techniques discussed to this point have all utilized equal-frequency laser beams propagating in opposite directions. In this case the nonlinear resonances are observed when the laser frequency ω equals the atomic transition frequency ω_0 and both laser beams interact with the $v = 0$ velocity group. A more generalized situation is one in which oppositely propagating beams of *unequal* frequency are used. Let ω_1 be the frequency of the strong saturating beam and ω_2 be the frequency of the probe beam. In this case also the nonlinear resonance occurs when both laser beams interact with the same velocity group of atoms, but $v \neq 0$. Instead, the beams interact with atoms for which the longitudinal velocity is

$$v = \left(\frac{\omega_0 - \omega_1}{\omega_1} \right) c \approx \frac{\omega_0 - \omega_1}{\omega_0} c. \qquad (10.3.2)$$

The importance of this fact is that by making measurements with different values for ω_1 one can interact with atoms having different rms velocities. Thus, for instance, if collision broadening is the dominant homogeneous broadening mechanism in an experiment, this technique allows one to study collision broadening as a function of velocity.[7]

Since saturation spectroscopy yields narrow resonances centered at atomic or molecular resonance frequencies, it is natural that these techniques be considered for use in precision frequency stabilization of lasers and in frequency standards work. In this regard it is of interest to mention mechanisms besides collisions that also lead to broadening or shifts of the observed resonances. These include power broadening, divergence and noncollinearity of the laser beams, transit time of the atoms through the laser beams, second-order Doppler effects, and recoil.

Several techniques have been devised to increase the sensitivity of saturation spectroscopy. Two of these methods are intermodulated fluorescence[8] and polarization spectroscopy.[9] The basic concept of intermodulated fluorescence is that when the absorption is reduced by saturation so also is the fluorescence. Therefore one can monitor the saturated absorption resonances by detecting changes in the fluorescence from the atomic

[6] T. W. Hansch, I. S. Shahin, and A. L. Schawlow, *Phys. Rev. Lett.* **27**, 707 (1972).

[7] A. T. Mattick, A. Sanchez, N. A. Kurnit, and A. Javan, *Appl. Phys. Lett.* **23**, 675 (1973); also see A. Javan, in "Laser Spectroscopy" (S. Haroche *et al.*, eds.), p. 440. Springer-Verlag, Berlin and New York, 1975.

[8] M. S. Sorem and A. L. Schawlow, *Opt. Commun.* **5**, 148 (1972).

[9] C. Wieman and T. W. Hänsch, *Phys. Rev. Lett.* **36**, 1170 (1976).

or molecular vapor.   In particular one modulates the two traveling waves at different frequencies and then records that component of the fluorescence that is modulated at either the sum or the difference frequency. Examination of Eq. (10.3.1) shows that the standard technique of saturation spectroscopy has a signal-to-background ratio in the limit $\alpha l \ll 1$, which is proportional to αl, where α is the absorption coefficient and l the length of the gas sample.   The fluorescence technique, on the other hand, has a signal to background ratio that is independent of αl since both the total fluorescence and the change in fluorescence due to saturation are each proportional to αl.   Hence the intermodulated fluorescence technique can have significantly better sensitivity for very weak absorption lines.   A typical experimental setup using this technique is shown in Fig. 3.

Another method of increasing the signal-to-background ratio is polarization spectroscopy.[9]   In this case the saturating beam's polarization is chosen such that the partial saturation of the transition results in a birefringent or optically active material.   As a result, the probe beam polarization is modified upon passage through the sample.   This modification results from both the polarization-dependent change in attenuation $\Delta\alpha$ due to the saturating beam given in Eq. (10.3.1) and the polarization-dependent change in the index of refraction Δn, which is also caused by the saturating beam.   The index change can be simply related to the absorption change with the Kramers–Kronig relationships.   The total resulting modification of the probe beam polarization is then de-

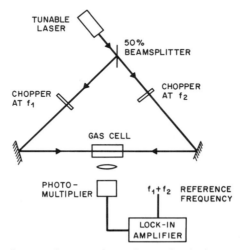

FIG. 3. Schematic diagram of an experimental setup for the intermodulated fluorescence technique of saturated absorption spectroscopy.

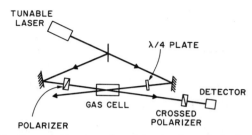

FIG. 4. Schematic diagram of an experimental setup for polarization spectroscopy.

tected with an analyzing polarizer. Figure 4 illustrates the scheme of such a polarization spectrometer. The saturating beam is circularly polarized and the probe beam is linearly polarized. The probe beam can be decomposed into two circularly polarized components, rotating in the same (+) and opposite (−) sense of the circularly polarized saturating beam. The saturating beam produces a saturation that is different for the two probe beam components. The difference in the attenuation $\Delta\alpha^+ - \Delta\alpha^-$ makes the probe beam elliptically polarized, while the difference $\Delta n^+ - \Delta n^-$ rotates the axis of polarization. If these changes in the polarization are small, the intensity of the probe beam that passes through the analyzing polarizer is given by

$$I = I_0 \left[\theta^2 + \frac{\omega}{c} (\Delta n^+ - \Delta n^-) \frac{l\theta}{2} + (\Delta n^+ - \Delta n^-)^2 \frac{l^2}{4} \right.$$

$$\left. + (\Delta\alpha^+ - \Delta\alpha^-)^2 \frac{l^2}{16} \right], \quad (10.3.3)$$

where θ is the small angle by which the analysis polarizer is rotated away from the perfectly crossed position. It has been shown[9] that

$$\frac{\Delta\alpha^-}{\Delta\alpha^+} = d = \begin{cases} 1 - 5/(4J^2 + 4J + 2), & \text{for } J = J', \\ (2J^2 + 5rJ + 3)/(12J^2 - 2), & J = J' + 1, \end{cases} \quad (10.3.4)$$

where $r = (\gamma_J - \gamma_{J'})/(\gamma_J + \gamma_{J'})$, J and J' are the angular momenta of the involved states, which have decay rate γ_J and $\gamma_{J'}$. Spontaneous reemission into the lower state has been ignored.

At the crossed position, $\theta = 0$, the combined effects of $\Delta\alpha$ and Δn produce a Lorentzian-shaped signal on a small background that is determined by the extinction ratio of the polarizer. This signal is, however, proportional to $(\Delta\alpha)^2$ and small signals can be better detected at a finite value for θ. In the limit that the last two terms in Eq. (10.3.3) can be ignored one finds a signal-to-background ratio given by

$$(\omega/2c)(\Delta n^+ - \Delta n^-)l\theta/(\theta^2 + \phi^2),$$

where ϕ^2 represents the finite extinction ratio of the polarizers. This signal-to-background ratio is maximum at $\theta = \phi$ and is improved over the normal technique of saturation spectroscopy by a factor $(1 - d)/8\phi$; values for this quantity in excess of 100 have been obtained.[9]

One rather annoying aspect of saturation spectroscopy is the existence of additional, spurious resonances known as crossover resonances. These signals occur halfway between any two resonance lines that share either a common upper or lower level. At this frequency the common level participates in both resonance transitions with one transition induced by the saturating beam and the other by the oppositely propagating probe beam. The effect of the saturating beam on the common level produces a change in the absorption of the probe beam and hence gives rise to the spurious resonances.

In this section we have used the "hole-burning" model to describe the results obtained in saturation spectroscopy. It must be cautioned, however, that for large intensities of the saturating beam this picture ceases to be valid. In this regime effects such as ac Stark splitting become important and the simple nonlinear resonance lines seen at low intensity become more complex and can even be observed to split.[10] Further discussion of these effects is beyond the scope of this chapter.

10.3.3. Two-Photon Absorption without Doppler Broadening

The first two-photon absorption experiment in an atomic vapor[11] came soon after the laser's discovery. However, it was not until recent years that the availability of widely tunable dye lasers made two-photon absorption a useful spectroscopic technique. In 1970, Vasilenko *et al.*[12] pointed out that it is possible to obtain two-photon absorption spectra without Doppler broadening. Such spectra are obtained with oppositely propagating laser beams. Consider an atom or molecule in a gas that is irradiated by two laser beams of frequency ν_1 and ν_2 propagating in opposite directions. In the rest frame of this atom the laser frequencies are Doppler shifted and appear as ν_1' and ν_2', where

$$\nu_1' = (1 + v/c)\nu_1, \qquad \nu_2' = (1 - v/c)\nu_2, \tag{10.3.5}$$

[10] S. Haroche and F. Hartmann, *Phys. Rev. A* **6**, 1280 (1972).

[11] I. D. Abella, *Phys. Rev. Lett.* **9**, 453 (1962).

[12] L. S. Vasilenko, V. P. Chebotaev, and A. V. Shishaev, *JETP Lett. (Engl. Transl.)* **12**, 161 (1970).

and v is the atom's velocity along the longitudinal direction. The sum frequency that determines the two-photon transition is given by

$$\nu = \nu_1' + \nu_2' = \nu_1 + \nu_2 + (\nu_1 - \nu_2)v/c. \qquad (10.3.6)$$

If $\nu_1 = \nu_2$, the sum frequency is independent of atomic velocity and the first-order Doppler effect has been eliminated; all atoms in the Maxwellian velocity distribution can be made simultaneously resonant, in distinct opposition to the case of saturation spectroscopy. Most Doppler-free two-photon experiments to date have utilized equal-frequency photons absorbed from oppositely propagating beams. With the use of equal-frequency photons, however, there generally is present another two-photon line, which is fully Doppler broadened, caused by absorbing the two photons needed to make the transition from the same beam. For this case only a small fraction of the atoms in the velocity distribution are simultaneously resonant and consequently the strength of this two-photon line is much weaker than that of the Doppler-free line. Quantitatively, the area under the Doppler-free line is twice that under the Doppler-broadened line. In some cases polarization selection rules that restrict the absorption to one photon from each beam can be used to eliminate the Doppler-broadened background.[13]

In some cases it may be advantageous to use counterpropagating beams of unequal frequency since that will allow one to greatly enhance the two-photon transition rate by tuning one laser near an allowed intermediate-state transition. In this case, of course, there is residual Doppler broadening since the factor $(\nu_1' - \nu_2)$ in Eq. (10.3.6) is nonzero. The two-photon transition rate W for transitions from the ground state g to the excited state f is proportional to

$$W \propto I_1 I_2 \left| \sum_n \mu_{gn}\mu_{nf}[(E_n - h\nu_1)^{-1} + (E_n - h\nu_2)^{-1}] \right|^2, \qquad (10.3.7)$$

where I_1 and I_2 are the intensities of the two counterpropagating beams with frequencies ν_1 and ν_2, respectively. The summation is over all possible intermediate states n having energies E_n, and μ_{gn}, μ_{nf} are the dipole matrix elements between the intermediate state and the ground or excited state. The resonance denominators give rise to the increase in transition rate that is observed[14] when one laser frequency is tuned near an intermediate-state resonance. The maximum increase occurs at exact intermediate-state resonance and this situation, which may be called

[13] B. Cagnac, G. Grynberg, and F. Biraben, *J. Phys.* (*Paris*) **34**, 845 (1973).
[14] J. E. Bjorkholm and P. F. Liao, *Phys. Rev. Lett.* **33**, 128 (1974).

two-photon spectroscopy with a resonant intermediate state or optical double resonance, is discussed in the next section. In sodium vapor a maximum resonant enhancement of 6×10^9 was demonstrated for the on-resonance case[15]; with the laser tuned several Doppler widths off resonance with the ground-state to intermediate-state transition frequency, enhancements of up to 10^6 were observed.[14] In many cases the residual Doppler broadening (reduced resolution) is a small price to pay for the increased signal strength. In the next section we point out that for the on-resonance case Doppler-free spectra are obtained even though unequal-frequency photons are employed.

Because the transition rate is proportional to the product of the two beam intensities one can attempt to increase the transition rate by focusing tightly. However, the focus of a Gaussian beam can only be held over approximately the distance of the confocal parameter $b = 2\pi w_0^2/\lambda$, where w_0 is the radius of the focal spot. The region that effectively contributes to the two-photon absorption can be thought of as a cylindrical volume of length b and cross-sectional area πw_0^2. The total transition rate is therefore approximately proportional to

$$W_{\mathrm{T}} \propto \frac{P_1}{\pi w_0^2} \cdot \frac{P_2}{\pi w_0^2} \cdot b \cdot \pi w_0^2 \propto P_1 P_2 \frac{2}{\lambda}, \qquad (10.3.8)$$

where we have assumed the length l of the vapor cell is equal to or greater than b, and the two beams with powers P_1 and P_2 have the same focal spot sizes and wavelengths. Since the result is independent of w_0, in most cases one will want to focus such that b is not shorter than l. Tighter focusing will only give rise to loss in resolution due to the increased divergence of the beams and increased transit time broadening. In addition, tight focusing may give rise to optically induced energy level shifts, as discussed later.

A typical experimental setup is shown in Fig. 5. The output from a cw dye laser is focused into a vapor cell and reflected back on itself with a mirror. To eliminate feedback of the reflected light into the laser cavity,

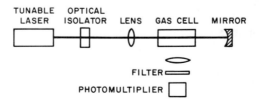

FIG. 5. Experimental setup for Doppler-free two-photon spectroscopy.

[15] J. E. Bjorkholm and P. F. Liao, *Phys. Rev. A* **14**, 751 (1976).

which would disturb the frequency stability of the dye laser, an optical isolator must be placed into the optical path as illustrated. The effects of feedback can also be eliminated by mounting the mirror on a vibrating loudspeaker.[16] Although this mirror retroreflects the light, the phase of the back-reflected light is rapidly modulated and apparently does not upset the dye laser. In the case of unequal-frequency beams the mirror is replaced by the second laser and a focusing lens. This second laser is usually held at a fixed wavelength while the first is scanned. Both pulsed[17] and cw[16] dye lasers have been used, with highest resolution usually obtained with cw lasers.

The two-photon transitions are usually monitored by observing the fluorescence that occurs as a result of the excitation of the final state. This technique provides an extremely sensitive method of detection since the fluorescence occurs at a different wavelength than the excitation and is usually observed only at the two-photon resonance. In some cases, however, the observation of fluorescence may be difficult. The sample may itself fluoresce strongly as, for example, in a discharge, or the fluorescence might be extremely weak or difficult to detect. In these cases it may be advantageous to monitor the transitions by detecting the change in transmission for one of the beams due to the two-photon resonances. Recent two-photon spectroscopy of infrared transitions of CH_3F and NH_3 were done in this manner.[18] In order to make the change in transmission detectable, these experiments also made use of a nearly resonant intermediate state by using unequal-frequency photons. In spite of a residual Doppler broadening of about 1 MHz, the resolution was more than adequate to resolve the spectra.

A possible alternative approach for monitoring two-photon transitions is to use two-photon polarization rotation.[19] This technique is similar to polarization saturation spectroscopy in that one beam causes the polarization of the second beam to be modified and this polarization modification is then detected through a crossed or partially crossed polarizer. The description of the polarization modification is basically identical to the one used for saturation spectroscopy except that the induced changes, $\Delta\alpha$ and Δn, are due to the two-photon susceptibility. Because the two-photon absorption line is Doppler free, the changes $\Delta\alpha$ and Δn are Doppler free; hence the signal detected through the analyzing polarizer is also Doppler free. This technique has the advantages of high signal-to-

[16] T. W. Hänsch, K. C. Harvey, G. Meisel, and A. L. Schawlow, *Opt. Commun.* **11,** 50 (1974).

[17] M. D. Levenson and N. Bloembergen, *Phys. Rev. Lett.* **32,** 645 (1974).

[18] W. K. Bischel, P. J. Kelly, and C. K. Rhodes, *Phys. Rev. A* **13,** 1817 and 1829 (1976).

[19] P. F. Liao and G. C. Bjorklund, *Phys. Rev. Lett.* **36,** 584 (1976).

background ratio and the ability to discriminate against background fluorescence since it is possible to spatially and spectrally isolate the probe beam.

Another detection technique that does not rely on the observation of fluorescence is the detection of ions. The ions can be produced by Stark-ionizing the two-photon excited states.[20] Because the two-photon excited states often lie close to the ionization limit, the Stark ionization is easily accomplished. In fact in some experimental situations, the ions are observed even in the absence of external fields.[21] Recently, Harvey et al.[22] reported the use of a detector that measured the effect of the highly excited atoms on the current of a space-charge-limited diode. Such a detector had nearly 100% detection efficiency.

The extremely narrow linewidths that, in principle, can be obtained using Doppler-free two-photon spectroscopy make ultraprecise measurements an attractive possibility. In making such measurements, however, care must be taken to account for the possibility of ac Stark shifts and splittings of the resonance lines. These effects are due to virtual transitions between energy levels caused by nonresonant light and, as such, are intrinsic to multiphoton processes. For the simple situation shown in Fig. 6 the shift of the energy levels $\delta\nu$ is given by

$$\delta\nu = \frac{1}{4} \frac{|\mathbf{p}_{12} \cdot \mathbf{E}_0|^2}{h^2(\nu_0 - \nu)}, \qquad (10.3.9)$$

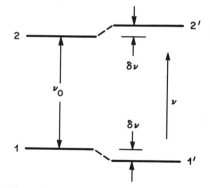

FIG. 6. Level diagram illustrating optically induced level shifts of magnitude $\delta\nu$ produced by light of frequency ν.

[20] T. Ducas, M. Littman, R. R. Freeman, and D. Kleppner, *Phys. Rev. Lett.* **35,** 366 (1975).

[21] P. Esherick, J. A. Armstrong, R. W. Dreyfus, and J. J. Wynne, *Phys. Rev. Lett.* **36,** 1296 (1976).

[22] K. C. Harvey, and B. P. Stoicheff, *Phys. Rev. Lett.* **38,** 537 (1977).

where $\mathbf{p}_{12}$ is the dipole matrix element connecting the two levels. This expression is valid for $|\delta\nu/(\nu_0 - \nu)| \ll 1$. The more general expression for the shift is

$$\frac{\delta\nu}{(\nu_0 - \nu)} = \frac{1}{2}\left(1 + \frac{|\mathbf{p}_{12}\cdot\mathbf{E}_0|^2}{h^2(\nu_0 - \nu)^2}\right)^{1/2} - \frac{1}{2}. \qquad (10.3.10)$$

It has been shown that ac Stark shifts can be significant even when using cw lasers to induce the two-photon transitions.[23] Saturation of the two-photon transition is *not* necessary for these shifts to be important.[23] In many cases it may be necessary to extrapolate measured resonance frequencies to zero ac Stark shift in order to determine precise values.[18]

10.3.4. Laser-Induced Line Narrowing

In saturation spectroscopy one observes the gain or loss profile experienced by a weak probe beam scanned over a Doppler-broadened transition in the presence of a strong collinear saturating beam tuned onto the same transition. Sharp resonances are observed when both beams interact with the same velocity group. Similar but more complicated behavior is observed if the collinear pump and probe beams are tuned onto different but *coupled* transitions. By the word coupled, we mean that the two transitions have a common energy level, as shown in Fig. 7a, where the (0–1) and (0–2) transitions are coupled by the common level 0. This situation in which the pump and probe beams are tuned onto different coupled Doppler-broadened transitions is often referred to as "laser-induced line narrowing." In order to illuminate the important qualitative difference between saturation spectroscopy and laser-induced line narrowing, consider the absorption experienced by the probe in Fig. 7a as its frequency is scanned over both the (0–2) and (0–1) transition frequencies. We assume that both transitions are absorbing, that the pump is a standing wave slightly detuned from ν_{02}, and, for simplicity, that levels 1 and 2 are closely spaced. The resulting absorption spectrum is shown in Fig. 7b. The absorption spectrum observed for probe frequencies in the vicinity of ν_{02} is the saturation spectrum. The two sharp saturation resonances are equally wide. The linewidth Γ_S is determined by the natural width (or, more generally, the homogeneous width) of the (0–2) transition. Let γ_i be the decay rate of level i; then, $\Gamma_S = \gamma_0 + \gamma_2$. In marked contrast, the two laser-induced line-narrowed resonances observed on the coupled transition for probe frequencies in the vicinity of ν_{01} are unequal. For the simple situation being considered here, the narrow

[23] P. F. Liao and J. E. Bjorkholm, *Phys. Rev. Lett.* **34**, 1 (1975); J. E. Bjorkholm and P. F. Liao, *in* "Laser Spectroscopy" (S. Haroche *et al.*, eds.), p. 176. Springer-Verlag, Berlin and New York, 1975.

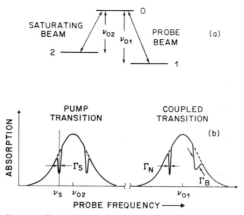

FIG. 7. (a) Level diagram for a coupled three-level system. The saturating and probe beams connect different levels to the common level 0. (b) Absorption spectrum seen by the probe light as a function of the probe frequency.

linewidth is $\Gamma_N = \gamma_1 + \gamma_2$ and the broad linewidth is $\Gamma_B = \gamma_1 + \gamma_2 + 2\gamma_0$. For the case shown in Fig. 7a, the narrow line is obtained when the pump and probe are copropagating and the wide line is obtained for oppositely propagating pump and probe. It should be cautioned, however, that this association is not a generality and depends upon the type of energy level configuration being considered.

The three different energy level configurations that are dealt with in laser-induced line-narrowing are shown in Fig. 8. The generalized theory applying to all three situations has been extensively treated.[24] We shall not attempt a thorough discussion of the generalized results here. Instead, later we shall present the qualitative results only for the case shown in Fig. 8c, the cascade configuration. We discuss the problem from the standpoint of on-resonance two-photon spectroscopy.[15]

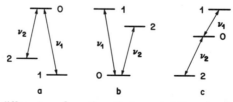

FIG. 8. The three different configurations that coupled three-level systems can take.

[24] For example, see M. S. Feld and A. Javan, *Phys. Rev.* **177,** 540 (1969); T. Hansch and P. Toschek, *Z. Phys.* **236,** 213 (1970); I. M. Beterov and V. P. Chebotaev, *Prog. Quantum Electron.* **3,** Part 1 (1974).

As in saturation spectroscopy, other techniques can be employed to observe laser-induced line-narrowed resonances. For instance, with reference to Fig. 7a, the earlier discussed techniques of intermodulated fluorescence or polarization rotation could be used as the probe frequency is scanned in the vicinity of ν_{01}. The Doppler-free resonances can be monitored without the need for a probe beam by monitoring along the direction of the pump beam the spectrum of the fluorescence given off spontaneously in the vicinity of ν_{01}.[25] This technique is called laser-induced fluorescent line narrowing. The narrow resonances are most easily detected when one level is the ground state and the other two levels are excited states as in Fig. 8c; in this case the fluorescence originating from level 1 is monitored as the probe frequency is scanned over the (0–1) transition frequency. This particular case will be discussed in more detail shortly.

Inherent in the technique of laser-induced line narrowing is the ability to interact with velocity groups having nonzero velocity along the direction of the beams. As previously discussed, this allows study of the velocity dependence of various line-broadening mechanisms.[7]

Laser-induced fluorescent line narrowing has also been employed to great advantage in solid-state spectroscopy. In solids, of course, the inhomogeneous broadening mechanism is not the Doppler effect since the atoms are stationary; instead the transitions are broadened because of the different local fields seen at different sites within the solid. Nonetheless, the same qualitative ideas apply. Instead of selecting a given velocity group, in a solid one selects a given local field group. The technique was first applied to the study of ions in a crystalline host[26] and later to glasses.[27] Thus the technique allows the selective study of the various classes of ion sites within the solid.

In order to further elucidate specific features of laser-induced line narrowing in gases we specifically consider the passage from the situation shown in Fig. 9, where $\Delta\nu \gg \Delta\nu_D$, to the cascade scheme shown in Fig. 8c, for which $\Delta\nu \ll \Delta\nu_D$; $\Delta\nu$ is the detuning of ν_2 from the g $\rightarrow$ i resonance frequency and $\Delta\nu_D$ is the Doppler width (FWHM) of that transition. In this way we shall be able to make contact with two-photon spectroscopy discussed in Section 10.3.3. Figure 10 illustrates lineshapes that can be expected in the limit of low optical intensities and no collisions; this problem is treated in detail in Bjorkholm and Liao.[15] The more gen-

[25] For example, see R. H. Cordover, P. A. Bonczyk, and A. Javan, *Phys. Rev. Lett.* **18**, 730, 1104 (E) (1967); W. G. Schweitzer, Jr., M. M. Birky, and J. A. White, *J. Opt. Soc. Am.* **57**, 1226 (1967); H. K. Holt, *Phys. Rev. Lett.* **20**, 410 (1968).

[26] A. Szabo, *Phys. Rev. Lett.* **25**, 924 (1970); **27**, 323 (1971).

[27] L. A. Riseberg, *Phys. Rev. Lett.* **28**, 786 (1972).

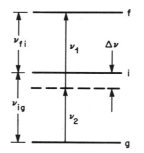

FIG. 9. A three-level system in which two-photon transitions from g → f are induced by two light beams with frequencies ν_1 and ν_2.

eral situations, which include collisions[28] and both collisions and arbitrary light intensities,[29] have also been treated. The spectra in Fig. 10 assume that ν_2 is fixed, that ν_1 is scanned over the resonance, and that the two-photon transitions are detected by monitoring the fluorescence from the final state f.

The generalized spectrum is seen to be composed of two types of lines. The first type predominates for the off-resonance case, $\Delta\nu \gg \Delta\nu_D$, and was discussed in Section 10.3.3. This two-photon line occurs where $\nu_1 + \nu_2 = \nu_{fg}$ (aside from small frequency-pulling effects), it has a lineshape

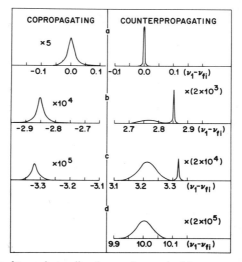

FIG. 10. Generalized two-photon lineshapes observed with copropagating and with counterpropagating light beams. The frequency ν_2 is fixed and ν_1 is scanned. Lineshapes are shown for various values of $\Delta\nu/\Delta\nu_D$ as follows: (a) 0, (b) -1.39, (c) -1.62, (d) -6.06.

[28] P. R. Berman, *Phys. Rev. A* **13**, 2191 (1976).

[29] R. Salomaa and S. Stenholm, *J. Phys. B* **8**, 1795 (1975); **9**, 1221 (1976).

described by the Voigt profile (a Gaussian convolved with a Lorentzian), and it has a Doppler-broadened linewidth [full-width at half-maximum (FWHM)] given approximately by

$$|(\nu_2 + \epsilon\nu_1)/\nu_{ig}|\Delta\nu_D + \Delta\nu_f, \qquad (10.3.11)$$

where $\Delta\nu_f$ is the natural linewidth of the final state. The parameter ϵ equals $+1$ or -1 depending upon whether the laser beams are copropagating or counterpropagating, respectively. In Fig. 10, this Doppler-broadened line is so broad and its peak intensity so low for the case of co-propagating beams that it does not appear visible in the figure.

For $\nu_2 = \nu_1$ and using counterpropagating beams, Doppler broadening is eliminated to first order. The reason for using unequal-frequency photons is that in many situations the two-photon transition rate can be increased many orders of magnitude by using resonant enhancement, making $\Delta\nu$ small. Maximum resonant enhancement occurs for $\Delta\nu \ll \Delta\nu_D$ and for this case the second type of two-photon line predominates. Importantly, this line is Doppler free; its lineshape is Lorentzian and its linewidth (FWHM) is given by

$$|(\nu_2 + \epsilon\nu_1)/\nu_{ig}|\Delta\nu_i + \Delta\nu_f, \qquad (10.3.12)$$

where $\nu_2 \approx \nu_{ig}$ and $\Delta\nu_i$ is the natural linewidth of the intermediate state. It is interesting to note that the linewidth of the two-photon spectrum for oppositely propagating beams is fairly insensitive to $\Delta\nu_i$. In particular, we can infer that it is only *slightly* dependent upon power broadening of the intermediate state. The exact sum frequency $\nu_1 + \nu_2$ at which this line appears depends upon ν_2, although $\nu_1 + \nu_2 \approx \nu_{fg}$. For the intermediate situation $\Delta\nu \approx \Delta\nu_D$ the two types of lines are of comparable magnitude and the two-photon lineshape is a more complicated combination of the two types of lines. The most important point to make here is that for the on-resonance situation $\Delta\nu \ll \Delta\nu_D$, maximum transition rates and Doppler-free spectra are simultaneously obtained.

In situations of practical interest the atom being investigated is usually not a simple three-level atom as shown in Fig. 9; in fact, often it is the fine or hyperfine structure of the various states that is to be determined. This multiplicity of the various levels makes the two-photon spectrum more complicated. The important point here is that the qualitative features of the spectrum, which we just discussed, remain basically the same. When the lasers are tuned far from intermediate-state resonance ($\Delta\nu \gg \Delta\nu_D$), the two-photon spectrum consists of several Doppler-broadened lines (full or residual Doppler broadening), the number and splittings of which are determined by the ground- and final-state level structure. The on-resonance case is more complicated and the form of the two-photon spectrum also depends upon the intermediate-state level structure. Recall that for the on-resonance case ($\Delta \nu \ll \Delta \nu_D$), the ν_2 beam has the effect of selecting the velocity group of atoms that is resonant with a ground-to-intermediate-state transition. If the ground and intermediate states have structure, a different velocity group will be selected by a beam at fixed ν_2 for each possible ground-to-intermediate-state transition. Hence when ν_1 is swept, two-photon resonances occur for each different velocity group. The difference in the Doppler shifts of the velocity groups causes a ground-state splitting $\delta \nu_g$ to appear in the spectrum as a splitting of $(\nu_1/\nu_2)\delta \nu_g$. Similarly, structure in the intermediate state will give rise to a splitting in the spectrum of $\delta \nu_i (\nu_2 + \epsilon \nu_1)/\nu_2$. For counterpropagating beams and $\nu_2 \sim \nu_1$, the resultant lines tend to merge together. For co-propagating beams the splitting is much larger and the on-resonance two-photon spectrum can yield Doppler-free information about the intermediate-state level structure. Thus, using on-resonance two-photon spectroscopy in a vapor, one can obtain Doppler-free spectra of the inter-mediate state with a resolution that rivals the best that can be obtained using single-photon absorption and well-collimated atomic beams. This has been experimentally demonstrated.[15]

The on-resonance situation has been investigated in both sodium vapor[15] and neon.[30] For the 3S $\rightarrow$ 4D two-photon transition in sodium vapor the signal strength for the on-resonance case was 6×10^9 larger than for the case of equal-frequency, counterpropagating photons with es-sentially no loss in resolution. In Ref. 15 the experimentally observed lineshapes were shown to be in good agreement with the theory.

The large resonant enhancements obtained for the on-resonant case also make it possible to observe extremely weak two-photon transitions. For instance, this technique has made it possible to observe the electric-dipole-forbidden 3S $\rightarrow$ 4F transition in sodium vapor.[31] This transition is about 10^{-6} weaker than an electric-dipole-allowed transition. The Doppler-free character of the spectra allowed the fine-structure split-ting of the 4F state to be precisely determined.

10.3.5. Conclusion

In this chapter we have attempted to illuminate the distinguishing quali-tative features of saturation spectroscopy, two-photon spectroscopy, and laser-induced line narrowing. All three techniques have found increasing use with the recent widespread availability of single-mode, continuously tunable lasers. They presumably will find even more application in the future.

[30] C. Delsart and J. C. Keller, *Opt. Commun.* **15**, 91 (1975).
[31] P. F. Liao and J. E. Bjorkholm, *Phys. Rev. Lett.* **36**, 1543 (1976).

10.4. Nonlinear Optical Effects*

10.4.1. Introduction

Nonlinear optics has long been one of the most important subfields in quantum electronics. Immediately after the laser was invented, it was recognized that the high field intensity of a laser beam could induce nonlinear responses in a medium. In 1961, Franken *et al.*[1] first demonstrated optical second-harmonic generation in a nonlinear crystal. Since then, many other nonlinear optical effects, such as optical mixing, parametric oscillation, stimulated light scattering, self-focusing, multiphoton transitions, and coherent transient phenomena, have been observed.[2] Although most of the nonlinear optical effects were discovered in the earlier years, the rapid progress in the field has not been interrupted since 1961. First, it is on the physical understanding of the various nonlinear optical effects, and then it is on the applications of these effects to various disciplines. More recently, the advent of tunable lasers, high-power lasers, and ultrashort laser pulses has brought new excitement to the field of nonlinear optics. As judged from the large number of papers on the subject in the recent literature, the field has apparently not yet lost its vigor in making advances.

It is clearly impossible for us to have any sort of comprehensible review of nonlinear optics in a short chapter. We should therefore be satisfied here with only a rather brief and qualitative description of the subject. Our emphasis is on physical concepts and potential applications. We shall not venture into serious discussion on any topic, nor shall we provide an extensive list of references. Readers are referred to other review articles or books for more details.[2]

In the following sections, we shall first give a brief summary on the fun-

---

[1] P. A. Franken, A. E. Hall, C. W. Peters, and G. Weinreich, *Phys. Rev. Lett.* **7**, 118 (1961).

[2] See, for example, the recent review article by Y. R. Shen, *Rev. Mod. Phys.* **48**, 1 (1976), and references therein.

---

* Chapter 10.4 is by Y. R. Shen.

METHODS OF EXPERIMENTAL PHYSICS, VOL. 15B

damentals of nonlinear optics, followed by a short discussion on nonlinear optical susceptibilities. We then describe as physically as possible the various nonlinear optical effects and their potential applications.

10.4.2. Fundamentals of Nonlinear Optics

As is well known, all electromagnetic phenomena, linear or nonlinear, are governed by the Maxwell equations or the resulting wave equations

$$\left[\nabla \times (\nabla \times) + \frac{1}{c^2}\frac{\partial^2}{\partial t^2}\right] \mathbf{E}(\mathbf{r}, t) = -\frac{4\pi}{c^2}\frac{\partial^2}{\partial t^2} \mathbf{P}(\mathbf{r}, t),$$

$$\nabla \cdot \mathbf{E}(\mathbf{r}, t) = -4\pi\nabla \cdot \mathbf{P}(\mathbf{r}, t),$$

(10.4.1)

where $\mathbf{P}(\mathbf{r}, t)$ is a generalized electric polarization that includes all multiple contributions.[3,4] This polarization $\mathbf{P}$ as a function of $\mathbf{E}$ describes the response of the medium to the incoming field. In the weak-field limit, $\mathbf{P}$ is linearly proportional to $\mathbf{E}$, but in general, $\mathbf{P}$ can be a complicated nonlinear function of $\mathbf{E}$. In some cases dealing with transients, $\mathbf{P}$ is an integral response function of $\mathbf{E}(\mathbf{r}, t)$ and obeys a certain equation of motion. Knowing $\mathbf{P}$ as a function of $\mathbf{E}$, one can in principle solve Eq. (10.4.1) and hence fully describe the corresponding optical phenomenon. Unfortunately, this is usually not the case in practice. First, it is sometimes difficult to write a simple correct expression for $\mathbf{P(E)}$. Second, it is often not possible to obtain a complete solution for Eq. (10.4.1) even with the largest computer available. One must resort to various approximations in order to circumvent the difficulties. This is actually the work of theoretical research on nonlinear optics.

In many cases of nonlinear optics, we can approximate the problem as one with several quasi-monochromatic light beams interacting in a nonlinear medium. It is then more convenient to decompose $\mathbf{E}$ and $\mathbf{P}$ into their Fourier components.

$$\mathbf{E}(\mathbf{r}, t) = \sum_i \mathbf{E}(\omega_i) = \sum_i \vec{\mathscr{E}}(\omega_i) \exp(i\mathbf{k}_i \cdot \mathbf{r} - i\omega_i t),$$

$$\mathbf{P}(\mathbf{r}, t) = \sum_i \mathbf{P}(\omega_i),$$

(10.4.2)

where $\vec{\mathscr{E}}(\omega_i)$ is a slowly varying spatial function. If the field is not excessively intense, then the polarization $\mathbf{P}(\omega_i)$ can be expanded into a power series

[3] L. D. Landau and E. M. Lifshitz, "Electrodynamics in Continuous Media," p. 251. Pergamon, Oxford, 1960.
[4] N. Bloembergen, "Nonlinear Optics." Benjamin, New York, 1965.

$$\mathbf{P}(\omega_i) = \chi^{(1)}(\omega_i) \cdot \mathbf{E}(\omega_t)$$
$$+ \sum_{j,k} \chi^{(2)}(\omega_i = \omega_j + \omega_k) : \mathbf{E}(\omega_j)\mathbf{E}(\omega_k)$$
$$+ \sum_{j,k,l} \chi^{(3)}(\omega_i = \omega_j + \omega_k + \omega_l) : \mathbf{E}(\omega_j)\mathbf{E}(\omega_k)\mathbf{E}(\omega_l)$$
$$+ \cdots . \tag{10.4.3}$$

In Eq. (10.4.3), $\chi^{(n)}$ are the nth-order susceptibility tensors. They actually govern the nth-order nonlinear optical effects. In the electric-dipole approximation, $\chi^{(n)}$ are independent of the wave vectors $\mathbf{k}$.

Written in terms of the Fourier components, Eq. (10.4.1) now splits into a set of nonlinearly coupled equations

$$\left[\nabla \times (\nabla \times) - \frac{\omega_i^2}{c^2} \right] \mathbf{E}(\omega_i) = \frac{4\pi\omega_i^2}{c^2} \mathbf{P}(\omega_i),$$
$$\nabla \cdot \mathbf{E}(\omega_i) = -4\pi \nabla \cdot \mathbf{P}(\omega_i), \tag{10.4.4}$$

where $\mathbf{P}(\omega_i)$ is given in Eq. (10.4.3). It is then the solution of these coupled equations one must strive for. As an example, let us consider the problem of three-wave interaction with $\omega_1 + \omega_2 = \omega_3$. Assuming all fields transverse, $\mathbf{E}(\omega_i) \perp \mathbf{k}_i$, we obtain from Eq. (10.4.4)[4]

$$[\nabla^2 + \omega_1^2\epsilon(\omega_1)/c^2]\mathbf{E}(\omega_1)$$
$$= -(4\pi\omega_1^2/c^2)\chi^{(2)}(\omega_1 = -\omega_2 + \omega_3) : \mathbf{E}^*(\omega_2)\mathbf{E}(\omega_3), \tag{10.4.5a}$$

$$[\nabla^2 + \omega_2^2\epsilon(\omega_2)/c^2]\mathbf{E}(\omega_2)$$
$$= -(4\pi\omega_2^2/c^2)\chi^{(2)}(\omega_2 = \omega_3 - \omega_1) : \mathbf{E}(\omega_3)\mathbf{E}^*(\omega_1), \tag{10.4.5b}$$

$$[\nabla^2 + \omega_3^2\epsilon(\omega_3)/c^2]\mathbf{E}(\omega_3)$$
$$= -(4\pi\omega_3^2/c^2)\chi^{(2)}(\omega_3 = \omega_1 + \omega_2) : \mathbf{E}(\omega_1)\mathbf{E}(\omega_2), \tag{10.4.5c}$$

where $\epsilon(\omega_1) = 1 + 4\pi\chi^{(1)}(\omega_i)$ are the linear dielectric constants. The above equations show explicitly that the three fields are nonlinearly coupled through the $\chi^{(2)}$ terms. With $\chi^{(2)}$ given, Eqs. (10.4.5) with appropriate boundary conditions can in principle be solved. The solution can be greatly simplified if depletion of the incoming pump fields is negligible, as is fairly often the case.

Physically, nonlinear coupling leads to energy transfer between waves. Then, as one would expect, the maximum rate of energy transfer should occur when both energy and momentum matching conditions are satisfied. In the steady- (or quasi-steady-) state case, energy matching $\omega_1 + \omega_2 = \omega_3$ is automatically satisfied. The momentum matching condition suggests that $\Delta\mathbf{k} = \mathbf{k}_1 + \mathbf{k}_2 - \mathbf{k}_3$ should vanish, where $\Delta\mathbf{k}$ is often called the phase mismatch between the interacting waves. This can be seen formally from Eqs. (10.4.5) if we transform, for example, Eq. (10.4.5c) into the form

$$(\mathbf{k}_3 \cdot \nabla)\mathscr{E}(\omega_3)$$
$$= (i2\pi\omega_3{}^2/c^2)\chi^{(2)}(\omega_3 = \omega_1 + \omega_2) : \mathscr{E}(\omega_1)\mathscr{E}(\omega_2) \exp(i\Delta\mathbf{k} \cdot \mathbf{r}), \quad (10.4.6)$$

assuming $\mathscr{E}(\omega_3)$ is a slowly varying function of distance. As Eq. (10.4.6) stands, we see clearly that the spatial growth of $\mathscr{E}(\omega_3)$ at the expense of $\mathscr{E}(\omega_1)$ and $\mathscr{E}(\omega_2)$ should be a maximum when $\Delta\mathbf{k} = 0$. This phase (or momentum) matching condition can be somewhat relaxed in an absorbing medium since $\mathbf{k}_i$ will then have an imaginary part.

We shall not proceed further on the discussion of nonlinear wave interaction here. We should, however, note that the coupled-wave approach described above is extremely powerful in helping our understanding of a large number of nonlinear optical phenomena. This is even more so if we generalize the waves to include also the excitational waves in the medium. In a later section, we shall discuss several well-known nonlinear optical effects along this line.

10.4.3. Nonlinear Susceptibilities

The role of nonlinear susceptibility tensors in nonlinear optics is the same as that of linear dielectric constants or refractive indices in linear optics. It is absolutely necessary to know $\chi^{(n)}$ in order to prescribe the relevant nonlinear optical effects. For many purposes, one would like to be able to predict $\chi^{(n)}$ theoretically even though $\chi^{(n)}$ can also be measured experimentally.

How do we find $\chi^{(n)}$ theoretically? As an nth-rank tensor, $\chi^{(n)}$ has 3^n elements, but governed by the symmetry of the medium, not all of them are independent or nonvanishing. In evaluating $\chi^{(n)}$, we should first find out from symmetry argument the nonvanishing and independent elements of $\chi^{(n)}$, and then from a microscopic theory, the magnitudes of those nonvanishing elements.

For a given medium with a certain class of symmetry, there is a set of symmetry operations such as inversion and rotation under which the properties of the medium remain invariant. Let S be such an operation. Then under the transformation of S, the tensor $\chi^{(n)}$ should remain unchanged. We have, for example,

$$(\mathbf{S}^\dagger \cdot \hat{\mathbf{i}}) \cdot \chi^{(2)} : (\mathbf{S} \cdot \hat{\mathbf{j}})(\mathbf{S} \cdot \hat{\mathbf{k}}) = \hat{\mathbf{i}} \cdot \chi^{(2)} : \hat{\mathbf{j}}\hat{\mathbf{k}}. \quad (10.4.7)$$

There is one such equation for each symmetry operation. These equations make some elements of $\chi^{(n)}$ vanish and some depend on others. As an example, Eq. (10.4.7) leads immediately to the familiar result that $\chi^{(2)}$

of a medium with inversion symmetry has no nonvanishing element. The reduced forms of $\chi^{(2)}$ and $\chi^{(3)}$ for various classes of materials have been tabulated in the literature.[5]

The formal microscopic expression for $\chi^{(n)}$ can always be obtained from the nth-order quantum-mechanical perturbation calculation. Numerical evaluation of $\chi^{(n)}$ from such an expression is difficult, however, since the transition frequencies and matrix elements of a given medium are usually not very well known. Thus, except in some simple cases such as alkali vapors, one would like to resort to simple theoretical models for evaluation of $\chi^{(n)}$.

One such simplifying model that has been quite successful in calculating $\chi^{(2)}$ and $\chi^{(3)}$ for molecules and crystals is the so-called bond model. This model was used in the early 1930s to calculate linear polarizabilities of molecules and linear dielectric constants of crystals.[6] According to the model, the induced polarization on a molecule or crystal is the vector sum of the induced polarizations on all the bonds between atoms in the molecule or crystal. This is known as the bond additivity rule.[6] We can write

$$\chi^{(n)} = \sum_i \beta_i^{(n)}, \tag{10.4.8}$$

where $\beta_i^{(n)}$ is the nth-order polarizability tensor of the ith bond and the summation is over all bonds in a unit volume. Thus, knowing the crystal structure, we need only to calculate $\beta^{(n)}$.

Let us consider here only $\beta^{(n)}$ due to electronic contribution. For optical frequencies far below the absorption bands, the calculation of $\beta^{(n)}$ is made simple by the recent development of the bond theory.[7] Following the Penn approximation,[8] we can write

$$4N(\beta_\parallel^{(1)} + 2\beta_\perp^{(1)})/3 = \hbar^2\Omega_p^2/4\pi\overline{E}_g^2, \tag{10.4.9}$$

where $\beta_\parallel^{(1)}$ and $\beta_\perp^{(1)}$ are the linear polarizabilities parallel and perpendicular to the cylindrically symmetric bond, respectively, and Ω_p is the plasma frequency of the N valence electrons per unit volume. The average energy gap $\overline{E}_g$ between bonding and antibonding states is composed of a homopolar gap E_h and a heteropolar gap C:

$$\overline{E}_g^2 = E_h^2 + C^2. \tag{10.4.10}$$

[5] P. N. Butcher, "Nonlinear Optical Phenomena," p. 43. Ohio State Univ. Press, Columbus, 1965.

[6] See, for example, K. G. Benbigh, *Trans. Faraday Soc.* **36,** 936 (1940).

[7] J. C. Phillips, "Covalent Bonding in Crystals, Molecules, and Polymers." Univ. of Chicago Press, Chicago, Illinois, 1969; "Bonds and Bands in Semiconductors." Academic Press, New York, 1973.

[8] D. R. Penn, *Phys. Rev.* **128,** 2093 (1962).

Approximate expressions of E_h and C in terms of the bond length, convalent radii, and valences of the bond atoms, etc., can be obtained from the bond theory.[7]

Knowing $\beta^{(1)}$, we can then calculate $\beta^{(n)}$ (and hence $\chi^{(n)}$) from the derivative:

$$\beta^{(n)} = \partial^{(n-1)}\beta^{(1)}/(\partial E)^{(n-1)}. \tag{10.4.11}$$

As suggested by Eq. (10.4.9), the effect of the external field E on $\beta^{(1)}$ is through $\overline{E}_g$. Two different models have been used to calculate the dependence of $\overline{E}_g$ on E. One is the bond-charge model.[9] It assumes that the bond charges along the bond can be approximated by a point charge located at a distance equal to the covalent radii from the two bond atoms, and that the external field E induces a shift in the position of this point charge. Using this model, Levine has successfully calculated $\chi^{(2)}$ and $\chi^{(3)}$ for a large number of liquids and crystals in good agreement with measured values. The other is the charge transfer model.[10] It assumes that E induces a transfer of a definite amount of valence charges from one bond atom to the other. Tang and co-workers[10] have calculated $\chi^{(2)}$ using the charge transfer model for a number of semiconductors. Their results also agree well with the measured values. Both models are in fact only crude approximations of the real picture. In reality, there is a broad distribution of valence electrons along and around the bonds. The external field simply induces a slight redistribution of the valence electron distribution.

The bond-theoretical study of nonlinear susceptibilities has led to the following conclusion. For highly nonlinear optical materials, the bond polarizabilities must have large nonlinearity. For materials with large $\chi^{(2)}$, the crystal structure should also be as asymmetric as possible so that in the summation of Eq. (10.4.8), the vectorial cancellation of $\beta^{(2)}$ from different bonds is as small as possible.

The approximations used in the bond theory discussed above does not hold when the optical frequencies are close to or in the absorption bonds. We must then use the full microscopic expression to calculate $\chi^{(n)}$. Recent empirical calculations[11] suggest that in order to obtain the correct dispersion of $\chi^{(2)}$, one cannot make any drastic approximation on the matrix elements and density of states. As in the linear case, the effect of critical points dominates the dispersion of $\chi^{(2)}$. In principle, the measurements of the dispersion of $\chi^{(2)}$ should compliment the linear optical measurements of $\chi^{(1)}(\omega)$ or $\epsilon(\omega)$ in the study of electronic structures of solids.

[9] B. F. Levine, *Phys. Rev. B* **7**, 2600 (1973).

[10] C. L. Tang and C. Flytzanis, *Phys. Rev. B* **4**, 2520 (1971); C. L. Tang, *IEEE J. Quantum Electron.* **qe-9**, 755 (1973); F. Scholl and C. L. Tang, *Phys. Rev. B* **8**, 4607 (1973).

[11] C. Y. Fong and Y. R. Shen, *Phys. Rev. B* **12**, 2325 (1975).

Unfortunately, limited by the tunability of lasers and by the usually poor signal-to-noise ratio of nonlinear optical experiments, such measurements cannot yet be considered as a useful technique to study band structures of solids.

10.4.4. Nonlinear Optical Processes

In this section, we describe briefly the various nonlinear optical processes. We can divide them into two groups: those in which a nonlinear susceptibility can be used to describe the nonlinear response of the medium laser and those in which it cannot. The former include harmonic generation, optical mixing, parametric amplification and oscillation, two-photon absorption, stimulated light scattering, optical-field-induced birefringence, and self-focusing. The latter includes coherent transient effects, multiphoton ionization, laser breakdown, and multiphoton dissociation of molecules. Because of limitation of scope, we shall not discuss here the highly nonlinear optical effects in plasmas, in the formation of laser-induced plasmas, and in laser-induced fusion.

10.4.4.1. Optical Mixing and Harmonic Generation.
Among all nonlinear optical effects, optical mixing is probably the most important and well understood. Its basic physics is simple. Incoming pump beams interact nonlinearly in a nonlinear medium. They beat with one another and induce in the medium a nonlinear polarization oscillating at the combination (sum and difference) frequencies. This nonlinear polarization then acts as a source in generating an output beam at the combination frequencies. The rate of energy transfer from the pump beams to the output beam is maximum when phase matching between input and output waves is realized. Optical harmonic generation is a special case of optical mixing where an incoming pump wave beats with itself and generates a harmonic wave.

The theoretical description of optical mixing follows Eq. (10.4.4). We discuss here sum-frequency generation by optical mixing as an example. Consider only the equation for the sum-frequency field $E(\omega_3)$. Assuming $E(\omega_3) \perp k_3$, we have

$$[\nabla^2 + \omega_3^2 \epsilon(\omega_3)/c^2]E(\omega_3) = -(4\pi\omega_3^2/c^2)P^{NL}(\omega_3), \quad (10.4.12)$$

which is the same as Eq. (10.4.5c) with the nonlinear polarization given by $P^{NL}(\omega_3) = \chi^{(2)}(\omega_3 = \omega_1 + \omega_2) : E(\omega_1)E(\omega_2)$. Noting that we can write $P^{NL}(\omega_3) = \mathscr{P}^{NL}(\omega_3) \exp(ik_s \cdot r - i\omega_3 t)$, we can transform the corresponding Eq. (10.4.6) into

$$(k_3 \cdot \nabla)\mathscr{E}(\omega_3) = (i2\pi\omega_3^2/c^2)\mathscr{P}^{NL}(\omega_3) \exp(i\,\Delta k \cdot r), \quad (10.4.13)$$

with $\Delta \mathbf{k} = \mathbf{k}_s - \mathbf{k}_3$ being the phase mismatch. Let $\Delta \mathbf{k}$ be along $\hat{\mathbf{z}}$. The solution of Eq. (10.4.13) can then be expressed in an integral form,

$$|\mathscr{E}(\omega_3)|^2 = (2\pi\omega_3{}^2/c^2 k_3 \cdot \hat{\mathbf{z}})^2 \left| \int_0^l dz [\hat{\mathscr{E}}(\omega_3) \cdot \mathscr{P}^{NL}(\omega_3)] \exp(i \, \Delta kz) \right|^2 \quad (10.4.14)$$

assuming $\mathscr{E}(\omega_3) = 0$ at $z = 0$. From Eq. (10.4.14), we see immediately that if depletion of the pump fields is negligible, then $P^{NL}(\omega_3)$ is independent of z and the sum-frequency output is given by[4]

$$|\mathscr{E}(\omega_3)|^2 = 2\pi\omega_3{}^2/c^2 k_3 \cdot \hat{\mathbf{z}})^2 |\hat{\mathscr{E}}(\omega_3) \cdot \mathscr{P}^{NL}(\omega_3)|^2 [\sin^2(\Delta kl/2)/(\Delta kl/2)^2], \quad (10.4.15)$$

which is a maximum at $\Delta k = 0$. The maximum output is therefore proportional to the square of the nonlinear susceptibility, the product of the pump field intensities, and the square of the crystal length l. Depletion of the pump fields is of course nonnegligible when the output conversion efficiency is high. Then in order to solve the integral in Eq. (10.4.14), we must first find the pump field variation as a function of distance. This must be done by solving the set of coupled equations in Eqs. (10.4.5). The theoretical description here can be easily extended to other cases of optical mixing when $\mathbf{P}^{NL}(\omega_3)$ is taken as the nonlinear polarization at the output frequency.

An obvious application of optical mixing is in constructing coherent sources at new frequencies. This is described in Chapter 9.4 in detail. So far, optical mixing has already been used to generate coherent radiation throughout the spectral range from microwave or far infrared below 1 cm^{-1}[12] to extreme ultraviolet at 512 Å.[13] In selecting nonlinear materials for such applications, one must take several factors into consideration.

(1) The material should have a large nonlinearity.

(2) The material should be more or less transparent to all frequencies involved.

(3) Phase matching should be achievable in the material.

(4) The material should have a sufficiently large dimension.

(5) The laser breakdown threshold should be sufficiently high.

For efficient optical mixing, we often need to focus the pump beams in

[12] Y. R. Shen, *Prog. Quantum Electron.* **4,** 207 (1976), and references therein; see also Y. R. Shen, ed., "Nonlinear Infrared Generation." Springer-Verlag, Berlin and New York, 1977.

[13] J. Reintzes, R. C. Eckardt, C. Y. She, N. E. Karangelen, R. C. Elton, and R. A. Andrews, *Phys. Rev. Lett.* **37,** 1540 (1976).

order to increase the pump field intensities and to use phase-matchable collinear beam geometry in order to increase the active interaction length. Research in this respect has been concentrated in finding materials that can have noncritical collinear phase matching for focused beams or in devising methods to circumvent the difficulty such as using waveguide or periodic structure. The noncritical collinear phase matching can occur, for example, in gas media and in crystals along a crystal axis.

The requirement of large nonlinearity is, of course, crucial for most nonlinear optical applications. We have already seen in Section 10.4.2 how a molecule or crystal can have large nonlinearity at frequencies well in the transparent region. Physically, a nonlinear susceptibility can also become much larger as a result of resonance enhancement when one or several optical frequencies approach resonances. The resonance is particularly strong in atomic vapors where the spectral lines are narrow. In addition, unlike solids, an atomic vapor can have wide transparent regions between sharp spectral lines, a very high laser breakdown threshold, and a cell length longer than we need. Thus, with a resonant nonlinear susceptibility and somewhat higher pump laser intensities, we can find high-order optical mixing in atomic vapors almost as efficient as the nonresonant lower-order optical mixing in solids.[14,15] Third-order optical mixing in atomic vapors has already been used to generate tunable radiation both in the infrared and in the ultraviolet.[14] In fact, because of the transparency requirement, atomic vapors are the only media one can use to generate extreme ultraviolet radiation by optical mixing.

The resonance enhancement of nonlinear susceptibilities can also be used as a spectroscopic means to study excitations in a medium.[16,17] This is especially useful for one-photon-forbidden but two-photon-allowed transitions. Such transitions can be easily detected from resonance enhancement in optical mixing when the sum or difference of two pump frequencies approaches these transitions. As an example, consider the use of four-wave mixing to study the Raman-allowed vibrations. The three pump fields are at ω_1, ω_2, and ω_3, and the output is at $\omega_1 - \omega_2 + \omega_3$. When $\omega_1 - \omega_2$ approaches a particular vibrational frequency, the output gets strongly resonance-enhanced. Since the output is coherent, it can be easily discriminated from incoherent scattering or fluorescence back-

[14] J. J. Wynne and P. P. Sorokin, in "Nonlinear Infrared Generation" (Y. R. Shen, ed.), p. 159. Springer-Verlag, Berlin and New York, 1977.

[15] R. B. Miles and S. E. Harris, IEEE J. Quantum Electron. qe-9, 470 (1973).

[16] N. Bloembergen, in "Laser Spectroscopy" (S. Haroche et al., eds.), p. 31. Springer-Verlag, Berlin and New York, 1975.

[17] Articles in "Course on Nonlinear Spectroscopy" (N. Bloembergen, ed.). North-Holland Publ., Amsterdam, 1976.

ground. This method therefore has great advantage over spontaneous Raman scattering technique in cases where sample fluorescence is a nuisance. With ultrashort pump pulses, the technique can also be used to measure the vibrational relaxation times. Usually, we can let $\omega_1 = \omega_3$. This four-wave mixing technique to probe low-frequency excitations is often called coherent anti-Stokes Raman spectroscopy (CARS). As is clear from our general discussion, the technique is certainly not limited to the probing of only low-frequency excitations.[17]

10.4.4.2. Parametric Amplification and Oscillation. In second-order sum-frequency mixing, two pump fields at ω_1 and ω_2 generate an output field at $\omega_3 = \omega_1 + \omega_2$. One would expect that the inverse process in which one pump field at ω_3 generates two output fields at ω_1 and ω_2 should also be possible. The latter is known as parametric generation.[18] The splitting of ω_3 into ω_1 and ω_2 is clearly not unique, but the output has a sharp maximum near phase matching. Thus, one can tune the frequencies of the output by tuning the phase-matching condition, e.g., by varying the orientation or the temperature of the nonlinear crystal. The difference between parametric amplification and oscillation is that the latter occurs in a cavity so that feedback effectively increases the active length. However, if the pump field is sufficiently intense, then the single-path parametric amplification from noise can already yield intense output. This is sometimes known as parametric superfluorescence and has been used to generate tunable ultrashort optical pulses.[19] The formal theoretical derivation of parametric amplification and oscillation follows the coupled equations in Eqs. (10.4.5) with proper boundary conditions. The solution is particularly simple if depletion of the pump field $E(\omega_3)$ is negligible, since then the equations of $E(\omega_1)$ and $E(\omega_2)$ become linearly coupled.

Parametric oscillators are important optical devices for generating tunable coherent sources, in particular in the infrared range.[18] They are discussed in detail in Chapter 9.3. Presently, with different nonlinear crystals, they have already been used to cover a tuning range from 0.42 to 10.5 μm.

10.4.4.3. Two-Photon Absorption. Two-photon processes were well-known even long before the laser was invented,[20] but they are more readily detectable with laser beams. The rate of two-photon absorption in a medium is given by

[18] See, for example, R. L. Byer, in "Quantum Electronics" (H. Rabin and C. L. Tang, eds.), p. 588. Academic Press, New York, 1975.

[19] A. Laubereau, L. Greiter, and W. Kaiser, *Appl. Phys. Lett.* **25**, 87 (1974); T. Kushida, Y. Tanaka, M. Ojima, and Y. Nakazaki, *Jpn. J. Appl. Phys.* **14**, 1097 (1975).

[20] M. Goeppert-Mayer, *Ann. Phys. (Leipzig)* [5] **9**, 273 (1931).

$$\frac{\partial |E(\omega_1)|^2}{\partial z} = -\left(\frac{4\pi\omega_1^2}{c^2 k_1}\right)$$

$$\cdot [\text{Im } \chi^{(3)}(\omega_1 = \omega_1 + \omega_2 - \omega_2)]|E(\omega_1)|^2 |E(\omega_2)|^2 \quad (10.4.16)$$

and a similar equation for $E(\omega_2)$. Here, the imaginary part of the third-order nonlinear susceptibility, Im $\chi^{(3)}$, is directly proportional to the two-photon absorption cross section derived by the golden rule.[21] It is seen that the absorption rate is proportional to the product of the two incoming beam intensities. Therefore with sufficiently intense laser beams, two-photon absorption can be quite appreciable.

Two-photon absorption is useful as a spectroscopic technique complimenting the single-photon absorption spectroscopy because of the different selection rules.[21] Recently, two-photon absorption with oppositely propagating beams has also been developed into a high-resolution spectroscopic technique for gases to beat Doppler broadening. This is discussed in detail in Chapter 10.3.

10.4.4.4. Stimulated Light Scattering. Light scattering is also a two-photon process in which one incoming photon at ω_1 is absorbed, one photon at ω_2 is emitted, and the material system makes a transition from the initial state $|i\rangle$ to the final state $|f\rangle$ with a transition frequency $\omega_{fi} \cong \omega_1 - \omega_2$ (see Fig. 1). The final state can correspond to any material excitation, namely, electronic excitation, vibration, rotation, magnon, plasmon, etc. Spontaneous light scattering is of course well known to

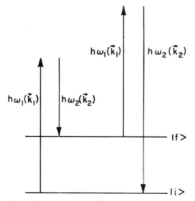

FIG. 1. Schematic drawing showing the Stokes ($\omega_1 > \omega_2$) and anti-Stokes ($\omega_1 < \omega_2$) Raman transition between states $|i\rangle$ and $|f\rangle$.

[21] J. M. Worlock, in "Laser Handbook" (F. T. Arecchi and E. O. Schulz-Dubois, eds.), p. 1323. North-Holland Publ., Amsterdam, 1972; H. Mahr, in "Quantum Electronics" (H. Rabin and C. L. Tang, eds.), p. 286. Academic Press, New York, 1975.

most scientists. If, however, the incoming pump field at ω_1 is very intense, then, the scattering probability can become so high that stimulated emission by the scattered photons dominates and leads to an avalanche build-up of the emitted radiation at ω_2. This is known as stimulated scattering. For each type of spontaneous scattering, there should be a corresponding stimulated scattering process.[22] Formally, Eq. (10.4.16) with ω_1 and ω_2 interchanged can be used to describe the growth of the Stokes field at ω_2 in the stimulated scattering. The quantity Im $\chi^{(3)}(\omega_2)$ in this case is negative and is directly proportional to the spontaneous scattering cross section.

The following stimulated scattering processes have been observed: phonon-Raman, electronic-Raman, spin–flip-Raman, polariton, Brillouin, Rayleigh, Rayleigh-wing, and concentration scattering. Among these, the Raman and polariton scattering have found useful applications in generation of tunable radiation, particularly in the infrared. Stimulated Raman scattering in atomic and molecular vapors pumped by a dye laser can generate tunable radiation in the range between 670 and $\sim 10,000$ cm^{-1} with maximum peak power as high as 60 MW.[14] Stimulated spin–flip Raman scattering pumped by a molecular laser can tune over the ranges 590–1100 and 1540–2000 cm^{-1}. Finally, stimulated polariton scattering pumped by a ruby laser can generate tunable far-infrared radiation between 14 and 200 cm^{-1}.[12]

10.4.4.5. Optical-Field-Induced Birefringence. A polarized laser beam of high intensity can induce a birefringence in a medium. Formally, this can be seen from the expression for the induced polarization $\mathbf{P}(\omega)$ at the frequency ω of a weak probing beam. Let the intense beam be at ω'. Then following Eq. (10.4.3), we can write

$$\begin{aligned}\mathbf{P}(\omega) &= [\chi^{(1)}(\omega) + \chi^{(3)}(\omega = \omega' - \omega' + \omega) : \mathbf{E}(\omega')\mathbf{E}^*(\omega')]\mathbf{E}(\omega) \\ &\equiv (\chi^{(1)} + \Delta\chi) \cdot \mathbf{E}(\omega),\end{aligned} \quad (10.4.17)$$

assuming $\chi^{(2)} = 0$ and higher order terms negligible. The induced susceptibility $\Delta\chi$ proportional to $|E(\omega')|^2$ has in general a different tensorial symmetry than $\chi^{(1)}$. In particular, $\Delta\chi$ is usually anisotropic even if the medium is originally isotropic.

Physically, the induced $\Delta\chi$ can arise from various different mechanisms. The intense optical field can perturb the electron distribution around atoms or molecules, modifies the density of the medium through electrostriction, induces molecular reorientation, redistribution, vibration, and population change among energy states, etc. All these mecha-

[22] Y. R. Shen, in "Light Scattering in Solids" (M. Cardona, ed.), p. 275. Springer-Verlag, Berlin and New York, 1975.

nisms can effectively cause a change in the linear response of the medium. The optical-field-induced $\Delta\chi$ gets resonantly enhanced when $\omega \pm \omega'$ approaches a two-photon allowed transition. This fact has recently been used as a high-resolution nonlinear optical spectroscopic technique to study Raman transitions in liquids[23] and optical transitions in gases.[24] The induced birefringence at resonance in an atomic vapor is so large that it may find application in the construction of a fast optical switching device.[25]

10.4.4.6. Self-Focusing. The optical-field-induced $\Delta\chi$, or equivalently the optical-field-induced refractive index Δn, changes the propagation velocity of a laser beam in a medium. In most media, Δn is positive. Then, if the beam has a Gaussian-shaped profile, its central portion is more intense and should propagate more slowly than the edge portion. Consequently, as described in Fig. 2a, the wavefront of the beam gets dis-

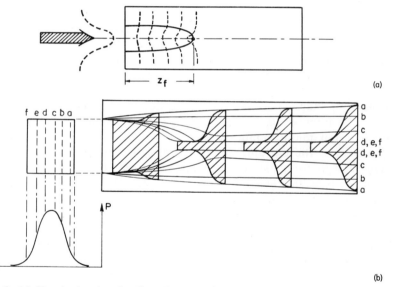

FIG. 2. (a) Sketch showing the distortion of the wavefront and self-focusing of a laser beam in a nonlinear medium. (b) Sketch showing transient self-focusing of a laser pulse in a Kerr Liquid. Different parts (a, b, c, etc.) of the pulse focus and defocus along different ray paths. The pulse first gets deformed into a horn shape and then propagates on without much further change.

[23] D. Heiman, R. W. Hellwarth, M. D. Levenson, and G. Martin, *Phys. Rev. Lett.* **36**, 189 (1976); J. J. Song, G. L. Eesley, and M. D. Levenson, *Appl. Phys. Lett.* **29**, 567 (1976).

[24] R. Teets, R. Feinberg, T. W. Hansch, and A. L. Schawlon, *Phys. Rev. Lett.* **37**, 683 (1976).

[25] P. F. Liao and G. C. Bjorklund, *Phys. Rev. Lett.* **36**, 584 (1976).

torted, and since ray propagation is always perpendicular to the wave-front, the beam tends to focus by itself. This of course happens only when the self-focusing action is strong enough to overcome the usual dif-fraction action.[26]

Relatively high intensity laser pulses (≥ 10 MW/cm^2) are often neces-sary for self-focusing experiments. The self-focusing dynamics varies ac-cording to the ratio of the laser pulse width T_p to the response time τ of Δn. If the ratio is much larger than 1, we have the quasi-steady-state case. The focusing depends only on the instantaneous incoming laser power and both focusing and subsequent diffraction are rather abrupt, leading to a sharp focal spot that moves along the axis as the input laser power varies with time. On a time-integrated photograph, it results in an intense filament of the order of 10 μm in diameter. This is the so-called moving-focus model. If the ratio T_p/τ is close to or smaller than 1, we have the transient case. The field-induced Δn is now a transient cumula-tive response to the laser intensity variation over a period of $\sim \tau$ in the past. Focusing and subsequent diffraction become less abrupt for the more transient case. In the extreme transient limit, the front part of the laser pulse hardly focuses, while the lagging part focuses more tightly, such that the longitudinal pulse profile first gets deformed into a horn shape and then propagates on without much further change (see Fig. 2b). As the horn-shaped pulse travels in the medium, its neck part again forms a track of intense filament on the integrated photograph. This is the so-called dynamic-trapping model.

Because of the extremely high intensity (≥ 10 GW/cm^2) in the focal region, self-focusing can readily initiate in the medium other nonlinear op-tical processes such as stimulated Raman and Brillouin scattering, multi-photon absorption, and optical breakdown. Optical breakdown is the most detrimental effect one must avoid in the design of high-power laser systems.

The field-induced Δn can sometimes be negative, leading to self-defocusing of light.[27] This often happens when the medium absorbs the laser beam and heating causes a decrease in n. Self-defocusing is extremely sensitive to the absorption coefficient. As a result, it has been used as a spectroscopic technique to study very weak absorption bands in materials.

10.4.4.7. Coherent Transient Effects. In magnetic resonance, it is well known that resonant coherent pulsed excitation leads to coherent

[26] Y. R. Shen, *Prog. Quantum Electron.* **4**, 1 (1975).

[27] F. W. Dabby, T. K. Gustafson, J. R. Whinnery, Y. Kohanzadeh, and P. L. Kelley, *Appl. Phys. Lett.* **16**, 362 (1970).

mixing of initial and final states with a transverse relaxation time T_2. If the width of the exciting pulse or pulses is shorter than or comparable with T_2, then coherent transient effects result. There is a close analog between magnetic resonance and optical resonance. A system under resonant excitation can often be approximated by a two-level system. Then, as shown by Feynman *et al.*,[28] any two-level system can be treated as a spin $-\frac{1}{2}$ system and hence its dynamic response to a rotating or cirularly polarized field can be described by the well-known Bloch equation. Physically, the Bloch equation is simply an equation of motion that describes the precession of a pseudodipole $\mathbf{p}$ around an effective field $\mathbf{E}$, subject to initial conditions and longitudinal and transverse relaxations. We define $\mathbf{p}$ and $\mathbf{E}$ as follows: $p_x\hat{\mathbf{x}} + p_y\hat{\mathbf{y}}$ is the real induced dipole; p_z is the product of the population difference and the transition matrix element γ between initial and final states; $E_x\hat{\mathbf{x}} + E_y\hat{\mathbf{y}}$ is the applied field; E_z is the quotient of the resonant energy and γ. Neglecting relaxation, the equation has the form

$$\partial\mathbf{p}/\partial t = -(\gamma/\hbar)\mathbf{E} \times \mathbf{p}. \tag{10.4.18}$$

In magnetic resonance cases, $\mathbf{p}$ and $\mathbf{E}$ correspond to the induced magnetic dipole and the applied magnetic field, respectively.

The Bloch equation coupled with the Maxwell equations essentially describe all the transient coherent phenomena in a two-level system.[29] There is however an important difference between optical and magnetic resonance cases. The former deal with propagating waves with a wavelength λ much smaller than the dimension d of the medium, while the latter deal with standing waves with $\lambda \gtrsim d$. Thus, there are a number of observed coherent transient effects that are common for both optical and magnetic resonances. These include one- and two-photon transient nutation and free-induction decay, photon or spin echoes, adiabatic following and inversion, and superradiance. There are also others that are unique for optical resonances. They include self-induced transparency, $n\pi$ pulse propagation, and some aspects of superradiance.

Applications of coherent transient effects are mainly in the measurements of relaxation times of excited states of a system. One hopes to learn how various interactions and external perturbations affect the relaxations. Recently, it has been proposed that adiabatic inversion[30] and π pulse propagation[31] can also be used for efficient isotope separation.

[28] R. P. Feynman, F. L. Vernon, and R. W. Hellwarth, *J. Appl. Phys.* **28,** 49 (1957).
[29] L. Allen and J. H. Eberly, "Optical Resonance and Two-Level Atoms." Wiley, New York, 1975.
[30] I. Nebenzahl and A. Szöke, *Appl. Phys. Lett.* **25,** 327 (1974).
[31] J. C. Diels, *Phys. Rev. A* **13,** 1520 (1976).

10.4.4.8. Multiphoton Ionization. Higher order nonlinear optical effects can also occur if the laser field intensity in a medium is sufficiently high. Thus, in low-pressure gases, n-photon ionization processes with $n \gg 1$ can readily be observed if the laser intensity is 10^{11} W/cm^2 or more. For example, Lecompte et al.[32] have observed 11-photon ionization of Xe, and Baravian et al.[33] have observed 16-photon ionization of Ne.

Theoretically, multiphoton ionization could presumably be described by a high-order perturbation calculation of transition probability. However, at the laser intensity necessary for multiphoton ionization, the optical-field-induced level shifts and level broadenings often become important. They can contribute significantly to the multiphoton ionization probability. This is particularly true when the laser combination frequencies are not far from intermediate resonances.[34] Being a very high-order nonlinear process, multiphoton ionization probability is very sensitive to the laser mode structure. For example, in the 11-photon ionization of Xe, it has been observed that a laser beam with 100 longitudinal modes yields an ionization probability $10^{6.9 \pm 0.3}$ times larger than a single-mode laser beam.[32]

Since ion detectors are extremely sensitive, two-photon ionization and two-photon transition with subsequent dc field-induced ionization have been used to study autoionization states[35] and high rydberg states of simple atoms.[36] Multiphoton ionization may also find potentially important applications in isotope separation[37] and in generating highly polarized electrons.[38]

10.4.4.9. Optical Breakdown. In the study of multiphoton ionization, one must keep the gas pressure low ($\leqslant 10^{-3}$ torr), the laser pulse short ($\leqslant 10^{-8}$ sec), and the focal dimension small ($\leqslant 10^{-2}$ cm), so that electron–atom and atom–atom collision effects can be neglected. At higher pressures, free electrons released from atoms by multiphoton ionization can absorb photons by inverse bremsstrahlung during their collisions with atoms. An electron can then acquire enough energy capable of

[32] C. LeCompte, G. Mainfray, C. Manus, and F. Sanchez, *Phys. Rev. A* **11**, 1109 (1975).

[33] G. Baravian, J. Godart, and G. Sultan, *Phys. Rev. A* **14**, 761 (1976).

[34] See, for example, C. S. Chang and P. Stehle, *Phys. Rev. Lett.* **30**, 1283 (1973); P. Agostini and C. LeCompte, *ibid.* **36**, 1131 (1976).

[35] P. Esherick, J. A. Armstrong, R. W. Dreyfus, and J. J. Wynne, *Phys. Rev. Lett.* **36**, 1296 (1976).

[36] M. G. Littman, M. L. Zimmerman, T. W. Ducas, R. G. Freeman, and D. Kleppner, *Phys. Rev. Lett.* **36**, 788 (1976).

[37] S. A. Tuccio, J. W. Dubrin, O. G. Peterson, and B. B. Snavely, *Int. Quantum Electron. Conf., 8th, 1974,* Postdeadline paper Q.14(1974).

[38] P. Lambropoulos, *Phys. Rev. Lett.* **30**, 413 (1973).

ionizing an atom to release another electron. This starts the electron multiplication process and leads to avalanche ionization. As soon as there is appreciable ionization, the incoming photons can be readily absorbed by electrons via free–free transitions in the field of ions. It causes intense heating of the electron plasma and consequently a rapid hydrodynamic expansion of the plasma in the form of a spherical shock wave with a visible spark. The phenomenon is known as optical breakdown. After the breakdown, the laser beam is preferentially absorbed by the part of the shock wave moving toward the laser, and hence the spark appears to propagate against the laser beam.[39]

The theory of optical avalanche ionization is similar to that of microwave avalanche ionization. However, in the optical case, the photon energy is not negligible compared to the electron energy. This is a quantum effect, which must be taken into account in the correct treatment of optical avalanche ionization.[40] Interaction of the laser beam with the expanding plasma is another difficult theoretical problem. Besides direct optical absorption, other nonlinear optical processes can occur in the plasma. How efficiently a high-power laser beam can heat up a plasma in a very short period is presently of great importance for the understanding and realization of laser-induced fusion.

Optical breakdown can also occur in solids and is an extremely important problem for laser optical technology. In transparent solids, the physical mechanisms for optical breakdown are again multiphoton and avalanche ionization.[41] However, in solids with small energy gaps or appreciable impurity concentrations, the number of free electrons in the conduction bands resulting from thermal ionization can be fairly high. Then, multiphoton ionization is no longer needed to provide the first few electrons to start the avalanche ionization. Multiphoton ionization becomes important only when the energy gap is much larger than kT, but not much larger than the laser photon energy. The above description also applies to surface optical damage even though the surface contamination may significantly lower the breakdown threshold.

10.4.4.10. Multiphoton Dissociation of Molecules. Collisionless multiphoton dissociation of polyatomic molecules is presently the most interesting topic in nonlinear optics.[42] It deals with the problem of how a single molecule can absorb more than 30 or 40 infrared photons and get

[39] S. A. Ramsden and W. E. R. Davis, *Phys. Rev. Lett.* **13**, 227 (1974).

[40] N. Kroll and K. M. Watson, *Phys. Rev. A* **5**, 1883 (1972).

[41] See, for example, N. Bloembergen, *IEEE J. Quantum Electron.* **qe-10**, 375 (1974).

[42] See, for example, articles on the subject, *in* "Tunable Lasers and Applications" (A. Mooradian, T. Jaeger, and P. Stokseth, eds.), references therein. Springer-Verlag, Berlin and New York, 1976.

dissociated, and what the subsequent dynamics of dissociation is. The process is technically very important because of its potential application to isotope separation. It has been demonstrated by Ambartzumian *et al*. and others[42] that multiphoton molecular dissociation under quasi-monochromatic laser excitation is isotopically selective. In the case of a cell filled with 0.18 torr natural SF_6 and 2 torr H_2 as the scavenger, the observed enrichment factor of $^{34}S/^{32}S$ after irradiation of 2000 2 J CO_2 laser pulses was as high as 2800. Isotope separation of heavy elements such as ^{181}Os and ^{192}Os by multiphoton molecular dissociation has also been observed.[43]

Scientifically, multiphoton dissociation of molecules is also an extremely interesting process. In normal circumstances, even the occurrence of a two- or three-photon absorption process would require a laser intensity of ~ 1 MW/cm^2. One would expect that for absorption of more than 30 or 40 photons by a single molecule, the laser intensity needed should be close to the optical breakdown threshold $\geqslant 10$ GW/cm^2. In reality, the observed threshold for multiphoton dissociation of SF_6 was only ~ 10 MW/cm^2. The reason for this low dissociation threshold is now qualitatively understood[42] and is briefly sketched here.

For a single molecule to absorb several tens of infrared photons with high probability from a reasonably intense laser beam, two conditions must be satisfied. First, each single-photon step in the absorption must be at resonance or near resonance. Second, the density of states must increase rapidly with increase of energy so that even with a thermal distribution, up-pumping can dominate over down-transitions. For a polyatomic molecule, the second condition is usually satisfied because of the large number of vibrational degrees of freedom. That the first condition can also be satisfied is however not obvious. Consider the case of SF_6. Its ν_3 vibrational mode at 948 cm^{-1} can be excited by a CO_2 laser beam tuned to the ν_3 absorption band. Around 29 photons must be absorbed by a single SF_6 molecule in order to reach the dissociation level of $SF_6 \rightarrow SF_5 + F$[44] at 27,600 cm^{-1}. If we treat the problem as interaction of monochromatic laser radiation with a ν_3 anharmonic oscillator having an anharmonic shift of $\Delta\nu_{anh} = 2.88$ cm^{-1}, then as shown in Fig. 3a, the laser frequency will quickly get out of step with the ν_3 ladder in the up-excitation process. This picture is clearly not realistic since an SF_6 molecule should be far more complicated than an anharmonic oscillator. Associated with each vibrational state, there is a set of rotational levels.

[43] R. V. Ambartzumian, Yu. A. Gorokhov, V. S. Letokhov, and G. N. Makarov, *JETP Lett.* (*Engl. Transl.*) **22**, 43 (1975).

[44] M. J. Coggiola, P. A. Schulz, Y. T. Lee, and Y. R. Shen, *Phys. Rev. Lett.* **38**, 17 (1977).

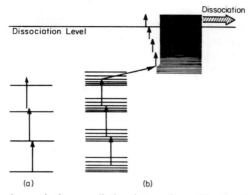

FIG. 3. (a) Monochromatic laser radiation interaction with an anharmonic oscillator. The laser excitation quickly runs out of steps with the ladder of excited states. (b) Schematic diagram showing near-resonant transitions over the discrete rotational–vibrational states followed by successive resonant transitions in a quasi-continuum to beyond the dissociation level.

Each vibrational transition now has three branches P ($\Delta J = -1$), Q ($\Delta J = 0$), and R ($\Delta J = +1$) with frequencies given by

$$\nu(v, j \to v + 1, J) \cong \nu_3 + v(\Delta\nu_{anh}) + 2\tilde{B}M,$$

where $M = -J$ for $\Delta J = -1$, $M = 0$ for $\Delta J = 0$, and $M = J + 1$ for $\Delta J = 1$. The constant $\tilde{B}$ is 0.027 cm^{-1} for SF$_6$. With a proper J value, the anharmonic shift can be nearly compensated by the rotational shift. Thus, in the up-excitation, if it starts with a proper rotational state in the ground-vibrational manifold ($J \sim 50$ in the case of SF$_6$), the first three or four single-photon steps can all be nearly resonant. The situation can be further improved when one considers in reality that the degeneracy of vibrational states is also split by Coriolis coupling and anharmonic coupling. Therefore, as shown in Fig. 3b, one can expect that at least the first three–six single-photon steps in the vibrational excitation are nearly resonant. Then, the density of states of a polyatomic molecule increases very rapidly and soon forms a quasi-continuum. In SF$_6$ for example, the density of states at 5000 cm^{-1} is already as high as 10^3/cm^{-1}. Therefore, after the first three–six near-resonant steps, the radiation field can now resonantly excite the molecule all the way through the quasi-continuum to above the dissociation level. This then qualitatively explains the observed high probability of multiphoton dissociation. While isotopic selectivity comes in through the initial near-resonant steps over the discrete states, absorption in the quasi-continuum simply puts enough energy into the molecule for subsequent dissociation. Clearly, the above theory is not valid for diatomic molecules and perhaps also triatomic molecules. This is the reason why no multiphoton dissociation of such small molecules has ever been reported.

An experiment using a molecular beam to study collisionless multiphoton dissociation[45] has shown that each SF_6 absorbs around 35 photons from a CO_2 TEA laser pulse. The absorbed energy appears to be randomly distributed in all vibrational degrees of freedom before dissociation. The dissociation rate is finite and increases very rapidly with the excitation energy. It is around 10^7 sec^{-1} at the excitation level of 35 photons absorbed per molecule. The power or/and energy threshold and the power or/and energy dependence of multiphoton dissociation are not yet clearly understood. The quantitative aspect of multiphoton excitation and also other characteristic features of the dissociation process are yet to be investigated. A full understanding of the process will be essential for its future application to isotope separation or other induced chemical processes.

10.4.5. Concluding Remarks. We have not discussed here a number of other important nonlinear optical effects. In particular, nonlinear optics in plasmas has developed into a major subfield in the study of laser interaction with plasmas and laser-induced fusion.[46] Nonlinear optics in optical waveguides is an important branch in the development of integrated optics. Applications of nonlinear optics to various disciplines have been growing rapidly in the recent years. We can anticipate that the field of nonlinear optics will certainly remain active for a number of years to come.

Acknowledgment

This work was supported by the Division of Materials Sciences, Office of Basic Energy Sciences, U. S. Department of Energy under contract No. W-7405-ENG-48.

[45] E. R. Grant, M. J. Coggiola, P. A. Schulz, Y. T. Lee, and Y. R. Shen, *in* "State-to-State Chemistry" (P. Brooks and E. Mayes, eds.), p. 72. ACS Symp. Ser. No. 56, Amer. Chem. Soc. (1977).

[46] See, for example, H. J. Schwarz and H. Hora, eds., "Laser Interaction and Related Plasma Phenomena," Vols. I, II, III, IV, 1969, 1971, 1973, and 1975 resp.

10.5. Laser-Selective Chemistry*

10.5.1. Introduction

The modern laser can provide extremely high intensities of light in relatively small volumes. Generally laser radiation is extremely monochromatic, often with frequency tunability. These factors make it possible to produce well-defined excited states of molecules and atoms, which may undergo specific chemical reactions. Since the energy available in a single photon may greatly exceed thermal energies, reactions may easily be induced that would not occur thermally. We may thus expect to be able to generate new chemical products that are characteristic of the selective nature of the excitation. Of practical importance is the selective reaction of particular nuclear isotopes. It is now clear that laser chemistry does provide viable methods for isotope separation and for the preparation of isotopically enriched samples. We may anticipate many other applications also. For instance, it is possible that chemical isomers (compounds that are structurally somewhat different but have the same composition) may be separated. It may be feasible to mass-produce and/or separate moderately expensive chemical products with laser excitation. Furthermore, in condensed media, we may expect the development of new electronic and optical devices based on selective chemical reaction.

The idea of using light to provide energy for specific chemical reactions is not new. We are all familiar with the fact that photochemical reactions are used quite effectively by nature. Chemists have long been interested in photochemistry; it is a most active area of research today, particularly among organic chemists. Several important books are available on the subject; some are listed at the end of this chapter. Many photochemical processes, including some described here, do not *require* laser sources. Some do, however. The use of laser sources provides a certain amount of control over the properties of the radiation, so that emphasis can be placed on chemical problems.

There are available at the present time several excellent and comprehensive reviews of achievements in laser chemistry, also listed at the end

---

* Chapter 10.5 is by **James T. Yardley**.

METHODS OF EXPERIMENTAL PHYSICS, VOL. 15B

of the chapter. Furthermore, results in this field are developing rapidly. This chapter is concerned primarily with providing some of the fundamental ideas and methods for developing laser chemistry. First some basic principles are outlined. An understanding of chemical reaction rates (kinetics) and excited state relaxation rates is crucial for successful selective chemistry and is considered in some detail. Finally we examine some well-documented examples of selective chemistry involving laser excitation of specific electronic states and of specific vibrational states.

10.5.2. Chemistry, Spectroscopy, and Laser Excitation

10.5.2.1. Principles for Laser-Selective Chemistry. Our object is to utilize the unique properties of laser radiation to generate a specific chemical reaction, yielding products that would be otherwise difficult to obtain. A little reflection (and some experience) leads us to a number of requirements that generally should be satisfied:

a.   The optical absorptions (or other optical properties) of the medium under investigation must allow for excitation of the appropriate excited state of the desired constituent.

b.   The light source must not produce too much extraneous excitation in the medium.

c.   A specific reaction (either externally induced or naturally occurring) must follow or result from the excitation process.

d.   Reactions or relaxation processes that compete with the reaction of interest or destroy the selectivity must be eliminated.

In order to satisfy the above requirements, it is clear that many considerations are required for a successful process. Let us consider excitation of a molecule AB to form AB*. Some of the many possibilities for selective chemical processes are

$$
\begin{aligned}
AB \xrightarrow{\hbar\omega_1(\uparrow)} AB^* &\longrightarrow A + B && \text{Dissociation}\\
&\longrightarrow \widetilde{AB} && \text{Isomerization}\\
&\xrightarrow{C} ABC && \text{Addition}\\
&\xrightarrow{C,CD} A + BC,\ ABC + D && \text{Abstraction}\\
&\xrightarrow{\hbar\omega_2(\uparrow)} A + B && \text{Two-step dissociation}\\
&\xrightarrow{\hbar\omega_2(\downarrow)} A + B && \text{Stimulated dissociation}\\
&\xrightarrow{(E,H)} A + B && \text{Field-induced predissociation}\\
&\xrightarrow{\hbar\omega_2(\uparrow)} A^+ + B^- && \text{Two-step ionization}\\
&\xrightarrow{(E)} A^+ + B^- && \text{Field-induced ionization}\\
&\xrightarrow{M} A + B && \text{Collision-induced predissociation}
\end{aligned}
$$

As illustrated above, selective chemistry can in principle be achieved simply by direct dissociation or isomerization of AB* or by simple reaction with an additional molecule C to form the desired products.   In practice, however, we may find it desirable to induce the desired reaction externally in order to eliminate competing processes or to provide sufficient energy for the reaction.   This might be done, for instance, through the absorption of additional quanta of radiation or through the application of an external electric or magnetic field.

10.5.2.2. Chemical Energy Surfaces. Chemists often consider the reactants and the products for some particular reaction to consist of two points on some set of multidimensional energy surfaces.   An understanding of these surfaces provides clues as to how the system may be moved from one point to the other.   The traditional starting point for the description of a molecular system is the Born–Oppenheimer approximation.[1]   In this approximation the electronic energy levels for a fixed nuclear configuration or geometry are determined.   In an ultrasimplistic view, each electronic energy level may be thought of as arising from an assignment of electrons to various *molecular orbitals,* which may be associated with several nuclei and which are somewhat analogous to atomic orbitals.   If the electronic energy levels are determined as a function of all variations in nuclear geometry, a set of energy surfaces is generated, each surface corresponding to a particular electronic state.   These states are often labeled with group-theoretical labels, which give an indication of how the wavefunction is transformed under the appropriate symmetry operations.   These surfaces define the potential energy for movement of the nuclei.   They determine the stable geometries for each electronic state of the molecule.   In many cases, the energy for nuclear motion is quantized, giving rise to discrete vibrational energy levels.   For unbound states, i.e., dissociative ones, the nuclear motion is described by one or more continua.

For a first approximation, certain terms in the total Hamiltonian for the system are ignored.   These usually include spin–orbit coupling, electronic–rotation interactions, vibration–rotation interactions, and those terms ignored in making the Born–Oppenheimer approximation. The total internal energy of the system is then just the sum of (1)   the minimum electronic energy for the appropriate state, (2)   the vibrational energy, and (3)   the rotational energy.   It should be emphasized, though, that the terms ignored above may give rise to energy shifts and splittings of degenerate energy surfaces.   In addition, they often play a key role in determining reaction paths by inducing transitions between surfaces.

Many typical features of chemical energy surfaces are demonstrated

[1] See, for instance, M. Bixon and J. Jortner, *J. Chem. Phys.* **48,** 715 (1968).

most easily for a diatomic molecule. In this case, there is only one nuclear degree of freedom, namely the internuclear distance R. As an example we may consider the Br_2 molecule. Figure 1 shows somewhat schematically some of the electronic energy surfaces. In this case the coupling of electron spin and electronic angular momentum is taken into account. For some electronic states, the surfaces have been experimentally mapped out. In other cases we must rely upon theoretical considerations. We see that two ground-state Br atoms (labeled $^2P_{3/2}$) may come together to produce three different states that are expected to be bound (i.e., stable with respect to separated atoms). The states $X(^1\Sigma_g^+)$ and $A(^3\Pi_{1u})$ have been experimentally investigated.[2] They are described by equilibrium internuclear distances of 2.284 and 2.695 Å, respectively, and by dissociation energies of 15,895.6 and 2077 cm^{-1}, respectively.[2a] There are also numerous unbound, or dissociative states about which, unfortunately, little is definitely known. The lowest-energy state (X) is strongly bound; thus Br_2 is stable at room temperature. Note that excited-state $B(^3\Pi_{ou}^+)$ is bound with respect to $Br(^2P_{3/2})$ + $Br(^2P_{1/2})$. However, it is sufficiently energetic to dissociate into two ground-state Br

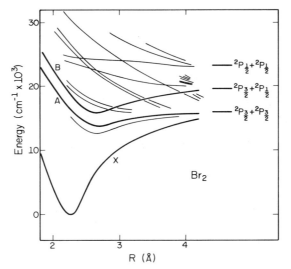

FIG. 1. Potential energy surfaces for Br_2. States A, B, and X are reconstructed from experimental data [J. A. Coxon, *J. Mol. Spectrosc.* **41**, 548 (1972); **51**, 428 (1974)]. Other states are sketched after the diagram of R. S. Mullikan [*J. Chem. Phys.* **55**, 288 (1971)].

[2] J. A. Coxon, *J. Mol. Spectrosc.* **41**, 548 (1972).

[2a] Wavenumber (cm^{-1}) is a spectroscopist's unit of energy. E (cm^{-1}) = E (erg)$/hc$. Thus, an energy given in cm^{-1} is multiplied by 1.986 × 10^{-16} to give energy in erg/molecule.

atoms. Because of electronic–rotation interaction, this state can interact with one of the dissociative states ($^3\Pi_1{}^-$). Thus, excitation of bound levels in the B state generally results in the production of two ground-state Br atoms.[3]

For a polyatomic molecule consisting of N atoms there are $3N - 6$ nuclear degrees of freedom ($3N - 5$ for a linear molecule). As a result, the potential surfaces become more difficult to visualize. Let us consider the system Br–H–Cl. If the system is constrained to be linear, there are only two nuclear degrees of freedom for each electronic state. Each potential surface may then be represented by a contour plot such as that shown in Fig. 2, in which the Br–H and H–Cl distances are used as nuclear coordinates to describe the ground state. It may be seen that, on this particular surface, both HBr and HCl are stable. However, HCl is more stable than HBr by 1.1×10^{-12} erg/molecule. The chemical reaction Br + HCl → BrH + Cl can be studied on this single potential surface using simple classical mechanics. Figure 2 gives an illustration of two reaction trajectories on this surface. For the trajectory shown in Fig. 2a, sufficient energy for the reaction is provided primarily in the form of HCl vibrational energy; in the other (Fig. 2b), the same amount of energy is entirely in the form of translational energy. For these particular examples, it may be seen that vibrational excitation is much more effective than translational energy in causing reaction to occur. This is not always the case, however. It must be emphasized that for nonlinear encounters the potential surface in this contour plot will be different. Extensive investigations of classical reactive trajectories such as those shown in Fig. 2 have been carried out.[4] They provide a great deal of insight concerning reactions on a single potential energy surface.

For more complicated molecular systems, chemists may find it convenient to assemble qualitative or semiquantitative energy "surfaces." In order to do this, the appropriate electronic energy levels of the products and reactants must first be obtained. There are many tools available for this. Spectroscopic measurements generally yield the most accurate information when they are available. A considerable amount of thermochemical data is available (heats of reactions, etc.), which may be of considerable use if properly manipulated. Unfortunately, such information is rarely of use for excited states. Sometimes simple concepts such as "bond energies" may be helpful. For instance, the breaking of a C–C bond requires about 5.8×10^{-12} erg/molecule; a C–H bond requires

[3] M. S. Child and R. B. Bernstein, *J. Chem. Phys.* **59**, 5916 (1973).

[4] R. N. Porter, *Annu. Rev. Phys. Chem.* **25**, 317 (1974); D. S. Perry, J. C. Polanyi, and C. W. Wilson, *Chem. Phys.* **3**, 317 (1974); J. C. Polanyi and W. H. Wang, *J. Chem. Phys.* **51**, 1439 (1969).

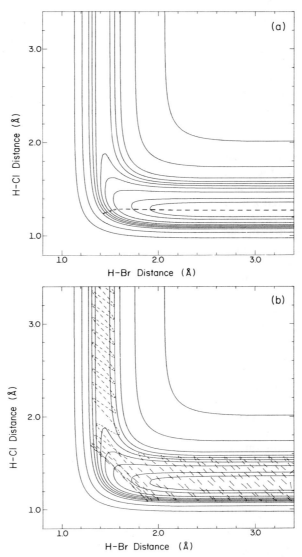

FIG. 2. Model ground-state potential surface for linear Br–H–Cl system. Energy contours in electron volts are 0.1, 0.2, 0.4, 0.55, 0.65, 0.75, 0.85, 1.0, 1.5, and 2.5, where zero of energy corresponds to ClH + Br. (a) Example of trajectory for reaction Br + HCl → BrH + Cl, where sufficient energy for reaction is present only in the form of Br–HCl relative kinetic energy. (b) Example of trajectory for same reaction except that sufficient energy for reaction is present in HCl vibration and only a small amount of translational energy is present. These trajectories were kindly provided by Lindley Specht and were carried out in a manner similar to those described in J. C. Polanyi and W. H. Wang [*J. Chem. Phys.* **51**, 1439 (1969)].

$\sim 7 \times 10^{-12}$ erg/molecule. A number of useful references for thermochemical calculations are provided at the end of the chapter.

After reactant and product energies are determined or estimated, it is useful to attempt to determine which of the product and reactant states "correlate," i.e., which reactant states are transformed into which product states if the reaction occurs gradually and smoothly on the same potential surface. For this purpose, a physical picture of the electronic configurations involved is very helpful. Group theory is also useful since certain symmetry elements must often be preserved during the transformation. In many cases, assumptions may be necessary concerning the geometry of the reaction and concerning the importance of various terms in the Hamiltonian for the system. These considerations have been discussed in detail by Herzberg, who provides many examples of correlation diagrams. These works are listed in the Bibliography.

As a specific example we might consider again the reaction Br + HCl → BrH + Cl. Figure 3 gives a simple pictorial representation of the molecular orbitals (as linear combinations of s and p atomic orbitals) for HCl and for linear Br–H–Cl. Schematic energies for the molecular orbitals are also indicated along with the electron occupancies for ground-state and low-lying excited states. Figure 4 shows a possible cor-

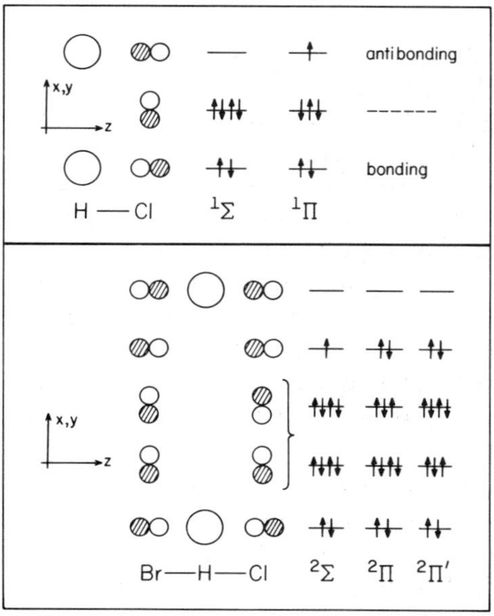

FIG. 3. Simple pictorial representations of molecular orbitals for HCl and for BrHCl as linear combinations of H(1s), Br(4p), and Cl(3p) atomic orbitals. Also shown are orbital occupancies for ground and low-energy excited states.

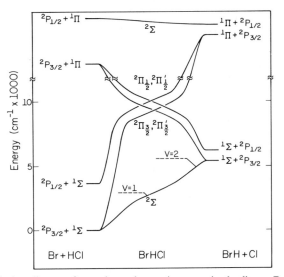

FIG. 4. Correlation diagram for various electronic states in the linear Br–H–Cl system [after S. R. Leone, R. G. Macdonald, and C. B. Moore, *J. Chem. Phys.* **63**, 4735 (1975)]. The dashed lines indicate energies of $[Br(^2P_{3/2}) + HCl\ (v = 1)]$ and $[Br(^2P_{3/2}) + HCl\ (v = 2)]$.

relation diagram for reactions on several distinct surfaces of the linear Br–H–Cl molecular system. We observe that given sufficient energy, ground-state Br atoms $(^2P_{3/2})$ may react with ground-state $HCl(^1\Sigma)$ to produce ground-state products HBr + Cl or to produce excited $HBr(^1\Pi)$ and ground-state Br. Since the $^1\Pi$ state of HBr is dissociative, the latter reaction would in fact produce Cl + H + Br. We observe also that in this linear geometry the excited-state Br $(^2P_{1/2})$ in reaction with HCl $(^1\Sigma)$ may not produce ground-state products. For a nonlinear reaction some restrictions are removed. For instance, Br $(^2P_{1/2})$ + HCl $(^1\Sigma)$ can yield HBr $(^1\Sigma)$ + Cl $(^2P_{1/2})$ but a significant *barrier* to that reaction would be expected. It should be emphasized that correlation schemes such as this one are approximate and that various interactions can induce transitions between surfaces.

If the electronic energies of the products and the reactants are known, it is energetically possible for the reaction of interest to occur if the *total* energy of the reactants (electronic, vibrational, rotational, and translational) exceeds that of the products. This does not *ensure* reaction, however. The object of laser selection is to provide the energy for reaction in the appropriate way.

10.5.2.3. Molecular Spectroscopy. For bound molecular states, where the energy levels are quantized, direct single-photon absorption of radiation can occur for light frequencies that satisfy the Bohr condition $\hbar\omega = \Delta E$. Here and throughout, we shall use angular frequency $\omega = 2\pi\nu = 2\pi c/\lambda$. It is also necessary that the energy levels involved obey spectroscopic selection rules. Associated with each transition is a lineshape, which may be characterized qualitatively by a "linewidth." Finite linewidths may arise from the existence of finite dephasing times for the two levels involved. In practical situations the observed lineshapes may also be influenced by the Doppler effect. Furthermore, line broadening may be induced by high-intensity radiation. Because of these finite linewidths, excitation of molecular states may occur even if the light frequency does not exactly satisfy the Bohr condition.

Optical excitation of quantized states is possible through processes other than single-photon absorption. For high excitation intensities, multiple-photon absorption can occur[5] and may induce dissociation or ionization directly. Various Raman and stimulated Raman processes also in principle may provide selective laser excitation.[6]

The existence in molecular systems of discrete absorption frequencies associated with specific excited states is important for laser chemistry because (1) it allows excitation of states leading to specific reactions, and (2) it makes possible the selection of a particular chemical species, isotope, or isomer for chemical reaction. Optical excitation to dissociative (continuum) states is possible, but because of the significant linewidths the excitation process tends to be nonselective and tends to occur with a relatively low probability.

Isotope effects in molecular spectra are particularly significant because of the practical importance of laser-induced enrichment of nuclear isotopes. In molecular spectroscopy, isotope effects are quite common. Important examples include the following:

10.5.2.3.1. VIBRATIONAL ENERGIES. Molecular vibrational frequencies depend upon the forces holding the nuclei together (which are isotope independent) and directly upon the nuclear masses. For a simple diatomic molecule the vibrational frequency is $\omega = (k/\mu)^{1/2}$, where μ is the reduced mass. Thus, the vibrational frequencies for HCl and for DCl are 5.635×10^{14} and 3.941×10^{14} sec^{-1}, respectively; for $^{79}Br^{79}Br$ and $^{81}Br^{81}Br$ they are 3.1593×10^{13} and 3.1201×10^{13} sec^{-1}. For many polyatomic molecules, Herzberg gives the isotope effects explicitly for

[5] W. L. Peticolas, *Annu. Rev. Phys. Chem.* **18**, 233 (1967).

[6] C.-S. Wang, *in* "Quantum Electronics: A Treatise" (H. Rabin and C. L. Tang, eds.), p. 447. Academic Press, New York, 1975.

each normal mode of vibration.[7]   In the general case they may be calculated using methods discussed by Wilson et al.[8] Computer programs to do this are readily available.[9]

10.5.2.3.2. ELECTRONIC ENERGY LEVELS. Direct isotope effects on electronic energy surfaces are exceedingly small.   In molecules the principal apparent isotope effect on electronic energy levels results from a difference in the zero-point vibrational energy.   As an example, we may consider the $A(^1A'') \leftarrow X$ absorption[10] in H_2CO for which the electronic origin is found to be 28,188.0 cm^{-1}. For HDCO the value is 28,243 cm^{-1} and for D_2CO it is 28,301 cm^{-1}. These energy differences translate into rather obvious spectral differences.

10.5.2.3.3. ROTATIONAL ENERGIES. Since rotational energy levels may usually be described in terms of three principal moments of inertia, they also exhibit isotope effects.   For a diatomic molecule the rotational levels are given by $(\hbar^2/2I)J(J + 1)$, where the moment of inertia $I = \mu R^2$ and μ is the reduced mass.   For HCl, $B = 10.59$ cm^{-1}, while for DCl $B = 5.445$ cm^{-1}.   For other molecules the effect is usually smaller.

10.5.3. Chemical Kinetics

10.5.3.1. Excitation Processes. For laser chemistry it is usually not sufficient to know whether or not a reaction is energetically possible.   In addition we need to understand chemical kinetics, that is, the *rates* of various processes.   These rates are usually described in terms of the time derivative of the concentration of a particular species, rate $= d[N]/dt$, where we shall express the concentration of N, [N], in molecules/cm^3. For excitation of a molecule in nondegenerate state a to a particular excited state b by one-photon absorption, the rate of excitation (at any point in a medium) is given by the Einstein equation $d[N_b]/dt = B([N_a] - [N_b])$.   The coefficient B may usually be written

$$B = 4\pi^2(3v\hbar^2)^{-1}|S||D_{ba}|^2 g(\omega),$$

where v is the velocity of light in the medium, S is the light intensity in the medium (erg sec^{-1} cm^{-2}), D_{ba} is the magnitude of the electric dipole matrix element between states a and b, and $g(\omega)$ is a normalized absorption lineshape function that peaks near the transition frequency $\omega_{ba} =$

---

[7] G. Herzberg, "Infrared and Raman Spectra of Polyatomic Molecules." Van Nostrand-Reinhold, Princeton, New Jersey, 1945.

[8] E. B. Wilson, J. C. Decius, and P. C. Cross, "Molecular Vibrations." McGraw-Hill, New York, 1955.

[9] Several programs are available from the Quantum Chemistry Program Exchange, Bloomington, Indiana.

[10] V. A. Job, V. Sethuraman, and K. K. Innes, *J. Mol. Spectrosc.* **30**, 365 (1969).

$\hbar^{-1}(E_b - E_a)$. For instance, in low-pressure gases $g(\omega)$ is usually dominated by the Doppler effect and is given by

$$g(\omega) = (v/\omega)(m/2\pi kT)^{1/2}\exp[-(mv^2/2kT)(\omega - \omega_{ba})^2\omega_{ba}^{-2}].$$

Where intermolecular interactions dominate the lineshape, $g(\omega)$ may often assume a Lorentzian form $(\pi\tau)^{-1}[(\tau)^{-2} + (\omega - \omega_{ba})^2]^{-1}$, where τ is a time characteristic of the interaction.

If the medium is uniformly illuminated and if the net absorption is small, the excitation may be expected to be reasonably uniform. Otherwise, the spatial behavior must be taken into account. For pulsed excitation the total concentration of excited molecules at any time may be determined by integrating $d[N_b]/dt$. If $[N_a] > [N_b]$ then $[N_a]$ is approximately constant, and, if relaxation processes can be ignored, the integration is straightforward.

With the high radiation intensity available from lasers, the excitation kinetics can become much more complicated. For a proper description, the interaction of radiation with matter must be considered in detail. For different situations many interesting phenomena may result. Examples include *saturation,* in which the populations of a and b may become equalized, and *multiphoton absorption.*

10.5.3.2. Unimolecular Decay. We need to consider not only excitation processes, but also the decay of excited molecules. It is often found that an excited molecular system can decay unimolecularly, i.e., without the need of any intermolecular interaction. In this case, the kinetics should obey the first-order decay law $d[N_b]/dt = -k[N_b]$, where k is the first-order *rate constant.* Since energy must be conserved, such decay must involve the flow of excitation into some continuum or near-continuum lying at the same energy. There are many important examples of unimolecular decay. For spontaneous emission of radiation k is just the Einstein A coefficient: $A = 4\omega_{ba}^3|D_{ba}|^2/3\hbar vc^2$. For electronically excited molecules, typically $A \sim 10^6$–10^8 sec^{-1}; for vibrational excitation $A \sim 10^3$–10^0 sec^{-1}. Of more chemical interest, it has been found that molecules excited to bound molecular states may undergo unimolecular decay into a dissociative continuum, a process known as "predissociation." For example, as mentioned previously Br_2 excited into the B state (see Fig. 1) predissociates into ground-state Br atoms. Large polyatomic molecules may in addition undergo unimolecular radiationless decay into a near-continuum of vibrational levels belonging to another bound state. This is called internal conversion if the electron spin angular momentum is conserved and intersystem crossing if it is not.

For decay according to a simple first-order rate law with no excitation present for times $t > 0$, direct solution of the differential equation for the decay of $[N_b]$ gives exponential decay: $[N_b(t)] = [N_b(0)] \exp[-kt]$. Often this behavior is observed even when the decay is not truly unimolecular. In this case, the derived constant k is called a pseudo-first-order rate constant.

10.5.3.3. Bimolecular Relaxation. Many chemical reactions or decay processes require the encounter of *two* species. In this case, the expected rate law is $d[N_b]/dt = -k[N_b][X]$ in which X is the designated partner, assuming access of available species in proportion to their concentrations. Here k is a second-order rate constant and has units of cm^3 molecule^{-1} sec^{-1}. For a gas phase bimolecular reaction, simple theories show that $k = \int \sigma(v)vf(v)\, dv$ where v is the relative speed, $\sigma(v)$ the "reaction cross section," and $f(v)$ the relative speed distribution function. Therefore, it is sometimes convenient to describe a reaction with a known second-order rate constant k in terms of an average cross section $\langle\sigma\rangle \equiv k/\int vf(v)\, dv$. For hard spheres the cross section is πd^2, where d is the collision diameter. Thus, we could also think of a reaction probability per collision $P = \langle\sigma\rangle/\pi d^2$. Note that the rate constant for hard-sphere bimolecular encounters is just $\pi d^2 \int vf(v)\, dv = \pi d^2 (8kT/\pi\mu)^{1/2}$. As an example, the thermal rate constant for the reaction Cl + HBr $\rightarrow$ HCl + Br is[11] 7.2×10^{-12} cm^3 molecule $^{-1}$ sec^{-1}. This translates into a cross section of 1.4 Å^2 or a probability per collision of 0.028, assuming a collision diameter of 4 Å.

If $[X]$ is constant, it is seen that second-order decay of N_b is pseudo-first order. If $d[N_b]/dt$ obeys the rate law $d[N_b]/dt = -k_1[N_b] - k_2[N_b][X]$ the pseudo-first-order rate constant is just $k_1 + k_2[X]$. Note that in this case a plot of pseudo-first-order rate constant against $[X]$ is linear and allows determination of both k_1 and k_2. This is called Stern–Volmer plot. Of course, in general $[X]$ may not be constant for a reactive system. In this case, the coupled differential equations for $[N_b]$ and $[X]$ must be solved to obtain the complete time behavior.

10.5.3.4. Termolecular Reactions. An exothermic chemical reaction such as AB + C $\rightarrow$ A + BC can occur as a true two-body reaction because the energy released may be partitioned between A–BC relative translational energy and BC vibrational energy. However, if two atoms come together [for instance, Br($^2P_{3/2}$) + Br($^2P_{3/2}$)], a bound diatomic molecule [$Br_2(X)$] can be formed only if some energy is removed from the two-body system. This may occur during interaction with a third body. Gas phase atomic recombination is thus one example of a termolecular reaction. The rate of termolecular reaction of N_b with species X and M is $d[N_b]/dt = -k[N_b][X][M]$. The third-order rate constant k has units cm^6

[11] K. Bergmann and C. B. Moore, *J. Chem. Phys.* **63**, 643 (1975).

molecule^{-2} sec^{-1}. The quantitative theoretical description of such rate constants is rather tedious,[12] but elementary arguments suggest that $k \sim 10^{-32}$ cm^6 molecule^{-2} sec^{-1} if each three-body encounter is effective. For example, the recombination of Ar*(^{3}P) with ground-state Ar to produce ^{3}Ar$_2$* bound molecules has been found to proceed with a third-order rate constant[13] of 6.6×10^{-33} cm^6 molecule^{-2} sec^{-1}.

10.5.3.5. Diffusion-Controlled Reactions. In Section 10.5.3.3, it has been assumed that the rate of bimolecular encounters depends only upon the concentrations of the two species. However, in liquids and in high-pressure gaseous media (pressures above ~ 1 atm) the encounter rate between two dilute species may depend upon their ability to move through the medium. From simple arguments[14] the encounter rate between A and B in this situation will be limited to a quantity that may be written $k_D[A][B]$. The rate constant for bulk diffusion is approximately $k_D = 4\pi(r_A + r_B)(D_A + D_B)$ where r_A and r_B are the radii of A and B and where D_A and D_B are the diffusion coefficients for A and B in the medium. In aqueous solution at room temperature the diffusion-limited reaction rate constant is $\sim 10^{-11}$ cm^3 molecule^{-1} sec^{-1}. In 1 atm of buffer gas such as Ar, this limiting rate constant is $\sim 1.5 \times 10^{-10}$ cm^3 molecule^{-1} sec^{-1}. It is inversely proportional to buffer gas pressure.

10.5.3.6. Surface-Catalyzed Reactions. Some chemical reactions and excited-state relaxation processes are known to occur efficiently on the walls of the containing vessel. The kinetic behavior in these cases can be quite complicated and may require solution of some appropriate diffusion equation. For gases in which the pressure is sufficiently low, the mean free path λ exceeds the vessel dimension d_v, and an excited molecule encounters the wall with a frequency on the order of $\langle v \rangle / d_v$. For $d_v \sim 2.5$ cm and $\langle v \rangle \sim 3 \times 10^4$ cm/sec this is $\sim 10^4$ sec^{-1}. Note that this frequency is independent of pressure as long as $d_v < \lambda$. If a bimolecular encounter with a molecule adsorbed on the surface is required, the reaction rate will also depend on the amount of surface coverage. For higher pressure gases and for liquids, the motion of a molecule to the walls of the vessel is diffusional. For diffusional motion from the center of a cylindrical vessel of radius R, the mean rate constant for reaching the wall is[15] $\sim 3.6D/R^2$, where D is the diffusion coefficient. For gases, D is inversely proportional to pressure. It may be worth noting that atomic recombina-

[12] See, for instance, H. O. Pritchard, *Acc. Chem. Res.* **9**, 99 (1976).

[13] R. Boucigne and P. Mortier, *J. Phys. D* **3**, 1905 (1970).

[14] W. C. Gardiner, Jr., "Rates and Mechanisms of Chemical Reactions," pp. 165–170. Benjamin, Menlo Park, California, 1969.

[15] A. C. G. Mitchell and M. W. Zemansky, "Resonance Radiation and Excited Atoms," p. 246. Cambridge Univ. Press, London and New York, 1961.

tion, which requires termolecular interactions in the bulk, may occur readily on surfaces.

10.5.3.7. Equilibrium Considerations.

It must always be remembered
that for every reaction there is a reverse reaction. Thus, for the general
reaction aA $+$ bB $+$ cC $+ \cdots \rightleftharpoons x$X $+ y$Y $+ z$Z $+ \cdots$ there is a forward rate constant k_f and a reverse rate constant k_r. When all internal
states are at equilibrium, the net reaction rate must be zero, giving

$$k_f/k_r = [A]_e^a[B]_e^b[C]_e^c \cdots / [X]_e^x[Y]_e^y[Z]_e^z \cdots,$$

where the subscript e refers to the equilibrium value. The right-hand side
of this equation may be calculated from statistical mechanics[16]:

$$k_f/k_r = q_A{}^a q_B{}^b q_C{}^c / q_X{}^x q_Y{}^y q_Z{}^z \cdots,$$

where q_A is the partition function per unit volume for species A. This fact
has two important consequences for laser-induced chemistry.

(a)   If either k_f or k_r is known, the reverse rate constant can be calculated from statistical mechanics. Care must be taken when applying
these considerations when the internal degrees of freedom are not in equilibrium. For instance, if we know the rate constant for HCl ($v = 2$) $+$ Br
($^2P_{3/2}$) $\rightarrow$ HBr ($v = 0$) $+$ Cl ($^2P_{3/2}$), where v refers to vibrational quantum
number, we can calculate (assuming rotational and translational equilib-
rium) the rate constant for the reaction HBr ($v = 0$) $+$ Cl ($^2P_{3/2}$) $\rightarrow$ HCl
($v = 2$) $+$ Br ($^2P_{3/2}$). However, we cannot calculate the total rate constant for HBr ($v = 0$) $+$ Cl ($^2P_{3/2}$) $\rightarrow$ HCl $+$ Br without more information.

(b)   Regardless of how fast a reaction can occur, if the internal degrees
of freedom remain in equilibrium, there is a limit on the concentration of
products that can be formed. For example, suppose we wish to produce
the diatomic molecule AB starting with atoms A and B. The *most* AB we
can form is the equilibrium amount, given by[16]

$$[AB]_e \cong [A]_e[B]_e \left[\left(\frac{2\pi\hbar^2}{kT}\right)\left(\frac{m_{AB}}{m_A m_B}\right) \right]^{3/2} \left[\exp\left(\frac{D_e}{kT}\right) \right]$$
$$\left[\frac{kT}{B}\right]\left[\exp\left(\frac{-\hbar\omega}{2kT}\right)\right]\left[1 - \exp\left(\frac{-\hbar\omega}{kT}\right)\right]^{-1},$$

where D_e is the dissociation energy, $B = \hbar^2/2I$ the rotational constant of
AB, and ω the vibrational frequency of AB. In particular, consider the
formation of NaXe* ($D_e \sim 1770$ cm^{-1}) from sodium atoms produced initially in the ^{2}P state (Na*) and from Xe atoms at a temperature of 723 K.

[16] D. A. McQuarrie, "Statistical Mechanics," Chapter 9. Harper, New York, 1970.

The above formula gives $[NaXe^*] \leqslant 3 \times 10^{-2}[Na^*]p_{Xe}$ where p_{Xe} is the xenon pressure in atmospheres. Thus, ~ 100 atm of Xe will be needed if appreciable conversion from Na* to NaXe* is desired.

10.5.3.8. Chain Reactions. Since several chemical reactions can occur simultaneously and can be strongly coupled, complex behavior can be observed that, at times, can be quite misleading with regard to the mechanism of the reaction. We can consider the simple chain reaction

$$Cl_2 \longrightarrow Cl + Cl \qquad \text{Initiation}$$

$$\left.\begin{array}{l} Cl + H_2 \longrightarrow HCl + H \\ H + Cl_2 \longrightarrow HCl + Cl \end{array}\right\} \quad \text{Propagation}$$

The net effect of the above set of reactions is $Cl_2 + H_2 \rightarrow 2HCl$. However, in this case the reaction proceeds through formation of Cl atoms. Once formed, a single Cl atom can cause (or catalyze) the production of many HCl molecules. The chain can only be terminated by a reaction that destroys the chain carrier Cl. One example of a termination reaction is $Cl + Cl \rightarrow Cl_2$. From an experimental point of view, the mechanisms of chain reactions are often difficult to elucidate. A variety of techniques, such as the addition of known scavengers of chain carries, may be employed.[17] A yield of several product molecules per reactant molecule often is characteristic of a chain reaction mechanism.

10.5.4. Relaxation Processes

10.5.4.1. General Considerations. There are many processes that *relax* or change the state function for a molecular system following selective optical excitation. These relaxation processes may be detrimental for selective laser chemistry because they provide pathways for the removal of the excitation, thereby preventing reaction or for scrambling of the excitation removing selectivity. On the other hand, relaxation processes can be selective in some cases and thus can be of assistance in producing the desired chemical result.

For relaxation occurring during bimolecular encounters, a first-order semiclassical calculation for the probability per collision of a transition from internal states of the bimolecular system designated a to states b gives the Born approximation[18]:

$$P_{a \rightarrow b} \cong \hbar^{-2} \left| \int_{-\infty}^{\infty} V_{ba}(t') \exp(i\omega_{ba}t') \, dt' \right|^2 ,$$

[17] W. C. Gardiner, Jr., "Rates and Mechanisms of Chemical Reactions," Chapter 8. Benjamin, Menlo Park, California, 1972.

[18] A. Messiah, "Quantum Mechanics," Vol. II, pp. 722–727. Wiley, New York, 1962.

where V_{ba} is the matrix element of the collisional perturbation coupling b and a and ω_{ba} is the transition frequency. It may be seen from this that the collision must indeed couple states a and b before a transition may take place. It is also clear that the transition probability will be greatest when the perturbation is "resonant" with the transition frequency, i.e., if V_{ba} has significant Fourier components at frequency ω_{ba}. Thus, resonant transfer ($\omega_{ba} \sim kT$) may occur with high probability while nonresonant transfer ($\omega_{ba} \gg kT$) is often inefficient.

Relaxation can also occur in isolated molecules due to intramolecular interactions. In many instances the unimolecular rate constant may be expressed by Fermi's golden rule[19] $k = 2\pi\hbar^{-1}|V_{ba}|^2\rho$, where V_{ba} is the matrix element of the intramolecular hamiltonian between initial state b and the continuum a, and ρ is the number of states per unit energy in the continuum.

10.5.4.2. Relaxation of Electronic Excitation. In isolated molecules, intramolecular interactions can induce transitions from one electronic surface to another. If the final electronic surface in the transition is dissociative, we have already seen that "predissociation" can take place as in the decay of Br_2 excited to the $B(^3\Pi_{0u}^+)$ state. If the final electronic state is stable, then it must have a sufficiently high density of internal (vibrational) energy levels to provide an effective continuum into which excitation may flow, conserving energy within the restrictions of the uncertainty principle. For molecules composed of less than six atoms, the density of vibrational levels is usually too small for intramolecular electronic relaxation to occur. In these cases relaxation for isolated molecules can be radiative or predissociative. For molecules of six atoms or more, radiationless intramolecular decay processes are relatively common. It may be noted that, in principle, the matrix element V_{ba} may couple some vibrational levels more effectively than others. In this case, the radiationless process can be somewhat selective. In large organic molecules, it has been found empirically that radiationless relaxation to the lowest energy excited singlet and triplet states is usually extremely efficient. Experimental observations and theoretical considerations concerning radiationless electronic relaxation processes have been discussed in several review articles.[20]

Molecular collisions and other forms of intermolecular interaction can

[19] A. Messiah, "Quantum Mechanics," Vol. II, pp. 732–736. Wiley, New York, 1962.

[20] J. Jortner, S. A. Rice, and R. M. Hochstrasser, *Adv. Photochem.* **7**, 149 (1969); K. F. Freed, *Top. Curr. Chem.* **31**, 105 (1972); J. T. Yardley, *in* "Chemical and Biological Applications of Lasers" (C. B. Moore, ed.), Vol. 1, p. 231. Academic Press, New York, 1974; P. Avouris, W. M. Gelbart, and M. A. El-Sayed, *Chem. Rev.* **77**, 793 (1977).

induce radiationless electronic relaxation, often with great facility. Here we give a few of the more important processes.

10.5.4.2.1. ELECTRONIC EXCITATION TRANSFER (E → E). During an intermolecular interaction it is possible for electronic excitation (denoted with an asterisk, *) to be transferred from one molecule to another: A* + B ⇄ A + B*. This is a reasonably common occurrence in gas phase collisions involving atoms and diatomic molecules. The decay of the initially excited molecule or atom usually follows Stern–Volmer behavior, although if the acceptor B* is metastable, an equilibrium may be established and double-exponential decay observed. These processes have been examined extensively both theoretically and experimentally.[21] When dipole–dipole interactions dominate, collision cross sections as high as ~500 Å^2 have been found when the electronic states are nearly resonant in energy. In molecules, it may be noted that electronic exchange can bring about apparent excited-state vibrational relaxation: A*† + A → A† + A*, where † denotes vibrational excitation. This has been suggested, for example, in the case of NO ($A^2\Sigma^+$).[22] Excitation transfer can also take place in condensed phases, where in many cases it is known as "Förster Transfer." Here the observed kinetics for decay of an excited molecule A* undergoing excitation transfer to B are complicated since the transfer efficiency depends upon the relative orientations and intermolecular distances.[23]

10.5.4.2.2. ELECTRONIC DEACTIVATION (E → T). In atoms, it is possible for intermolecular interaction to quench the electronic excitation: A* + M → A + M. The cross sections for such processes are expected to be small since this process is generally nonresonant. In molecules, other effects usually dominate.

10.5.4.2.3. INTERMOLECULAR ELECTRONIC–VIBRATIONAL EXCITATION TRANSFER (E → V). Processes can occur in which electronic excitation in one molecule or atom is converted into vibrational excitation of another molecule: A* + B → A + B†. These processes may be particularly important if A is an atom so that intramolecular E → V transfer is not possible. Many examples are known[24,25] including transfer from Hg (6^3P) to CO ($v = 2$–9) and from Br($^2P_{1/2}$) to HCN ($v_3 = 1$).

[21] E. E. Nikitin, *Adv. Chem. Phys.* **28**, 317 (1975).

[22] L. A. Melton and W. Klemperer, *J. Chem. Phys.* **55**, 1468 (1971).

[23] T. Förster, *in* "Comparative Effects of Radiation" (M. Burton, J. Kirby-Smith, and J. Magee, eds.), p. 300. Wiley, New York, 1960; J. B. Birks, "Photophysics of Aromatic Molecules," p. 567ff. Wiley (Interscience), New York, 1970.

[24] G. Karl, P. Kruus, and J. C. Polanyi, *J. Chem. Phys.* **46**, 224 (1967).

[25] A. Hariri, A. B. Petersen, and C. Wittig, *J. Chem. Phys.* **65**, 1872 (1976); D. S. Y. Hsu and M. C. Lin, *Chem. Phys. Lett.* **42**, 78 (1976).

10.5.4.2.4. INTRAMOLECULAR ELECTRONIC–VIBRATIONAL ENERGY TRANSFER (E → V). Relaxation of the form A* + M → A† + M is probably the most common form for electronic relaxation of polyatomic molecules in the gas phase as well as in condensed phases. The role of the intermolecular interaction is either to provide a matrix element coupling A* to the final energy surface or to provide for conservation of energy through additional degrees of freedom. Some simple theories for this type of process exist,[26] although successful general theories would appear to be lacking.

10.5.4.3. Vibrational Relaxation. Since chemical reactivity is intimately related to nuclear motion, it is not surprising that vibrational energy transfer is important in laser chemistry. The net *removal* of vibrational energy during intermolecular interaction usually proceeds through transfer to the translational and rotational degrees of freedom (V → T,R): A† + M → A + M. Average collisional efficiencies at room temperature range from $\sim 10^{-13}$ for N_2 to ~ 1 for H_2O. Generally the coupling of translational and rotational motion to vibrational motion is thought to be induced by the repulsive part of the intermolecular potential and thus the efficiency increases sharply with increasing temperature.[27] At low temperatures or in chemically interacting systems, however, this is not always the case, and inverse-temperature dependencies have been noted.[28] In condensed phases, vibrational energy relaxation is rapid because of the close coupling with the nuclear motions of the medium,[29] although some exceptions have been noted, particularly at very low temperatures.[30]

The transfer of vibrational excitation from one form to another as a result of intermolecular interaction (V → V transfer) can occur with much greater efficiency than V → T,R transfer because of the possibilities for resonance. The available vibrational energy may be partitioned in a number of ways between the interacting molecular systems.[31] Because of the rapidity of V → V transfer relative to V → T,R transfer, it is possible on appropriate timescales for the vibrational energy levels to attain a distribution significantly different from an equilibrium one. The resulting

[26] C. A. Thayer and J. T. Yardley, *J. Chem. Phys.* **57**, 3992 (1972).

[27] T. L. Cottrell and J. C. McCoubrey, "Molecular Energy Transfer in Gases," Chapter 6. Butterworth, London, 1961.

[28] D. J. Miller and R. C. Millikan, *J. Chem. Phys.* **53**, 3384 (1970).

[29] A. Laubereau and W. Kaiser, *Annu. Rev. Phys. Chem.* **26**, 83 (1975).

[30] D. S. Tinti and G. W. Robinson, *J. Chem. Phys.* **49**, 3229 (1968); L. E. Brus and V. E. Bondybey, *ibid.* **63**, 786 (1975); W. F. Calaway and G. E. Ewing, *ibid.* p. 2842.

[31] C. B. Moore, *Adv. Chem. Phys.* **23**, 41 (1973).

distribution has been discussed by Treanor[32] and by Brey[33] for diatomic molecules. Such a distribution may favor excitation of specific isotopes. For instance, V $\to$ V transfer following excitation of ^{12}CO in an Ar matrix near 10 K results in a considerable flow of excitation into ^{13}CO and $C^{18}O$.[34] In condensed phases at room temperature V $\to$ V transfer should occur on picosecond timescales and can be both intramolecular and intermolecular.

In polyatomic molecules nuclear motion is usually thought of in terms of "normal modes of vibration," which are natural motions of the system for small displacements from equilibrium. In the small-displacement limit, excitation of a natural vibration in an isolated molecule should not result in excitation of other vibrations. However, if excitation involves larger amplitudes of vibration, energy may flow into other vibrational modes when this is energetically possible. Experiments on molecules with as few as seven atoms and 4200 cm^{-1} of vibrational energy have shown almost statistical distribution[35] of the energy among the various normal modes on timescales of ~ 1 msec. Calculations suggest[36] that this intramolecular relaxation may generally occur on nanosecond or faster timescales in many situations. Whether or not intramolecular V $\to$ V transfer occurs and on what timescale will depend upon the nature and amount of the excitation and upon the density of vibrational energy levels into which excitation can flow.

10.5.4.4. Rotational Relaxation. Rotational energy transfer processes occur readily in molecular collisions because of the strong angular dependence in intermolecular interaction potentials. This is particularly true if both collision partners have permanent dipole moments; cross sections for rotational energy transfer greater than 250 Å^2 have been observed in such cases.[37] Even for collisions between He and relatively spherical Li$_2$ molecules, cross sections ~ 20 Å^2 are reasonable.[38] Exceptions apparently include high rotational levels in diatomic hydrides[39] (such as HCl) where rotational relaxation processes can be relatively nonresonant ($\Delta E \sim 522$ cm^{-1} and $J = 26 \to 27$ in HCl).

[32] C. E. Treanor, J. W. Rich, and R. G. I. Rehm, *J. Chem. Phys.* **48**, 1798 (1968).

[33] K. N. C. Bray, *J. Phys. B* **1**, 705 (1968); **3**, 1515 (1970).

[34] H. Dubost, L. Abouaf-Marguin, and F. Legay, *Phys. Rev. Lett.* **29**, 145 (1972); H. Dubost and R. Charneau, *Chem. Phys.* **12**, 407 (1976).

[35] J. F. Durana and J. D. McDonald, *J. Chem. Phys.* **64**, 2518 (1976).

[36] J. D. McDonald and R. A. Marcus, *J. Chem. Phys.* **65**, 2180 (1976).

[37] S. L. Coy, *J. Chem. Phys.* **63**, 5145 (1975).

[38] K. Bergmann and W. Demtröder, *J. Phys. B* **5**, 2098 (1972).

[39] J. C. Polanyi and K. B. Woodall, *J. Chem. Phys.* **56**, 1563 (1972).

10.5.5. Selective Chemistry by Electronic Excitation

10.5.5.1. Introduction. It should be readily apparent at this point that selective electronic excitation can be used to produce specific chemical events through many types of reactions. Here no attempt will be made to review the great number of such experiments that have been carried out. Instead a few examples of selective chemistry through excitation will be given in an effort to illustrate the general principles involved.

10.5.5.2. Addition of Br_2* to Olefins. The chemistry of halogen atoms is of considerable interest for several reasons. Halogen atoms are readily produced by photolysis (exposure to light) of a number of compounds. Furthermore, halogen atoms react with almost any organic compound; the net photochemistry has been extensively examined using traditional techniques. Finally, halogen atom reactions often result in products with considerable internal excitation and therefore are suitable for use in chemical lasers.

Among the first experiments in laser photochemistry were those pioneered by Tiffany.[40] In these experiments a ruby laser operating near 6940 Å was temperature-tuned to excite specific transitions of various isotopes in Br_2. In this wavelength range excitation is primarily to the $A(^3\Pi_{1u})$ state, some 500–800 cm^{-1} *below* the dissociation limit (Fig. 1). Laser irradiation of mixtures of Br_2 and selected olefins $\left(\begin{array}{c} \diagup \\ \diagdown \end{array} C{=}C \begin{array}{c} \diagdown \\ \diagup \end{array} \right)$ in the gas phase resulted in *addition* of the Br_2 to the double bond, as in

$$Br_2 + (CF_3)FC{=}C(CF_3)F \rightarrow (CF_3)FBrC{-}CBr(CF_3)F.$$

This reaction is exothermic by typically 6300 cm^{-1}. It is energetically and mechanistically possible that such a reaction could proceed by direct bimolecular addition of Br_2* to the olefin. However, the experimental observations are not consistent with this hypothesis. A clear indication of this is the fact that the amount of product yield was found to increase linearly in proportion to $[Br_2]$! The reaction is also inhibited by the addition of a small amount of NO. Furthermore, excitation of specific isotopes does *not* result in products labeled with those isotopes as would be expected for simple addition. The experimental observations *can* be explained with the following mechanism, which is based, in part, on earlier classical observations, where the superscript ^{i} denotes isotope selection and X denotes an olefin:

---

[40] W. B. Tiffany, *J. Chem. Phys.* **48**, 3019 (1968).

$$^1Br_2 + \hbar\omega \longrightarrow {}^1Br_2^* \qquad \text{Excitation} \qquad (1)$$
$$^1Br_2^* + M \longrightarrow {}^1Br + {}^1Br + M \qquad \text{Dissociation} \qquad (2)$$

$$^1Br_2^* + M \longrightarrow {}^1Br_2 + M \qquad \text{E} \longrightarrow \text{V Transfer} \qquad (3)$$

$$\left.\begin{array}{l} ^1Br + X \rightleftharpoons {}^1BrX \\ {}^1BrX + Br_2 \longrightarrow {}^1BrXBr + Br \end{array}\right\} \quad \text{Propagation} \quad \begin{array}{l}(4)\\(5)\end{array}$$

$$^1Br + Br_2 \longrightarrow {}^1BrBr + Br \qquad \text{Exchange} \qquad (6)$$

$$^1Br_2^* + Br_2 \longrightarrow {}^1Br_2 + Br_2^* \qquad \text{E} \longrightarrow \text{E Transfer} \qquad (7)$$

$$\left.\begin{array}{l} Br(\text{wall}) \longrightarrow \tfrac{1}{2}Br_2 \\ Br + Br + M \longrightarrow Br_2 + M \end{array}\right\} \quad \text{Termination} \quad \begin{array}{l}(8)\\(9)\end{array}$$

A complete solution of the kinetic equations including all of the above reactions would be a difficult task indeed. For a steady-state approximate solution, it is assumed that the net rates of production of intermediates (such as Br, BrX, and Br_2^*) are zero. In this case, the quantum yield for the reaction defined here as $Q = -d[XBr_2]/dt \div d[Br_2]/dt$ is readily shown to be

$$Q = [k_2/(k_2 + k_3)]2k_4[Br_2][X](k_4[X] + k_8)^{-1}([Br_2] + k_{-4}/k_5)^{-1},$$

if k_9 the rate constant for reaction (9) is neglected. If k_8 is neglected, the yield is roughly

$$Q = [k_2/(k_2 + k_3)](k_4[Br_2][X])([Br_2] + k_{-4}/k_5)^{-1}[M]^{-1}(k_8[Br_2^*]k_2)^{-1/2}.$$

It may be seen that in either case, the yield is proportional to $[Br_2]$ for small $[Br_2]$ as observed. The factor $k_2/(k_2 + k_3)$ denotes the competition between E $\rightarrow$ V relaxation and collision-induced dissociation. The small experimental product yields suggest that $k_2/k_3 \sim 10^{-2}$. The experimental results also indicate that the rates of reaction (4) are rapid compared with reaction (5) under the reported experimental conditions. Finally, the lack of measurable isotope selection in the products suggests the importance of reaction (6) or (7). In these experiments, reaction (6) appears to be destroying the isotopic selection.

Now, it would be absurd to propose the commercial synthesis of halogen-substituted hydrocarbons by ruby-laser catalyzed addition of Br_2 to double bonds. After all, these reactions are readily carried out with ordinary light sources. The particular advantage of the laser in these experiments is its monochromaticity and tunability, which allow selection of specific spectral features. The importance of this experiment lies in the demonstration of laser-induced chemistry by electronic excitation. This experiment also illustrates some of the experimental difficulties involved in isotopically selective chemistry.

10.5.5.3. Reaction of Br* with HI. As we have already seen (Figs. 2 and 4) displacement reactions between a halogen atom and a hydrogen halide molecule can occur quite readily. Two questions of concern are (1) What is the role of electronic excitation in determining chemical reaction rates? (2) To what extent are correlation diagrams such as those shown in Fig. 4 a reasonable guide to chemical reactivity?

In order to examine these questions, Bergmann, Leone, and Moore[41] have studied the exothermic reaction of HI with ground-state $Br(^2P_{3/2})$ and with excited $Br*(^2P_{1/2})$. The diagram shown in Fig. 4 is relevant for linear encounters if Cl and Br are replaced by Br and I, respectively. The $^2P_{3/2}-^2P_{1/2}$ splitting is 3685 cm^{-1} for Br and 7603 cm^{-1} for I. The correlation diagram suggests that Br will be reactive while Br* will not.

In the experiments, a doubled Nd:YAG laser producing 0.5 mJ in 0.5 μsec at 4730 Å is used to dissociate Br_2 in a flowing mixture of Br_2, HI, and Ar. Light at this wavelength is sufficiently energetic to directly dissociate Br_2 into one $Br(^2P_{3/2})$ atom and one $Br*(^2P_{1/2})$ atom; this is apparently the predominant dissociation path. Experimentally, the luminescence from Br* or from HBr† (where † indicates vibrational excitation) is observed. There are several possible reactions that concern production and decay of the particular excited states under investigation:

$$Br_2 + \hbar\omega(21{,}142 \text{ cm}^{-1}) \longrightarrow Br + Br* \qquad \text{Excitation}$$

$Br + HI \longrightarrow HBr\dagger \ (v = 0,1,2) + I$		(1)
$Br* + HI \longrightarrow HBr\dagger \ (v = 0,1,2,3) + I$ Reaction		(2a)
$Br* + HI \longrightarrow HBr\dagger \ (v = 0) + I*$		(2b)

$$Br* + HI \longrightarrow Br + HI\dagger \qquad E \longrightarrow V \text{ Transfer} \qquad (3)$$

$Br* + HI \longrightarrow Br + HI$	$E \longrightarrow T$ Transfer	(4)
$Br* + Br_2 \longrightarrow Br + Br_2$		(5)

$HBr\dagger(v) + HI \longrightarrow HBr \ (v - 1) + HI\dagger$	$V \longrightarrow V$ Transfer	(6)
$Br_2 + HI \longrightarrow BrI + HBr$	Reaction	(7)

$HBr\dagger(v) \longrightarrow HBr(v - 1) + \hbar\omega \ (2100{-}3000 \text{ cm}^{-1})$	Luminescence	(8)
$Br* \longrightarrow Br + \hbar\omega \ (3685 \text{ cm}^{-1})$		(9)

Processes involving Ar are not expected to be of any importance under the conditions of these experiments. The "dark" reaction, (7), affects the concentrations of the reagents, forcing some corrections to be made. The luminescence reactions (8) and (9) contribute negligibly to the kinetic behavior of the system since $k_8 \sim 10v$ sec^{-1} and $k_9 \sim 0.9$ sec^{-1}. They do allow the time decay measurements, however. Experimentally the formation of HBr† as a function of time is found to be described by a double exponential curve. A solution to the kinetic equation describing the production of HBr luminescence takes the form

[41] K. Bergmann, S. R. Leone, and C. B. Moore, *J. Chem. Phys.* **63**, 4161 (1975).

$$A \exp\{-k_1[\text{HI}]t\} + B \exp\{-[(k_2 + k_3 + k_4)[\text{HI}] + k_5[\text{Br}_2]]t\},$$

where $k_2 = k_{2a} + k_{2b}$. The luminescence decay from Br* is proportional to only the second term. It may be seen that k_2, k_3, and k_4 cannot be distinguished by the experimental data. The experimental results give $k_1 = (1.0 \pm 0.3) \times 10^{-11}$ cm³ molecule⁻¹ sec⁻¹, which corresponds to a reactive cross section of ~3 Å² or a reactive probability per collision of 1/16. The results also show that $k_1/(k_2 + k_3 + k_4) = 4 \pm 1$ or that $k_1/k_2 > 4$. The emission intensities indicate that k_2 contributes less than 30% to the observed $(k_2 + k_3 + k_4)$, thus suggesting that $k_2 \ll k_1$. These results would indicate that the correlation diagram (Fig. 4) is of some use in understanding chemical reactivity in this system. The conclusiveness of these experiments has been questioned by Houston,[42] who points out that the reaction

$$\text{I*} + \text{Br}_2 \rightarrow \text{IBr} + \text{Br*} \tag{11}$$

can occur rapidly. If reaction (11) occurs more readily than the reaction to produce ground state Br

$$\text{I*} + \text{Br}_2 \rightarrow \text{IBr} + \text{Br} \tag{12}$$

or the quenching of I*,

$$\text{I*} + \text{Br}_2 \rightarrow \text{I} + \text{Br}_2, \tag{13}$$

then the decay of Br* by reaction (2b) will be masked since the net result of reactions (2b) and (11) is no destruction of Br*. The relative importance of (11)–(13) is not known at present.

Laser excitation, in these chemical experiments, allows the production of a large concentration of atomic reagents in a short period of time. The rapid kinetic processes that follow may thus be observed very conveniently.

10.5.5.4. Photodissociative Selective Chemistry: Formaldehyde.

The atoms H, H, C, and O can be assembled to form a number of stable systems, the energies of some of which are indicated in Fig. 5. Other possibilities include HC(OH) and $\text{CH}_2 + \text{O}$. It may be seen that it is energetically possible for the formaldehyde molecule (H_2CO) to decompose into H_2 and CO but that a large barrier normally hinders such dissociation. Electronic excitation into either the singlet or triplet excited states may provide sufficient energy to overcome this barrier and may open the possibility of other decay channels such as H + HCO or H + H + CO.

Formaldehyde exhibits a large number of relatively weak, sharply defined absorptions in the spectral region 3550–2500 Å. These involve ex-

[42] P. L. Houston, *Chem. Phys. Lett.* **47**, 137 (1977).

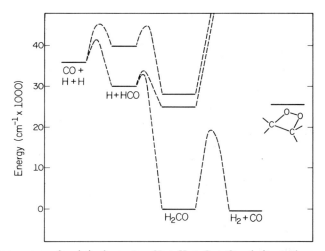

FIG. 5. Some energy levels in the system H + H + C + O, relative to the ground state of H$_2$CO. The excited states of H$_2$CO shown are ^{1}A'' at 28,188 cm^{-1} and ^{3}A'' at 25,194 cm^{-1} as determined from spectroscopic measurements. The energy of H$_2$ + CO may be determined from standard enthalpies of formation at 0°K as tabulated in the "JANAF Thermochemical Tables" (US Govt. Printing Office, Washington, D.C., 1971). The barrier for H$_2$CO → H$_2$ + CO is uncertain, but kinetic measurements provide a value ~19,000 cm^{-1} [C. J. M. Fletcher, *Proc. R. Soc. London, Ser. A* **146**, 357 (1934)]. Other barriers are represented only schematically. The energy for H + HCO is estimated from the photochemical studies of R. D. McQuigg and J. G. Calvert [*J. Am. Chem. Soc.* **91**, 1590 (1969)]. It is an upper limit. The energy of the dioxetane ĊH$_2$—O—O—ĊH$_2$ relative to *two* ground-state H$_2$CO molecules is also shown. This may be estimated to be ~21,600 cm^{-1} from thermochemical tables [S. W. Benson, *Chem. Rev.* **69**, 279 (1969)] assuming a ring strain energy of 26.4 kcal/mole as has been found for cyclobutane and for oxetane.

citation of various rovibronic levels within the lowest-energy excited singlet state, generated by excitation of an electron from a nonbonding molecular orbital (n) centered on the O atom to an antibonding π orbital (π^*) involving the CO group. The sharpness of the absorptions allows excitation of specific isotopes. Experimentally Yeung and Moore[43] have found that excitation of a 1:1 mixture of D$_2$CO and H$_2$CO at 3.0 torr pressure with a doubled ruby laser producing 8 MW of power in 15 nsec at 3742 Å results in production of D$_2$ and H$_2$ in the ratio of 6:1. Using narrow-band tunable sources, much higher isotopic selectivity has been obtained. Yeung and Moore[44] have also examined the decay of initially excited H$_2$CO as a function of excitation energy and gas pressure. The rate constant for intramolecular decay was found to be rapid ($k > 3.5 \times$

[43] E. S. Yeung and C. B. Moore, *Appl. Phys. Lett.* **21**, 109 (1972).
[44] E. S. Yeung and C. B. Moore, *J. Chem. Phys.* **58**, 3988 (1973).

10^7 sec^{-1}) and was found to increase significantly with increasing internal excitation. These observations suggested that the intramolecular decay involved E $\rightarrow$ V relaxation to the ground state S_0. The excess vibrational energy in S_0 should be sufficient to overcome the barrier to decomposition into H_2 + CO. Thus at first glance, isotopic selective chemistry appears to result from a single excitation step.

More recently, however, Houston and Moore[45] have examined the rate of *appearance* of CO in various vibrational states. Their results show that this rate of appearance is *slow* and directly proportional to gas pressure ($k \sim 1.65 \times 10^6$ sec^{-1} at $p = 1$ torr). They also show that the CO is produced with little vibrational excitation. Thus the chemistry of CO production must involve a more complicated chemical mechanism. The following mechanism is consistent with the above observations.

$$H_2CO + \hbar\omega \longrightarrow H_2CO(S_1) \qquad \text{Excitation}$$

$H_2CO(S_1) \longrightarrow I$	Intermediate	(1)
$H_2CO(S_1) + M \longrightarrow I$	formation	(2)
$H_2CO(S_1) \longrightarrow H_2CO(S_0)$	E $\longrightarrow$ V Transfer	(3)
$H_2CO(S_1) + M \longrightarrow H_2CO(S_0) + M$		(4)
$I \longrightarrow CO\ (v = i) + H_2$	CO formation	(5)
$I + M \longrightarrow CO\ (v = i) + H_2$		(6)
$I \longrightarrow$ other products	Intermediate	(7)
$I + M \longrightarrow$ other products	decay	(8)

Here an intermediate I has been postulated. Also the production of CO in particular vibrational states has been noted. The work of Yeung and Moore gives values for ($k_1 + k_3$) and ($k_2 + k_4$). For excitation of H_2CO at 3371 Å, Houston and Moore have determined $\Sigma_i\, k_{6,i}^F = 4.0 \times 10^{-11}$ cm^3 molecule^{-1} sec^{-1}, $k_8^F = 1.1 \times 10^{-11}$ cm^3 molecule^{-1} sec^{-1}, and $[(\Sigma_i\, k_{5,i}) + k_7] < 2.6 \times 10^5$ sec^{-1}, where the superscript F refers to M = formaldehyde. The nature of the intermediate I is not known at present. Possibilities include the triplet electronic state (T_1) and vibrationally excited S_0 states, although neither of these appears very attractive. Perhaps more palatable would be a collision complex [at least for reaction (2)] such as an elementary dioxetane $\lfloor\ CH_2-O-O-CH_2\ \rfloor$

Marling[46] has recently discussed in detail isotope separation schemes in formaldehyde. Several additional studies of formaldehyde have been reported.[46a]

[45] P. L. Houston and C. B. Moore, *J. Chem. Phys.* **65**, 757 (1976).

[46] J. Marling, *J. Chem. Phys.* **66**, 4200 (1977).

[46a] J. H. Clark, C. B. Moore, and N. S. Nogar, *J. Chem. Phys.* **68**, 1264 (1978); R. G. Miller and E. K. C. Lee, *ibid.* p. 4448.

10.5.6. Selective Chemistry by Vibrational Excitation

10.5.6.1. Enhancement of Bimolecular Reaction Rate. The existence of relatively large isotope effects in vibrational spectra, coupled with the fact that many chemical reactions require only a few vibrational quanta for activation, point out clearly that vibrational excitation is a fruitful direction for selective laser chemistry. Furthermore, the paths for degradation of selective excitation by various vibrational relaxation processes are reasonably well defined. As we have seen, elementary dynamical considerations would suggest that vibrational excitation is much more effective than translational excitation for catalyzing the endothermic reaction $Br + HCl \rightarrow BrH + Cl$ along the lowest energy potential surface. An examination of Fig. 3 shows that excitation of HCl to $v = 2$ provides sufficient energy for reaction. This idea has been employed by Leone $et\ al.$[47] and by Arnoldi $et\ al.$[48] in a scheme for separation of Cl isotopes. The essential steps in the scheme are

$$
\begin{array}{lll}
H^iCl + \hbar\omega_1 \longrightarrow H^iCl\ (v = 2) & \text{Selective excitation} & (1) \\
H^iCl\ (v = 2) + Br \longrightarrow HBr + {}^iCl & \text{Specific reaction} & (2) \\
{}^iCl + Br_2 \longrightarrow {}^iClBr + Br & \text{Product formation} & (3)
\end{array}
$$

These reactions can be carried out by laser excitation of $H^iCl\ (v = 2)$ in flowing mixtures of Br_2, Br, and HCl. An inert buffer gas such as He or Ar is used to act as a carrier for the reagents as well as to thermalize translational and rotational degrees of freedom and to help limit diffusion to the walls of the reaction vessel. For $[Br] \sim 3 \times 10^{15}$ molecules/cm^3 the rate of reaction (3) should be reasonably large since $k_3 \sim 1.2 \times 10^{-10}$ cm^3 molecule^{-1} sec^{-1}. This provides an effective trap for ^iCl. Numerous additional processes are possible including

$$
\left.
\begin{array}{l}
H^iCl\ (v = 2) + M \longrightarrow H^iCl\ (v = 0,1) + M \\
H^iCl\ (v = 2) + HCl \longrightarrow H^iCl\ (v = 1) + HCl\ (v = 1) \\
HCl\ (v = 2) + HCl \longrightarrow H^iCl + HCl\ (v = 2)
\end{array}
\right\}
\begin{array}{l}
\text{Vibrational relaxation}
\end{array}
\begin{array}{l}
(4) \\
*(5) \\
*(6)
\end{array}
$$

$$
\left.
\begin{array}{l}
{}^iCl + HCl \longrightarrow H^iCl + Cl \\
{}^iClBr + HCl \longrightarrow H^iCl + BrCl \\
{}^iClBr + HCl(wall) \longrightarrow H^iCl + BrCl
\end{array}
\right\}
\begin{array}{l}
\text{Exchange reactions}
\end{array}
\begin{array}{l}
*(7) \\
*(8) \\
*(9)
\end{array}
$$

$$
\left.
\begin{array}{l}
{}^iClBr + HBr \longrightarrow {}^iClH + Br_2 \\
{}^iClBr + HBr(wall) \longrightarrow {}^iClH + Br_2
\end{array}
\right\}
\begin{array}{l}
\text{Loss of product}
\end{array}
\begin{array}{l}
(10) \\
(11)
\end{array}
$$

[47] S. R. Leone, R. G. Macdonald, and C. B. Moore, $J.\ Chem.\ Phys.$ **63**, 4735 (1975).
[48] D. Arnoldi, K. Kaufmann, and J. Wolfrum, $Phys.\ Rev.\ Lett.$ **34**, 1597 (1975).

In the above, reactions marked with an asterisk result in scrambling of the isotopic selectivity. In order for interactions of HCl ($v = 2$) with Br ($k_2 + k_4^{Br} = 1.8 \times 10^{-13}$ cm³ molecule⁻¹ sec⁻¹) to compete effectively with vibrational relaxation [principally via reaction (5), $k_5 = 3.3 \times 10^{-12}$ cm³ molecule⁻¹ sec⁻¹] it is required that [Br] $\geqslant$ [HCl]. Also, since $k_4^{Br} \sim$ 1.3×10^{-12} cm³ molecule⁻¹ sec⁻¹ and $k_2 \sim 0.5 \times 10^{-12}$ cm³ molecule⁻¹ sec⁻¹ the maximum product quantum yield is ~ 0.3. Under typical experimental conditions, reactions (6)–(8) are not expected to be of importance.

Leone et al.[47] have attempted to separate Br³⁷Cl by exciting H³⁷Cl ($v = 2$) with an optical parametric oscillator. This was unsuccessful, presumably because of reaction (11). However, Arnoldi et al.[48] did observe transient enrichment in Br³⁵Cl in a similar system excited by predominant H³⁵Cl chemical laser transitions. This latter experiment demonstrates that vibrational excitation of specific isotopes can produce isotopically selected products. However, it is also clear that collection of the products can be made difficult by subsequent chemical reaction. It may be noted that lasers are virtually the only type of light source capable of producing chemically useful amounts of narrow-band infrared radiation on timescales that are short compared to vibrational relaxation times.

10.5.6.2. Dissociation with Intense Infrared Radiation. Vibrational photochemistry is possible in which the direct excitation of a particular molecular vibration may be followed by a radiationless dissociation (or predissociation) process. The resulting fragments could react with other reagents to produce the desired products. One problem with this scheme is that the vibrational selection rules usually allow absorption of only one or two quanta, providing insufficient energy to break any of the chemical bonds in many molecules. If one or two vibrational quanta are sufficient for dissociation, then the thermal accessibility of these excited vibrational levels and considerations of equilibrium may strongly favor thermal dissociation, removing the selectivity. As an example, consider a reagent for which the fundamental vibrational frequency is 1000 cm⁻¹ and suppose that 1000 cm⁻¹ is exactly the amount of energy needed to dissociate the molecule. From the elementary equilibrium constant mentioned in Section 10.5.3.7, we may calculate from an initial pressure of 1 torr, only four molecules in 10⁶ will remain undissociated for equilibrium at 300 K.[1] For this same example, a dissociation energy of ~ 4000 cm⁻¹ (4 quanta) would be necessary for the equilibrium to favor the reactants at 1 torr.

Recently a considerable amount of evidence has accumulated that supports the idea that very intense infrared radiation can directly dissociate even relatively strong chemical bonds in the absence of molecular collisions.[49-52] This is of particular importance since infrared photons may be produced with high efficiency and are thus relatively inexpensive. Furthermore, through the use of appropriate excitation wavelengths, specific nuclear isotopes may be dissociated and chemically isolated. Already macroscopic enrichment has been carried out for compounds containing specific isotopes of silicon, osmium, sulfur, boron, molybdenum, and many others. Many of the experiments thus far have been carried out for excitation of SF_6 near the very strong ν_3 absorption that occurs at 948 cm^{-1}. The SF bond dissociation energy is $\sim$40,000 cm^{-1}, which corresponds to more than 40 CO_2 laser quanta at 940 cm^{-1}. Some of the experimental observations may be summarized as follows.

(1) For excitation of SF_6 at pressures $\sim$100 μm by a single CO_2 laser pulse of intensity $\sim$5 $\times$ 10^8 W/cm^2 and 0.5 nsec duration the dissociation probability per molecule approaches unity.

(2) For pressures such that the time between collisions exceeds the excitation time, the dissociation probability decreases with increasing pressure.

(3) The amount of dissociation is relatively large only when the excitation intensity exceeds a threshold value, typically $\sim$3 $\times$ 10^7 W/cm^2.

(4) The average number of quanta absorbed per molecule is $\sim$16 at threshold and reaches $\sim$40 at considerably higher intensities.

(5) The optimum frequency for dissociation is slightly *below* the frequency for resonant excitation of the ν_3 mode.

The detailed chemistry following SF_6 dissociation has still not been fully examined. Subsequent to dissociation these reactive species may participate in a complicated series of chemical events. In pure SF_6, the apparent result is removal of the initially excited molecules from the gas phase. This removal can occur with impressive isotopic selectivity. In

[49] R. V. Ambartzumian, *in* "Tunable Lasers and Applications" (A. Mooradian, T. Jaeger, and P. Stokseth, eds.), p. 150. Springer-Verlag, Berlin and New York, 1976.

[50] N. Bloembergen, C. D. Cantrell, and D. M. Larsen, *in* "Tunable Lasers and Applications" (A. Mooradian, T. Jaeger, and P. Stokseth, eds.), p. 162. Springer-Verlag, Berlin and New York, 1976; N. Bloembergen, *Opt. Commun.* **15**, 416 (1975).

[51] K. L. Kompa, *in* "Tunable Lasers and Applications" (A. Mooradian, T. Jaeger, and P. Stokseth, eds.), p. 177. Springer-Verlag, Berlin and New York, 1976; V. S. Letokhov and C. B. Moore, *in* "Chemical and Biological Applications of Lasers" (C. B. Moore, ed.), Vol. 3, p. 1. Academic Press, New York, 1977; R. V. Ambartzumian and V. S. Letokhov, *ibid.* p. 167.

[52] A. L. Robinson, *Science* **194**, 45 (1976); N. Bloembergen and E. Yablonovitch, *Phys. Today* May, p. 23 (1978).

at least one case the enrichment factor for $^{34}SF_6$ following excitation of $^{32}SF_6$ has exceeded 2800.[53]

Numerous theories have been advanced to explain the dissociation processes discussed above.[50] In most pictures, moderately excited vibrational states ($\nu_3 \sim 4-5$) are excited with isotopic selectivity by multiple-photon absorption. Because of vibrational anharmonicity, such absorption would occur at lower frequencies than the fundamental absorption, although partial compensation is allowed by rotational selection rules. The selectively excited molecules may be further excited in a nonselective manner since *all* of the vibrational states of SF_6 at an energy of 4–5 quanta in ν_3 form a near-continuum. Although these theoretical pictures would seem to explain many of the observations, many questions remain to be answered and many new aspects of these fascinating processes remain to be uncovered.

It should be emphasized that the chemistry just described is unique to laser excitation. Conventional light sources simply cannot provide the instantaneous intensity necessary for the multiphoton absorption processes. This chemistry is likely to have significant practical value and to provide many exciting and challenging experimental and theoretical investigations.

Bibliography

General Photochemistry
G. H. Brown, ed., "Photochromism." Wiley (Interscience), New York, 1971.
J. G. Calvert and J. N. Pitts, Jr., "Photochemistry." Wiley, New York, 1966.
J. K. Kochi, "Free Radicals." Wiley, New York, 1973.
D. C. Neckers, "Mechanistic Organic Photochemistry." Van Nostrand-Reinhold, Princeton, New Jersey, 1967.
R. Srinivasan, ed., "Organic Photochemical Syntheses," Vols. 1 and 2. Wiley (Interscience), New York, 1976.

Chemical Kinetics
S. W. Benson, "The Foundations of Chemical Kinetics." McGraw-Hill, New York, 1960.
W. C. Gardiner, Jr., "Rates and Mechanisms of Chemical Reactions." Benjamin, Menlo Park, California, 1969.
K. J. Laidler, "The Chemical Kinetics of Excited States." Oxford Univ. Press, London and New York, 1955.
K. J. Laidler, "Theories of Chemical Reaction Rates." McGraw-Hill, New York, 1969.
R. E. Weston and H. A. Schwarz, "Chemical Kinetics." Prentice-Hall, Englewood Cliffs, New Jersey, 1972.

[53] R. V. Ambartzumian and V. S. Letokhov, *Laser Focus* **11**, 48 (1975); R. V. Ambartzumian, Yu. A. Gorokhov, V. S. Letokhov, and G. N. Makorov, *Zh. Eksp. Teor. Fiz., Pis'ma Red.* **21**, 375 (1975).

Spectroscopy and Thermodynamics
S. W. Benson, *Chem. Rev.* **69,** 279 (1969).
J. D. Cox and G. Pilcher, "Thermochemistry of Organic and Organometallic Compounds." Academic Press, New York, 1969.
G. Herzberg, "Infrared and Raman Spectra of Polyatomic Molecules." Van Nostrand-Reinhold, Princeton, New Jersey, 1945.
G. Herzberg, "Spectra of Diatomic Molecules." Van Nostrand-Reinhold, Princeton, New Jersey, 1950.
G. Herzberg, "Electronic Spectra of Polyatomic Molecules." Van Nostrand-Reinhold, Princeton, New Jersey, 1966.
G. Herzberg, "Spectra and Structures of Simple Free Radicals." Cornell Univ. Press, Ithaca, New York, 1971.
G. J. Janz, "Thermodynamic Properties of Organic Compounds." Academic Press, New York, 1967.
D. R. Small, E. F. Westrum, and G. C. Sinke, "The Chemical Thermodynamics of Organic Compounds." Wiley, New York, 1969.

Laser Chemistry and Isotope Separation
J. P. Aldridge, J. H. Birely, C. D. Cantrell, and D. C. Cartwright, *Phys. Quantum Electron.* **4,** 57 (1976).
M. J. Berry, *Annu. Rev. Phys. Chem.* **26,** 259 (1975).
S. Kimel and S. Speiser, *Chem. Rev.* **77,** 437 (1977).
E. K. C. Lee, *Acc. Chem. Res.* **10,** 319 (1977).
V. S. Letokhov, *Science* **180,** 451 (1973).
V. S. Letokhov and C. B. Moore, *Sov. J. Quantum Electron. (Engl. Transl.)* **6,** 129 (1976).
V. S. Letokhov and C. B. Moore, *Sov. J. Quantum Electron. (Engl. Transl.)* **6,** 259 (1976).
C. B. Moore, *Acc. Chem. Res.* **6,** 323 (1973).
C. B. Moore, ed., "Chemical and Biological Applications of Lasers," Vol. 1, Vols. 2–4. Academic Press, New York, 1974, 1977 resp.

AUTHOR INDEX FOR PART A

Numbers in parentheses are footnote reference numbers and indicate that an author's work is referred to although his name is not cited in the text.

AUTHOR INDEX FOR PART B

SUBJECT INDEX FOR PART A

SUBJECT INDEX FOR PART B

A

Absorption spectra, picosecond continuum monitoring of, 203
Ac Stark shifts, 242–243
ADA, *see* Ammonium dihydrogen arsenate crystals
ADP, *see* Ammonium dihydrogenate phosphate
Alkali halides, color centers in, 1–2
Ammonium dihydrogen arsenate crystals, in SFG and SHG, 177–180
Ammonium dihydrogen phosphate crystals, in SFG and SHG, 177–180
Ammonia laser, 76–77
Anion vacancy, in color center, 3
  H ion trapped in, 17
Argon fluoride lasers, 125–126
Arrested relaxation technique, in chemical lasers, 98
Associative ionization, 227
Atom exchange reactions, vibrational distributions and, 98
Auger recombination, in germanium, 208

B

Bacteriorhodopsin photochemistry, 206–207
Bathobacteriorhodopsin, 207
Bimodal rotational distributions, in hydrogen halide reaction products, 117
Bimolecular reaction rate, enhancement of, 294
Bimolecular relaxation, 280
Binary collision rate, molecular diameter and, 66
Biological processes, picosecond pulses in study of, 206–207
Birefringence, optical-field-induced, 260–261
Birefringence tuner, 31
Bond energies, concept of, 273–274
Born–Oppenheimer approximation, 271

Brewster's angle, 26, 30, 90
Brillouin scattering, 260
Bromine
  potential energy surfaces for, 272
  reaction of with halide, 289–291
Bromine atom, addition of to olefins, 288
Bromine–hydrogen–chlorine system
  correlation diagram for, 276
  ground state potential surface for, 274

C

Carbon dioxide
  electrical discharge in, 70–71
  in vibrational energy transfer, 72
Carbon dioxide laser, 83, 91
Carbon dioxide laser amplifiers, 91
Carbon monoxide–carbon monoxide collisions, 106
Carbon monoxide chemical laser, 105–111
Carbon monoxide laser, 83
CARS, *see* Coherent anti-Stokes Raman spectroscopy
Chain reactions, in chemical kinetics, 283
Chemical energy surfaces, 271–276
Chemical kinetics
  bimolecular relaxation in, 280
  chain reactions in, 283
  diffusion-controlled reactions in, 281
  excitation processes in, 278–279
  in laser-selective chemistry, 278–283
  surface-catalyzed reactions in, 281
  termolecular reactions in, 280–281
  unimolecular decay in, 279–280
Chemical lasers, 95–96, *see also* Chemically pumped lasers; Purely chemical lasers
  in double-resonance experiments, 118
  efficient visible, 139–140
  new, 120–140
  operation of at short wavelengths, 121–125
  photodissociation in, 111
  purely, *see* Purely chemical lasers